FEMINIST CLIMATE POLICY IN INDUSTRIALISED STATES

Feminist Climate Policy in Industrialised States explores ways in which policymakers can overcome institutional barriers and conventions in pursuit of the radical changes necessary for a gender-just climate emergency response.

In 2021, the Intergovernmental Panel on Climate Change acknowledged that addressing the climate emergency must involve social justice and equality. Feminist approaches to decision-making, policy-making, community organising and their underpinning methodologies can enable this. The authors draw critically on case studies, research and interviews with feminist practitioners, legislators and leaders who have implemented significant changes, to signal how change might be achieved and ask what lessons can be drawn. The book posits that we need to ultimately move beyond the gender mainstreaming and gender equality issues which have been integrated into existing – and failing – structures, to more transformative feminist approaches. It concludes by identifying key strands of feminist-oriented praxis that offer the potential to expedite responses to climate change across multiple levels of governance.

With industrialised states shifting rightwards to a politics which diminishes the importance and urgency of gender equality, diversity, human rights and the need for climate action, this volume will inspire, guide, and provide tools for policymakers, politicians, community activists, academics, and students to take transformative action to address the climate emergency.

Susan Buckingham is a writer, researcher, consultant, campaigner and activist. She edits the Routledge series on Gender and Environments, and her work develops the understanding of links between gender and environment and applies this to different contexts. Most recently, this has been in the UN

Decade of Ocean Science for Sustainable Development and through consultancies with the European Commission, including EIGE. She has edited and written extensively, is on the editorial board of the environmental justice journal *Local Environment*, and is currently writing a book on Ecofeminism. As an activist-academic, Susan has worked with women's organisations, and was a trustee and collaborator with Women's Environmental Network from 2000–2012. Susan co-founded Friends of the Cam in 2020 which campaigns against destructive, masculinist planning and water pollution practices which are destroying the chalk streams of SE England. She is also an activist in climate and social justice campaigns.

Martin Hultman is Professor in the Department of Sociology and Work Science, Gothenburg University, Sweden. Hultman leads three research groups analysing 'masculinities and environment, 'rights of nature' and 'climate change denial'. He heads the global research network *Center for Studies of Climate Change Denial* (CEFORCED) and was appointed the most influential academic in Gothenburg 2019 and awarded Linköping University alumnus of the year 2021. As part of his academic work he publishes chronicles in a wide range of newspapers and gives public lectures commenting on contemporary politics. Recent books include *Ecological Masculinities* (2018), *Men, Masculinities and Earth* (2021), *Climate Obstruction* (2022) and the forthcoming *Survival: Rights of Nature, Degrowth and Ecological Masculinities at the end of Anthropocene*.

Gunnhildur Lily Magnusdottir is Associate Professor in Political Science at the Department of Global Political Studies at Malmö University. Her main field of research is climate policy-making, in particular, how governmental authorities in Scandinavia and the European Union understand and work with gender and other issues related to the social dimensions of climate change and climate justice. Magnusdottir has published extensively on climate authorities, their practices, institutional norms and policy-making at different levels of governance. She currently leads a comparative research project on Gendered Norms and Practices in Nordic and Baltic Climate Policy Institutions: Implication for the Climate Transition (Nordforsk), exploring Nordic and Baltic governmental authorities and their institutional practices. She has previously been involved in various research projects financed by the Swedish Research Council for Sustainable Development and the Swedish Energy Agency, which circle around justice in climate and energy policy-making.

Karen Morrow has been Professor of Environmental Law at Swansea University (2007–date), and has formerly worked at Leeds University, Durham University, the Queen's University of Belfast and the University of Buckingham. Her research interests centre on public participation in environmental law and policy-making, and in particular on gender in the global climate governance regime. Her work spans theory and practice and multiple levels of law and governance, from the global/international to national and subnational levels. She also works on environmental law and policy in cross-border contexts in the UK. She is a member of the Earth Systems Governance Tipping Point subgroup. She was a founding co-editor of the *Journal of Human Rights and the Environment* and the IUCN e-journal. She sits on the editorial boards of the *Journal of Human Rights and the Environment*, the *Environmental Law Review* and the *University of Western Australia Law Review* and on the International Advisory Board for the Gender and Environment book series (Routledge).

Routledge Studies in Gender and Environments

With the European Union, United Nations, UN Framework Convention on Climate Change, and national governments and businesses at least ostensibly paying more attention to gender, including as it relates to environments, there is more need than ever for existing and future scholars, policy makers, and environmental professionals to understand and be able to apply these concepts to work towards greater gender equality in and for a sustainable world.

Comprising edited collections, monographs and textbooks, this *Routledge Studies in Gender and Environments* series incorporates sophisticated critiques and theorisations, including engaging with the full range of masculinities and femininities, intersectionality, and LGBTQI+ perspectives. The concept of 'environment' is drawn broadly to recognise how built, social and natural environments intersect with and influence each other. Contributions are especially welcomed from global regions and contexts which are not yet well represented in gender and environments literature, in particular Russia, the Middle East, and China, as well as other East Asian countries such as Japan and Korea.

Series Editor: Susan Buckingham, an independent researcher, consultant and writer on gender and environment related issues.

Titles in this series include:

Gender and the Social Dimensions of Climate Change
Rural and Resource Contexts of the Global North
Edited by Amber J. Fletcher and Maureen G. Reed

Feminist Climate Policy in Industrialised States
A Gender-Just Climate Emergency Response
Edited by Susan Buckingham, Martin Hultman, Gunnhildur Lily Magnusdottir and Karen Morrow

For more information about this series, please visit: www.routledge.com/Routledge-Studies-in-Gender-and-Environments/book-series/RSGE

FEMINIST CLIMATE POLICY IN INDUSTRIALISED STATES

A Gender-Just Climate Emergency Response

Edited by Susan Buckingham, Martin Hultman,
Gunnhildur Lily Magnusdottir and Karen Morrow

Routledge
Taylor & Francis Group

LONDON AND NEW YORK

Designed cover image: Getty Images

First published 2026
by Routledge
4 Park Square, Milton Park, Abingdon, Oxon OX14 4RN

and by Routledge
605 Third Avenue, New York, NY 10158

Routledge is an imprint of the Taylor & Francis Group, an informa business

We are also indebted to the Swedish Research Council for Sustainable Development, FORMAS, and research grant 2018-01704, which has financed a part of the time invested in the book and a part of the open access fees.

British Library Cataloguing-in-Publication Data
A catalogue record for this book is available from the British Library

ISBN: 978-1-032-59033-2 (hbk)
ISBN: 978-1-032-59029-5 (pbk)
ISBN: 978-1-003-46100-5 (ebk)

DOI: 10.4324/9781003461005

Typeset in Sabon
by SPi Technologies India Pvt Ltd (Straive)

CONTENTS

List of Figures   *x*
List of Tables   *xi*
List of Contributors   *xii*
Acknowledgements   *xviii*

Introduction   1
Susan Buckingham, Gunnhildur Lily Magnusdottir,
Karen Morrow and Martin Hultman

PART I
Global   **17**

1 To Practice What You Preach: Sweden's Feminist
  Foreign Policy in Diplomatic Work   19
  Malena Rosén Sundström and Ole Elgström

  Interview 1: Catherine McKenna, Former Minister
  of Environment and Climate Change, Canada:
  International and National Role in Climate Policy   37
  Interviewed by Dory Reeves and Julie MacArthur

2 A Globe of One's Own: The Inverse Effect of Women's
  Political Representation on GHG Emissions   48
  Laura Winther Engelsbak

3 To What Extent Can the European Union Contribute to a Feminist Climate Policy?   71
Gill Allwood

4 The Ocean We Want: A Feminist Approach to the Ocean Decade   89
Susan Buckingham, Mariamalia Rodríguez-Chaves, Ellen Johannesen, Renis Auma Ojwala, Zhen Sun, Momoko Kitada, Francis Neat and Ronán Long

5 Ensuring Justice through Good Practice: Establishing the Context for Change Across Organisational Scales   105
Seema Arora-Jonsson

Interview 2: The Hon Marama Davidson, co-Leader of the Green Party of Aotearoa New Zealand: The Importance of Grassroots and Community Action   122
Interviewed by Dory Reeves and Julie MacArthur

PART II
Initiatives   **133**

6 Gender Smart Mobility for All: Lessons Learned from Encounters with Danish Municipalities   135
Hilda Rømer Christensen and Michala Hvidt Breengaard

Interview 3: Ada Colau, Mayor of Barcelona 2015–2023: Addressing the Climate Emergency in Collaborative Ways at the City Level   152
Interviewed by Inés Novella Abril

7 What Does Degrowth say about Gender Equality and Social Justice?   159
Bipasha Baruah and Andrea Burke

Interview 4: Guðmundur Ingi Guðbrandsson, ex-Leader of The Left-Green Movement, and Minister of Social and Labour Affairs in Iceland   176
Interviewed by Gunnhildur Lily Magnusdottir

8 Climate Change Policies and Gender Equity: What are
   the Views of Women Who Work in Construction?   181
   Coralie Guedes, Vivian Price and Linda Clarke

9 Applying Intersectionality in Climate Policy and
   Planning: Experiences from Gothenburg and Malmö   200
   Nanna Rask, Angelica Lundgren and Annica Kronsell

   Interview 5: Marianne Borgen, Two-term Mayor of
   Oslo between 2015 and 2023   218
   Interviewed by Susan Buckingham

PART III
Methodologies   **225**

10 Young People and Old Trees: Posthuman
    Intersectionality in Swedish Climate Litigation   227
    Marie Widengård

11 Participatory Assessment Workshops as a Guiding Tool
    Towards Just and Inclusive Energy Strategies   244
    Gunnhildur Lily Magnusdottir and Anders Melin

12 Theatre and Stories that ReConnect: Embodiment
    Practices That Ecologise Masculinities   260
    Paul M. Pulé, Ilaria Olimpico and Uri Noy Meir

13 Photovoice: A Tool for Countering Social Path
    Dependencies in Climate Institutions?   277
    *Heidi Walker, Amber J. Fletcher, Maureen G. Reed
    and Nicholas Antonini*

14 Feminist Climate Approaches: How, Why and What?
    Why We Need Feminist Climate Approaches More
    than Ever, What Would they Look Like and How do
    we Get There?   296
    *Martin Hultman, Karen Morrow, Gunnhildur Lily
    Magnusdottir and Susan Buckingham*

Index   *305*

FIGURES

2.1   Marginal effects of women's political representation on
      climate footprint   59
6.1   Gender Smart Mobility dimensions   137
6.2   Gender Smart Mobility indicator   138
6.3   VEU project development model   140
6.4   Gender differences in transport   146
12.1 Theatre and Stories that Reconnect maps   264
12.2 Diagram of deep listening as it is used in TStR   266
12.3 Image of Milan workshop participants striking embodied
      poses of industrial/breadwinner masculinities   267
12.4 Image of Milan workshop participants striking embodied
      poses of ecomodern masculinities   268
12.5 Image of Milan workshop participants striking embodied
      poses of ecological masculinities   268

TABLES

2.1   Description of included variables   54
2.2   Main and sub-hypotheses   57
2.3   Primary model: Women's political representation and GHG
      emissions   58
2.4   Models with alternative dependent and primary
      independent variables   60
2.5   Fixed effect models with alternative dependent and primary
      independent variables   61
14.1 Common qualities identified by feminist climate leaders   297

CONTRIBUTORS

Inés Novella Abril is a researcher and coordinator at the UNESCO Chair on Gender, Universidad Politécnica de Madrid and Urban Planning lecturer at Universitat Politécnica de València, Spain. She is a consultant for the Spanish Government and Regional Administrations on gender in housing and spatial planning policies.

Gill Allwood is Professor of Gender Politics at Nottingham Trent University, UK and Visiting Professor in the Department of International Relations and Diplomacy Studies at the College of Europe in Bruges, where she delivers a course on Gendering EU External Affairs.

Nicholas Antonini is a Sociology Graduate Student at the University of Regina, Canada. His research is based in theories of practice and everyday life. Nicholas is a member of Barrier Free Saskatchewan and Mother Earth Justice Advocates.

Seema Arora-Jonsson is Professor of Rural Development at the Swedish University of Agricultural Sciences. She works on sustainability and justice in relation to environmental governance, climate politics and rural development, informed by feminist thinking and the need to decolonise development and environmental governance.

Bipasha Baruah is Professor and Western Research Chair in the Department of Gender, Sexuality and Women's Studies at Western University, Canada. She conducts interdisciplinary research on gender, economy, environment, and development.

Michala Hvidt Breengaard is a post-doctoral researcher at the University of Copenhagen, Denmark. Her research lies in the intersection of gender and diversity studies and research on climate and sustainability, including on transportation.

Susan Buckingham is a writer, researcher, consultant, campaigner and activist based in Cambridge, UK, and edits the Routledge series on Gender and Environments. Her work develops the understanding of links between gender and environment and applies this to different contexts. Most recently this has been in the UN Decade for Ocean Sustainability and through consultancies with the European Commission, including EIGE. She has edited and written extensively and is currently writing a book on Ecofeminism. As an activist-academic, Susan has worked with UK women's organisations, and was a trustee and collaborator with Women's Environmental Network from 2000 to 2012. Susan co-founded Friends of the Cam in 2020 which campaigns against destructive, masculinist planning and water pollution practices which are destroying the chalk streams of SE England. She is also an activist in climate and social justice campaigns

Andrea Burke is a PhD candidate at Western University, Canada. Her doctoral research identifies the challenges experienced by racialised and immigrant women who were essential workers during the COVID-19 pandemic as well as policy recommendations to address them.

Hilda Rømer Christensen is Associate Professor and Head of Coordination for Gender Studies at University of Copenhagen, Denmark. She is a coordinator and vice PI of research projects on smart mobilities and gendered innovations, such as TINNGO, GILL, gendered innovation living labs, and green transport, technology and diversity in Danish municipalities.

Linda Clarke is Emeritus Professor in the Centre for the Study of the Production of the Built Environment, University of Westminster, UK, where she has carried out comparative research on vocational education and training, employment, industrial relations, gender, equality, climate change, and labour in the construction sector.

Ole Elgström is Professor Emeritus of Political Science at Lund University, Sweden. He has published on internal and external negotiations involving the EU, and notably on external perceptions of EU foreign policy, in a large number of journals.

Laura Winther Engelsbak has been a member of the Danish Council for Human Rights, and has worked on the links between gender equality and climate change at the Danish UN-mission in New York and at the Danish Ministry of Foreign Affairs.

Amber J. Fletcher is Professor of Sociology and Academic Director of the Community Engagement and Research Centre at the University of Regina, Canada. She has been a contributing author to climate change reports from the Intergovernmental Panel on Climate Change and the Government of Canada.

Coralie Guedes is a research associate in the school of Organisation, Economy and Society at the University of Westminster, UK. She recently completed her PhD on the relevance of Global Framework Agreements for labour agency in processes of transnational environmental regulation.

Martin Hultman Why don't we act when we have known the existential challenges of climate change for more than thirty years? Which are the most effective levers to break free from our fossil fuelled modernity? Such are the questions Professor Martin Hultman, Department of Sociology and Work Science, Gothenburg University, Sweden deals with in his scholarship when doing in-depth empirical studies analysed and informed by interdisciplinary knowledge. Hultman leads three research groups analysing 'masculinities and environment, 'rights of nature' and 'climate change denial'. He heads the global research network *Center for Studies of Climate Change Denial* (CEFORCED) and has been appointed the most influential academic in Gothenburg 2019 and awarded Linköping University alumnus of the year 2021. As part of his academic work he publish chronicles in a wide range of newspapers and give public lectures commenting on contemporary politics. Recent books include *Ecological Masculinities* (2018), *Men, Masculinities and Earth* (2021), *Climate Obstruction* (2022) and the forthcoming *Survival. Rights of Nature, Degrowth and Ecological Masculinities at the end of Anthropocene*.

Ellen Johannesen is a researcher at the Norsk Institutt for Vannforskning, Denmark. She earned her PhD as part of the Empowering Women for the UN Decade of Ocean Science for Sustainable Development programme, World Maritime University, Sweden while working at the International Council for the Exploration of the Sea, where she led the development of the gender equality plan.

Momoko Kitada is a former seafarer, now holding the Nippon Foundation Professorial Chair in Gender and Innovation at the World Maritime University in Malmö, Sweden. She collaborates with the International Maritime Organization (IMO) on women's integration in the maritime sector.

Annica Kronsell, Professor of Political Science at Lund University and Chair of Environmental Social Science at the School of Global Studies, Gothenburg University, in Sweden is interested in how public institutions can govern climate and sustainability issues.

Ronán Long is the Director of World Maritime University – Sasakawa Global Ocean Institute in Malmö, Sweden where he holds the Nippon Foundation Professorial Chair in Ocean Governance and the Law of the Sea.

Angelica Lundgren is a former Vice President of a UN Environmental Group and a researcher at Gothenburg University, Sweden and Center for Social Studies of the University of Coimbra, Portugal. She has conducted research on climate policy, crisis communication, and nature-based solutions.

Gunnhildur Lily Magnusdottir is Associate Professor in Political Science and the Deputy Head of the Department of Global Political Studies at Malmö University. Her main research field is climate policy-making and how governmental authorities in Scandinavia understand and work with the social dimensions of climate change. Magnusdottir has published extensively on climate authorities, their practices, institutional norms and policy-making at different levels of governance. She is currently involved in a comparative research project on gendered norms and practices in Nordic and Baltic Climate Policy Institutions: Implication for the Climate Transition, (Nordforsk), exploring Nordic and Baltic governmental authorities and their institutional practices. She has previously been involved in research projects financed by the Swedish Research Council for Sustainable Development and the Swedish Energy Agency, on justice in climate and energy policy-making.

Julie MacArthur is an associate professor and the Canada Research Chair in Reimagining Capitalism at Royal Roads University in Victoria, Canada. She is also a resident Fellow in Energy Systems Transformation at the Cascade Institute.

Uri Noy Meir is a facilitator and trainer, integrating social arts into diverse contexts. Uri harnesses the power of theatre and embodied arts to drive meaningful change. He works as a social innovation consultant, coordinates activities for the ImaginAction network, and teaches at the Technical University of Applied Sciences Würzburg-Schweinfurt, Germany.

Anders Melin is Associate Professor in Ethics at the Department of Global Political Studies at Malmö University, Sweden. His main field of research is environmental ethics and he has led several interdisciplinary projects on environmental issues and justice.

Karen Morrow has been Professor of Environmental Law at Swansea University since 2007. Her research interests focus on theoretical and practical aspects of human rights as applied to environmental law and policy, and gender and the environment. She has published extensively in these areas. She sits on the editorial boards of the *Journal of Human Rights and the Environment*, the *Environmental Law Review* and the *University of Western Australia Law Review* and the International Advisory Board for the Gender and Environment book series (Routledge) and a member of the Global Network for the Study of Human Rights and the Environment.

Francis Neat is Professor on Sustainable Fisheries Management, Ocean Biodiversity (Nippon Foundation Chair) at the World Maritime University, Sweden. His current work is focused on interdisciplinary problems of illegal fishing and ocean governance particularly concerning gender equality and global equity.

Renis Auma Ojwala is a post-doctoral research associate at the World Maritime University, Sweden researching how to foster crew sustainability in the maritime industry. She was recently awarded a PhD from WMU which investigated gender equality in ocean science education in Kenya.

Ilaria Olimpico is a facilitator, trainer, and artist. She collaborates with CISP Interdisciplinary Centre Sciences for Peace in Pisa (Italy) and the Technical University of Applied Sciences of Würzburg-Schweinfurt (Germany). She created the methodology Stories That Reconnect.

Vivian Price is Professor at California State University, Dominguez Hills, USA and former union electrician, a researcher and filmmaker on labour and climate justice. Her films include *Transnational Tradeswomen*, *Harvest of Loneliness* and *Talking Union, Talking Climate*, centering experiences of gendered and racialised identity and climate.

Paul M. Pulé is a social and environmental justice scholar and consultant educator. Paul is currently exploring the links between personal trauma recovery and global crises through Somatic Experiencing, along with developing embodied practices and educational pedagogies through Theatre of the Oppressed, Social Presenting Theatre, Theatre and Stories that Reconnect.

Nanna Rask is a PhD student in Environmental Social Science at the School of Global Studies at the University of Gothenburg, Sweden, and was a visiting Fellow at Environmental Humanities South at the University of Cape Town, South Africa in autumn 2024.

Maureen G. Reed is Distinguished Professor and shares a UNESCO Chair in biocultural diversity, sustainability, reconciliation and renewal at the School of Environment and Sustainability, University of Saskatchewan, Canada. Her research focuses on the social dimensions of climate change, biocultural diversity, and eco-restoration.

Dory Reeves currently lives in Scotland and is a freelance urban planner and gender specialist. Between 2008 and 2019 she was Professor of Planning at Auckland University, New Zealand. She is currently working on a project to develop a more community-based approach to equality impact assessments to make places and spaces more gender sensitive.

Mariamalia Rodríguez-Chaves has a PhD from the National University of Ireland and is a Senior Post-doctoral Fellow at the World Maritime University-Sasakawa Global Ocean Institute, Sweden. She also works for the Interamerican Association for Environmental Defense and is a member of the Group of Experts of the UN Regular Process and the Legal and Technical Committee of the International Seabed Authority.

Zhen Sun is Associate Professor at the World Maritime University-Sasakawa Global Ocean Institute, Sweden. Her research focuses on the law of the sea, international regulation of shipping, gender equality in ocean governance, climate actions and the protection of the marine environment.

Malena Rosén Sundström is Associate Professor in Political Science at Lund University, Sweden. She is the PI of the research project "Sweden as a Norm Entrepreneur: The Case of the Feminist Foreign Policy" and has published on Sweden's feminist foreign policy, norm entrepreneurship and external perceptions.

Heidi Walker's research applies intersectional and decolonial lenses to build inclusive impact assessment processes, responses to climate hazards, and climate change adaptation. She currently works as a research coordinator with Narratives Inc., where she supports community-led impact assessment projects and residential school truth-telling initiatives.

Marie Widengård is a researcher at the School of Global Studies, Gothenburg University, Sweden. She works in Sweden, as well as in Africa, Asia, Latin America, and the Caribbean., focusing on the politics of nature and resource governance, including land, forests, water, mining, fisheries, and energy.

ACKNOWLEDGEMENTS

We would like to thank all the contributors to this book who have met deadlines and responded to our requests with good spirit, despite the increasingly difficult academic environments in which they work. We thank our interviewees, who took time out of their pressured schedules to talk about their work, and we are especially grateful for their frankness, and to their interviewers for enabling this to be shared.

We are also indebted to the Swedish Research Council for Sustainable Development, FORMAS, and research grant 2018-01704, which has financed part of the time invested in the book and a share of the open access fees.

We thank our various editors, under the initial guidance of Annabelle Harris, for responding to all these pressures with patience.

INTRODUCTION

Susan Buckingham, Gunnhildur Lily Magnusdottir,
Karen Morrow and Martin Hultman

Contexts

As we write this introduction, the world is in a perilous state; it will be even more so by the time you are reading this. At the time of writing, climate emergency declarations have been adopted in 2,364 national, regional, and local jurisdictions, covering 1 billion citizens (Climate Emergency Declaration 2025). Climate scientists weep on live TV in the USA (witness John Morales on BBC in October 2024) when describing the devastating impacts of climate change-infused Hurricane Milton, an indication of the increasing intensity of weather activity in central America. A much publicised academic paper writes about us 'stepping into a new phase of the critical and unpredictable climate crisis' (Ripple et al. 2024), while The *Copernicus Global Climate Highlights Report 2024* confirms 2024 as the warmest year on record and the first to exceed 1.5°C above pre-industrial levels for the annual global average temperature (European Union-Copernicus 2025). The combined climate science knowledge of the IPCC has, for many years now, warned that climate change has evolved into a climate catastrophe requiring 'deep, rapid and sustained reductions in greenhouse gases' (IPCC 2023: 12). And yet, the newly elected President Trump withdrew the USA from the Paris Agreement.

The year 2025 marks the sixtieth anniversary since Frank Ikard, then president of the American Petroleum Institute, informed the organisation that the combustion of fossil fuels would 'cause marked changes in climate' (Franta 2018). Fifty years has gone by since Exxon's internal research reports confirmed global warming and accurately projected the catastrophic effects of burning their products, acknowledging the reality of anthropogenic

DOI: 10.4324/9781003461005-1

global warming early on (Supran and Oreskes 2017). Forty years have passed since scientists and presidents (from all parts of the political landscape) worked together to form the Intergovernmental Panel on Climate Change, in the hope that the knowledge from fossil fuel companies and universities would be turned into action, lowering the risk for catastrophic change on Earth (Ekberg et al. 2022). But, since this acknowledgement by international leaders, confirmed by numerous agreements, the emissions accumulated have more than doubled and continue to rise every year along with global temperature. The question that needs to be asked is, *why have we not acted* (Stoddard et al. 2021)?

If general cognisance of a changed climate as the new, and increasingly problematic, normal, and specific experience of its impacts has grown rapidly in recent times, knowledge of its deeper social systemic nature has been with us for much longer. Scholarship since the early 2000s, clearly pointing to the gendered nature of climate change and its consequences, has been at the vanguard of this (see, for example, Nelson et al. 2002, Maskia 2002, Denton 2002, Lambrou and Laub 2004). However, as is clear from the contributions to this volume, the systemic factors which shape human aspects of climate change and responses to it often remain under appreciated. This is a significant factor stymying global responses to the climate crisis at a worldwide level while, at the same time, engaging effectively with systemic concerns is prompting significant progress at national and local levels. On the whole though, the way in which individual states (and the international organisations of which they are members), are governing is taking us further and further from mitigating emissions and dealing with the climate catastrophe (Stoddard et al. 2021, Magnusdottir and Kronsell 2024). Scholarly and NGO research has made it ever more evident that that the real decision-makers in global climate governance are the companies who emit the most greenhouse gases. The finance sector, with the so-called petro-states (notably Azerbaijan, Qatar, Russia, Saudi Arabia, the United States, and Venezuela), are decisively shaping processes and outcomes in numerous ways (Ekberg et al. 2022). Petro-states in particular, exert influence by hosting some of the high-profile United Nations Framework Convention on Climate Change (UNFCCC) Conference of Parties (CoP) events. It is salutary, but unfortunately unremarkable in this context, to note that 1700 representatives from the oil and gas industry turned up to COP29, a number that dwarfed almost all other participants, and certainly those representing populations most vulnerable to climate change (Frost 2024, Morrow 2022). With a clear focus on the embodiments of this hegemony, masculinities' studies are paying more attention to the organisation, values, practices, structures and money of cis-gendered males (Anshelm and Hultman 2014, Mehrabi et al. 2024a) that characterise these actors. Cara Daggett (2018) has, insightfully, coined the phrase 'petro-masculinities' to define both the nature of the fossil fuel industry (in which a mere 22% of oil and gas employees and 17% of oil and gas senior management are women) and its practices. The industry, the governments that facilitate them, and the global system of climate governance that appears to be powerless in the face of them, are all petro-masculinised in a deeply hegemonic way and empowered by the 80% share of fossil fuel (oil, coal and gas) that makes up the global trade of energy.

When the previous book in this series, *Gender, Intersectionality and Climate Institutions in Industrialised States* (Magnusdottir and Kronsell 2021) was published, we were cautiously optimistic that climate institutions and climate relevant sectors had the potential to integrate social equity, equality, and justice issues into their work. Global grassroots movements had recently progressively pushed the global agenda towards climate, equity, and gender policy work. That book utilised feminist institutionalism and intersectionality as analytical tools to examine how far climate (and climate-related) institutions in several industrialised states and at the international level, recognised and understood the relevance of gender and other social differences with which gender is imbricated. The gestation of the previous anthology coincided with growing momentum on climate activism in the Global North, inspired by a wide range of diverse actors and innovative campaigns, such as: Leave it in the Ground (LINGO)'s anti-extractivism initiative from 2015 (LINGO n.d.); the youth-driven Fridays for Future school strikes which started in 2018 (Euronews Green 2023); Extinction Rebellion's occupation of central London and subsequent high-profile activities in 2018 (Extinction Rebellion n.d.); and We Stay on the Ground's anti-flying campaign from 2019 (We Stay on the Ground n.d.). At the same time, the spectacular growth in momentum of the MeToo movement since 2006 (me too. Movement n.d.), illuminating violence against women and bringing some of the worst perpetrators to court for rape and sexual harassment, appeared to give cause for optimism about real progress on combatting sexism and misogyny. These social movement mobilisations were also translated into a shift towards at least nominally feminist political developments, including the climate change-focused actions which are interrogated in this volume. The previous book established that dominant path dependencies reinforced ecological modernisation and gender stereotypes, which hindered the development of inclusive and just climate-related policies. It ended by posing the question of whether it is possible for policymakers – even assuming that they are sympathetic – to overcome institutional barriers and conventions in pursuit of the radical changes necessary to achieve gender and climate justice. It found that while gender mainstreaming remains potentially transformative, it had not had sufficient traction to influence gender relations, let alone approaches to climate change. This anthology pursues what happened when women, often propelled into power by social movement support, tried to take on some of the path dependencies and find ways to progress beyond ecomodernism.

The optimism and ambition of the 2010s

In several industrialised countries during the mid-2010s, a number of self-declared feminist leaders came to power, seeking to promote and enable ambitious and innovative legislation that would address the climate and other environmental emergencies. A number of these key feminist leaders agreed to talk to us about their experiences during this period and interviews with them are a central element of this collection. At the international scale, Christina Figueras was taking gender equality seriously at the UNFCCC, and the Wellbeing Government Alliance had brought together women leaders (Jacinda Ardern, Nicola Sturgeon, Sanna Marin and Katrín Jakobsdóttir) at COP26 to stress the need to focus on wellbeing rather than growth. Domestically in New Zealand, Jacinda Ardern had given climate change an unprecedented profile in her administration and introduced groundbreaking policies at the national level (Ricketts 2023). Meanwhile, at city level, when Mayor of Barcelona, Ada Colau, known for her encouragement of women's and grassroots community groups, cited her feminist principles and her promotion of collective action with local citizens and networks of cities, as influential in defending cities against the interests of big business, such as the automotive industry lobby and Airbnb. Marianne Borgen, when Mayor of Oslo was known for her ability to pursue effective cooperation and under her leadership the city won environmental awards for achievement in sustainable transport (Borgen and Colau have been interviewed for this volume). Anne Hildalgo, Mayor of Paris, used her position as chair of the international C40Cities network between 2016 and 2019 to establish the 'Women4Climate' Programme, designed to encourage young women to develop careers in climate policy. She has also achieved what has been called the 'most remarkable and successful programme to reduce car use in cities' (Clark 2024). The interviews contained in this volume provide important insights into the real-world experience and impacts of feminist leadership and feminist informed policies – and of the obstacles encountered in attempting to make change.

Climate change and a climate of change in Europe in the 2020s

Since our previous book was published in 2021, another kind of urgency has emerged. The adverse impacts of climate change are no longer to be deflected as concerns for 'elsewhere' and/or 'the future' – people, places, and infrastructures in Europe, North America, Australia, and New Zealand have also been badly hit by its effects. Climate-related risks have escalated rapidly, and the timescale set under the Paris Agreement for reducing CO_2 and other greenhouse gas emissions is now recognised by scientists as untenable. It is widely understood that cumulative emissions will heat the planet to an average of 2.1 degrees before we can succeed in beginning to reduce temperatures.

(UNEP 2024) A 3.1 degree increase in global average temperature is regarded as a realistic prospect if we do not transform our societies today and stop using fossil fuels (UNEP 2024). Even if we do succeed in bringing the trajectory of global temperature rise under control, the assumed ability to reverse the change wrought is at best debatable (Malm and Carton 2024).

The developments alluded to above have coincided not only with a price spike in gas and electricity attributed to Russiás war against Ukraine and its impact on Germany's ability to source cheap fossil gas from Russia, but also with an increasingly polarised political landscape. One impact of the latter has seen the rise of far-right political parties which reject climate science and a simultaneous rise in climate disinformation distorting public discourse in many Global North countries (Hultman et al. 2019, Vowles 2024). The countries (including at the time of writing for example, Sweden, The Netherlands, Austria, and Italy) that have been tilted further right, sometimes under the leadership of women premiers, outnumber those which have ostensibly moved in the opposite direction, such as Australia, France, and the United Kingdom. It is, however, salutary to note that in the latter two instances, both governments have 'moderated' previous climate – and other – commitments to appease voters wooed by right-wing populist parties (Brulle et al. 2024). While the IPCC's sixth assessment report prioritises climate equity and justice, the mechanisms for achieving this are unclear (and indeed beyond the remit of this scientific advisory body). It is, however, clear that influential actors can obstruct or facilitate the realisation of such ideas depending on their political preferences, thus ideological shifts, not least those in government, are centrally important in this area. In an overlapping ideological register, we see mobilisation against gender equality in countries exhibiting shifts to the right, such as Hungary and the United States (Mehrabi et al. 2024b). With all this in mind, it is telling that one of the first acts of the incoming centre-right government in Sweden in 2022 was to abolish its groundbreaking feminist foreign policy which is discussed in this collection. This is the political landscape the feminist leaders, interviewed for this anthology, have had to navigate and the contentious ground in which future climate action must somehow succeed.

All these factors influence climate and social policies, in turn affecting international law and institutions. In the face of such volatility, it is therefore instructive to turn to 'good practice' examples, often found at the local level, which have shifted the focus towards more constructive approaches founded on 'wellbeing economy' and inclusion and justice principles, many of which chime with feminist approaches and values. Nevertheless, it is clear to us that focusing solely on gender is insufficient to explain power relations and behaviour in climate politics. In consequence, we revert here to an emphasis on feminism: identifying that what is needed is a broad feminist perspective rather than a gender equality approach which, alone, has not been sufficient.

We understand this feminist perspective to be a holistic and non-combative approach, based on inclusive justice and on building broad collaborations.

Feminist leadership and approaches

Feminist approaches have for many years shaped grassroots movements, often focused on local environmental concerns but since 2017 we have seen a feminist approach increasingly embraced on larger structural scales, apparent in the remits of many of the leaders interviewed in this book and in networks between them. One promising development is the Wellbeing Economy Governments partnership (WEGo) initiated by feminist premiers (or recent premiers) of Finland, Iceland, New Zealand, Scotland and Wales (WEGo n.d.). This has promoted collaboration, expertise, and policy experience exchange on a range of pressing societal issues, including the environment. Feminist approaches to leadership are also becoming more apparent at national levels, although the resignations of four of the five WEGo premiers and retirement of the fifth by the end of 2024 raise questions about how dependent such initiatives are on individuals rather than their offices, as the aforementioned abandonment (for the moment at least) of Sweden's Feminist Foreign Policy exemplifies. At the local and regional levels, mayors self-identifying as feminists from Reykjavik and Oslo to Paris and Barcelona to Mexico City have introduced radical social and environmental policies. However, the volatility of politics warns us to be cautious in assuming that these gains can be secured, and we look for significant popular, structural and systemic change as evidence for more sustained, enduring responses.

This book examines some of these feminist initiatives in relatively rich countries from a multilevel governance perspective. The justification for this geographical perspective is multifaceted: rich countries – collectively – are influential on a global scale; they are also some of the highest contributors to global heating per capita. At the same time they are also already experiencing marked climate change impacts (EEA 2023). However, their achievements towards gender equality/gender justice (and other forms of social justice and equality) have not been consistent, and tend to have been conceived within a 'liberal feminist' framework which adjusts the individual to the prevailing system. The COVID-19 pandemic illustrated how a disruption to national economies undermines progress towards gender and other equalities, but equally, as Baruah and Burke explain in the Canadian context in this volume, introducing compensating social programmes for the duration of the pandemic showed that innovation and radical action is also possible. The COVID-19 pandemic also allowed people to experience life with much reduced pollution from aviation, road traffic, and shipping, which is what the majority science has been advocating (Le Quéré et al. 2020). Discouragingly though, the bounce back in emissions following COVID has been rapid (WMO 2021).

This book, then, interrogates the need for structural change and asks what lessons can be drawn from 'good practice' cases including avoiding the negative sides of mitigation which can entrench existing social inequalities and injustices if conscious efforts towards social justice are not made. We also explore the possibilities local communities demonstrate in leading the way towards more progressive climate strategies, since cities and grassroots initiatives often aim for more stringent climate strategies than can be reached at the national and international levels. The book also suggests that we need to move beyond gender and that multiple social inequalities can be eliminated by addressing climate justice – and other forms of environmental justice (and vice versa) – intersectionally. We do this in two ways: a) by identifying examples of what we consider to be good feminist practice in institutions with a climate remit (international, national, local/city/regional); and b) through methodologies which align with feminist principles which have had/have the potential to influence climate-related policy. The number of smaller states included in the book is not an accident, as these seem to be more agile, and have had elected leaders with more scope to be able to generate change than larger states.

At the time at which we publish this book, we feel that we are balancing on a 'pivot' point. This captures a moment of change in which feminists did achieve some governing influence in industrialised states, but which – for the moment at least – seems to have been short-lived. We believe that there are lessons to be learned at this point, including the vulnerability of these changes. We are particularly interested in how local and grassroots movements articulated with and inspired governments to achieve change. In order to explore examples of change through feminist leadership, we need first to establish what we mean by this.

Srilatha Batliwala has written extensively on the concept and elements, and describes feminist leadership as a:

> process of transforming ourselves, our communities and the larger world to embrace a feminist version of social justice. It's the process of working to make the feminist vision of a non-violent, non- discriminatory world, a reality. It's about mobilising others around this vision of change.
>
> *(The Feminist Leadership Project 2020)*

Feminist governance principles are designed to be transformative, reflective, caring, responsible, transparent, non-violent, inclusive, courageous and with zero tolerance for discrimination and the abuse of power (Action Aid n.d.; Fair Share of Women Leaders, n.d.; Centre for Feminist Foreign Policy, 2023). Before its website was taken down in 2024, as a result of the extreme online violence it was experiencing, the Centre for Feminist Foreign Policy had declared climate justice to be one of its five policy areas which prioritised advancing climate justice, held accountable those mainly responsible for the climate crisis, and supported those who had already experienced severe climate-induced changes in their everyday life. Further, in a briefing on climate-migration, it claimed that:

> The climate crisis grew out of a patriarchal system that is entangled with racism, white supremacy, and extractive capitalism. The unequal impacts of climate change only make it more difficult to achieve a gender-equal world.

Feminist approaches are also evident in the work of Elisa Morgera, the latest UN Special Rapporteur on Human Rights and the Environment, who threads intersectional analysis through her treatment of climate mitigation, adaptation and finance in her scene-setting report on the promotion and protection of human rights in the context of climate change (UNHRC 2024).

Aims and structure of the book

The main aims of this book are to:

- identify feminist-inspired good practice in policymaking at a variety of scales in industrialised countries which addresses the climate emergency, with a specific focus on strategies which prioritise justice;
- explore the feminist leadership attributes and practices which have motivated the prioritisation of greenhouse gas reduction practices;
- evaluate innovative and creative practices to initiate climate responsive policies;
- inspire, guide, and provide tools for policymakers, politicians, community activists, academics, and students to act on the climate crisis.

The book's starting point is a question of whether it is possible for policymakers – even assuming that they are sympathetic – to overcome institutional barriers and conventions in pursuit of the radical changes necessary for a gender-just climate emergency response. Below we explore different kinds of feminist leadership, practices and approaches, and use this as an organising framework to assess what it can offer communities, cities, countries, and the planet to address the climate emergency. We are aided in this by the inclusion of specially commissioned interviews with leading politicians whose feminist commitments have resulted in a fresh approach to creating policy which has the potential to address the climate emergency. These run alongside analytical and practice-oriented chapters based on research by both well-established and upcoming authors.

The book is organised first through scale, in that analyses, research and interviews with feminist climate leaders are addressed internationally and nationally/locally, by section. The final section considers methodologies which can be used in research, participation, and policymaking to enable more transparent, authentic, intersectional, and fair decision-making.

The climate leaders who self-identified as feminist who were interviewed for this book, referred without prompting to many of the qualities identified above which have emerged as themes throughout the chapters. The interviewees gave examples of their strategies and practices, which can be defined as feminist, when describing what distinguished their approach, and which will be considered in more detail in the closing chapter.

International

In the opening chapter, Rosen Sundstrom and Elgstrom focus on how feminist foreign policy (FFP) was practiced and implemented by officials at Swedish embassies around the world. Using a framework of practice theory, implementation theory, and feminist institutionalism, they ask how FFP was conducted in diplomatic work and to what extent it was implemented. Their in-depth interviews led them to the conclusion that there was a relatively high degree of successful implementation, largely guided by a pragmatic approach. Its practice was found to be sensitive to context, with different aspects of FFP being in focus in different locations. Sweden's previous work on gender equality in foreign policy helped officials to adjust their practices to the more ambitious framework of FFP. This research leads Rosen Sundstrom and Elgstrom to argue that that a feminist agenda such as a feminist climate policy (FCP) can be put into practice if the political will exists, if clear goals and instruments for a well-structured implementation process are provided, and if those doing the policy on the ground are closely involved in the process.

In Chapter 2, Laura Winther-Englesbak utilises a longitudinal panel dataset with data from 137 countries from 2009 to 2018 and employs fixed effects regression models to estimate the effect of women's political representation on countries' GHG (greenhouse gas) emissions. The chapter establishes that there is a clear trend that the climate footprint is lower in countries where more women are politically represented. The chapter's findings do not suggest that the responsibility for solving the climate crisis should be placed on women but instead that improving gender equality across societies can address challenges of social injustice and global challenges like the climate crisis.

In Chapter 3, Gill Allwood considers the achievements of the European Union – long held to be a leader in both gender equality and climate policies. However, there has been a failure to integrate the two, as the current European Green Deal does not seem to have been gender mainstreamed, nor to have valued care and caring. This misses an opportunity to make climate and other policy gender-fair, but also transformative more widely. Nevertheless, Allwood concludes that the EU yet has the potential to contribute to a feminist climate policy (FCP), although she argues that an EU FCP has to be more than a climate-focused version of FFP.

By observing the progress of gender equality – a key component of the UN Decade of Ocean Science for Sustainable Development – the Empowering Women project at the World Maritime University in Sweden finds examples of what the few women who have achieved leadership in one of the most masculinist research areas have been able to achieve, although these examples are sporadic and not necessarily sustainable. The team-authored research presented in Chapter 4 considers progress towards gender equality in ocean governance, using examples from selected intergovernmental organisations and non-governmental organisations and reflects on two research projects: specifically participatory action research with the International Council for the Exploration of the Sea (ICES) and research in Kenyan universities providing ocean-related education. These two projects provide nuanced insights into the barriers and challenges women face in the ocean sector and lead the authors to consider what is needed to achieve gender equality, and whether more systemic structural change is required through a feminist approach to the Ocean Decade.

In Chapter 5, Seema Arora-Jonsson examines how Swedish policymaking has addressed gender and power in environmental and climate initiatives. This is accomplished in relation to criteria for justice considered vital by feminist research, addressing structural inequalities such as ownership of land and collective organising as well as attention to intersecting dimensions of power, such as ethnicity, race, residence, age. 'Good practices' on gender equality are studied by focusing on organisations in Sweden that have sought to work towards gender equality in relation to the environment and climate in various ways. Arora-Jonsson argues that such projects on gender mainstreaming (albeit depoliticised) can contribute to making a space for, although not necessarily bringing about, change. By making dissonances apparent as well as providing a platform, gender mainstreaming projects can be a vital piece of the puzzle for establishing a context for change.

National and municipal

In Chapter 6, Hilda Rømer Christensen and Michala Hvidt Breengaard focus on the notion of Gender Smart Mobility, a concept developed during an EU project on gender and diversity in Smart Transport. The concept, and its practice, links the UN Sustainable Development Goals on green and climate friendly city developments, and gender and social equality. The chapter demonstrates how Gender Smart Mobility enables stakeholders, notably technical and planning staff in Danish municipalities, to address Smart Mobility in ways that go beyond the dominant notions of what is 'Smart', conventionally centred around economy, growth, and technology. It is a visionary, yet also complicated and multifaceted, concept to translate into practice. It shows how the visions of green mobility are met in daily municipal policymaking where transport and mobility have a low priority and where there is little tradition of addressing gender and diversity in local affairs. The chapter presents the lessons learned, the potential and resistances to embrace and practice the visions of Gender Smart Mobility and its aligned methods. It raises the question of how municipal and citizens' work on climate change politics can become an advantage not only for some groups but for the broader population, including how to collect and disseminate actions of linking green transport with social equality.

Bipasha Baruah and Andrea Burke consider in Chapter 7 how far Canada has been able to engage with de-growth, bearing in mind that Canada is an associate member of the Wellbeing Alliance, which aligns with de-growth. Arguably, both these strategies are necessary to address the climate crisis. While the authors thought that progress on de-growth had promise in the period marked by COVID, Canada has not been able to sustain this and policies have since regressed to growth.

Coralie Gueddes, Vivian Price and Linda Clarke, in Chapter 8, have taken the opportunity of a women's construction conference in London to explore how women employed in this male-dominated industry in Europe and North America approach their work differently. Their findings, that women appear to favour collaborative working and prosper through networks, as well as having a commitment to sustainability, suggest that bringing more women into these industries could change its gendered nature and also move it more towards sustainability.

In Chapter 9, Nanna Rask, Angelica Lundgren and Annica Kronsell apply a critical feminist theory of intersectionality to analyse the climate policy efforts of two Swedish cities: Gothenburg and Malmö. As cities committed to ambitious climate goals, they are regarded as frontrunners. Actors in both cities acknowledge the challenges posed by organisational structures as they have made efforts to integrate measures, attempting to work more holistically across sectors. Qualitative research reveals that while civil servants were committed to climate action, a dominant institutional logic rooted in a technocratic and de-politicised understanding of climate issues impeded the transition toward just and sustainable societies. Consequently, the chapter argues that persistent logics must be challenged and transformed by using approaches that better address the holistic character of climate change. Intersectionality serves as a valuable analytical lens, exposing prevailing logics and problematising seemingly apolitical interpretations in climate policymaking and implementation, thereby contributing to (re)politicising climate actions. It is concluded that municipalities must engage in ethical reflection, allowing time and space for introspection within their organisations. Fostering an intersectional ethics is found to be crucial for effectively tackling climate problems, leading to more equitable outcomes for both humans and the 'environment'.

Methodology

In Chapter 10, Marie Widengård explores a legal case, where a group of Swedish youths challenged their nation's climate policy through the lens of posthuman intersectionality. It examines the intertwined destinies of young individuals and natural forests against the backdrop of Swedish forestry practices, which favour young, even-aged production forests at the expense of biodiverse, old-growth forests. By arguing that such practices not only diminish the country's carbon sink capabilities but also infringe upon the younger generation's rights, the chapter underscores the need for legal recognition of the complex relationships between humans and nonhumans in the pursuit of more-than-human climate justice. The youths' legal challenge, calling for Sweden to uphold its climate commitments, epitomises a growing movement towards acknowledging the interconnected welfare of human communities and the natural environment. This analysis enriches the climate litigation literature, paving the way for legal strategies that tackle the complex challenges of our era by advocating for justice across all life forms.

Gunnhildur Lily Magnusdottir and Andres Melin offer reflections on how feminist-based participatory workshops can be used as decision support and increase understanding of justice in energy policymaking. Recent increased awareness of justice issues, including energy justice, does not automatically lead to radical changes in strategies, political prioritisations, or institutional practices at different levels of climate governance, as Magnusdottir and Melin explain in Chapter 11. This is particularly evident when it comes to energy policymaking, which is a field historically framed as a natural science topic often situated in a hegemonic masculine environment. In this chapter, the authors contribute empirically and methodologically to the energy justice debate by analysing, with the help of feminist literature, how a participatory assessment workshop can be designed to heighten awareness and understanding of energy justice and how different voices in society can be included, although they also use their example to highlight the obstacles that a feminist approach faces.

In Chapter 12, Paul Pulé, Ilaria Olimpico, and Uri Noy Meir argue that considerable changes in what we value and the systems that we create and maintain to support them are required if we are to create an era of life-sustaining futures. To this end, they discuss an emerging social arts practice – *Theatre and Stories that Reconnect* (TStR) – which they used in a pilot workshop in 2019 in Milan, Italy. In this, participants explored three forms of masculinities: *industrial/breadwinner*, *ecomodern*, and *ecological*. Using deep listening, creativity, and empathy, the method was designed to awaken greater care for the world. The chapter argues that the root causes of global crises are enmeshed with destructive masculinities' norms and calls for embodied pedagogies that prioritise masculinities that are relational and caring. The chapter argues that TStR is one method that can move participants from masculinist 'power-over' social constructs to empowered, life-sustaining, and ecologised futures of justice for all.

Heidi Walker, Amber Fletcher, Maureen Reed, and Nicholas Antonini also address masculine and Western cultural norms in Chapter 13. They see these as embedded in emergency response and climate adaptation planning which have contributed to an emphasis on technical, physical, and economic impacts, and solutions. Impacts on social and more intangible values – and the diverse experiences of people who respond to hazards – often garner significantly less attention. They propose that their role as feminist scholars is, in part, to build, model, and promote participatory tools that support change in policy, decisions, and practice within and across institutional levels. However, realising this potential will require more attention by researchers and other societal stakeholders and decision-makers to come together to facilitate meaningful dialogue and drive actions that break from path dependencies. They propose the use of a participatory research method – Photovoice – which involves participants taking photographs that reflect specific aspects of their lived experiences and subsequently use those photos for critical reflection and collective dialogue around salient community issues. Drawing on two empirical studies involving community experiences of wildfire and flooding in Canada, the authors discuss opportunities and challenges of Photovoice as a potential tool for countering gendered and hegemonic cultural path-dependence in climate institutions.

The book concludes with the editors reflecting on the challenges to feminist climate policy, practice, and leadership in industrialised states. It argues that, in the face of a populist politics whose leading figures are openly sexist and misogynist, an intensifying climate emergency, and the apparently growing global influence of the fossil fuel lobby and big-tech, we need a feminist climate approach more than ever. The examples given by the chapters and the interviews suggest the efficacy of feminist climate leadership and practice (for true feminist leadership is collective). The big question is how to we get (back) there.

References

ActionAid (n.d.). Ten Principles of Feminist Leadership https://actionaid.org/feminist-leadership; https://actionaid.org/feminist-leadership (Accessed 24 Jan 2024).

Anshelm, J., & Hultman, M. (2014). A green fatwā? Climate change as a threat to the masculinity of industrial modernity. *NORMA: International Journal for Masculinity Studies*, 9(2), 84–96.

BBC (2024). 'Meteorologist becomes emotional giving Hurricane Milton Update BBC 8 October 2024 https://www.bbc.co.uk/news/videos/cly5z4w388qo

Brulle, R. J., Roberts, J. T., & Spencer, M. C. (Eds.). (2024). *Climate Obstruction across Europe.* Oxford University Press.

Centre for Feminist Foreign Policy (2023). Feminist Foreign Policy Summit (Accessed 26 January 2024) NB This website has now been disabled due to its experience of online violence.

Clark, B. (2024). The Last Car Free Olympics? *Prospect Magazine* August 12, 2024.

Climate Emergency Declaration (2025). Climate emergency declarations in 2,364 jurisdictions and local governments cover 1 billion citizens - Climate Emergency Declaration https://climateemergencydeclaration.org/climate-emergency-declarations-cover-15-million-citizens/

Dagget, C. (2018). Petro-masculinity: Fossil Fuels and Authoritarian Desire. *Millennium: Journal of International Studies* 47.1: 25–44.

Denton, F. (2002). Climate change vulnerability, impacts, and adaptation: Why does gender matter? *Gender & Development*, 10.2: 10–20.

Ekberg, K., Forchtner, B., Hultman, M., & Jylhä, K. M. (2022). *Climate obstruction: How denial, delay and inaction are heating the planet.* Routledge.

Euronews Green (2023). From solo protest to global movement: Five years of Fridays for Future in pictures | Euronews (Accessed 07 Jan 2025).

European Environment Agency. (2023). Europe's changing climate hazards — an index-based interactive EEA report — European Environment Agency. (Accessed 7 January 2025).

European Union-Copernicus. (2025). *The 2024 Annual Climate Summary. Global Climate Highlights* European Union https://climate.copernicus.eu/global-climate-highlights-2024 (Accessed 21 Jan 2025)

Extinction Rebellion. (n.d.). About - Extinction Rebellion. UK (Accessed 07 Jan 2025).

Fair Share of Women Leaders (n.d.). Feminist Leadership. Heinrich Boll Stiftung. The Green Political Foundation https://www.boell.de/en/feminist-leadership (Accessed 24 Jan 2024).

Franta, B. (2018). Early oil industry knowledge of CO2 and global warming. *Nature Climate Change*, 8(12), 1024–1025.

Frost, R. (2024). *More than 1,700 oil and gas lobbyists at COP29: Which European delegations invited them?* | Euronews.

Hultman, M., Björk, A., & Viinikka, T. (2019). The far right and climate change denial: denouncing environmental challenges via anti-establishment rhetoric, marketing of doubts, industrial/breadwinner masculinities enactments and ethno-nationalism. In Forchtner, B. Ed *The far right and the environment* (pp. 121–135). Routledge.

IPCC (2023). *Summary for Policymakers. In: Climate Change 2023: Synthesis Report. Contribution of Working Groups I, II and III to the Sixth Assessment Report of the Intergovernmental Panel on Climate Change* [Core Writing Team, H. Lee and J. Romero (eds.)]. IPCC, Geneva, Switzerland, pp. 1–34, doi: 10.59327/IPCC/AR6-9789291691647.001

Lambrou, Y. and Laub, R. (2004). *Gender perspectives on the conventions on biodiversity, climate change and desertification.* United Nations: Gender and Population Division.

Le Quéré, C., Jackson, R. B., Jones, M. W., Smith, A. J., Abernethy, S., Andrew, R. M., ... Peters, G. P. (2020). Temporary reduction in daily global CO2 emissions during the COVID-19 forced confinement. *Nature Climate Change, 10*(7), 647–653.

Leave it in the Ground (LINGO) (n.d.). Home - LINGO Home - LINGO. (Accessed 07 Jan 2025).

Magnusdottir, G.L. and Kronsell, A. (eds) (2021). *Gender, Intersectionality and Climate Institutions in Industrialised States.* Routledge.

Magnusdottir, G.L. and Kronsell, A. (2024). "Climate institutions matter: The challenges of making gender-sensitive and inclusive climate policies" in *Cooperation and Conflict, 59*(3), 361–378.

Malm, A. and Carton, W. (2024). *Overshoot: How the World Surrendered to Climate Breakdown.* Verso.

Maskia, R. (ed.) (2002). *Gender, development, and climate change.* Oxford: Oxfam.

me too Movement. (n.d.). me too. Movement | Get To Know Us. (Accessed 07 Jan 2025).

Mehrabi, T., Hultman, M., Uldbjerg, S., & Xin, L. (2024a). Introduction: Gender and Climate Catastophe. *Women, Gender & Research, 37*(1), 24. https://tidsskrift.dk/KKF/article/view/150748

Mehrabi, T., Hultman, M., Uldbjerg, S., & Xin, L. (2024b). Gender and Climate Catastrophe. *Kvinder, Køn og Forskning* 1, 6–29.

Morrow, K. (2022). Cop26 and beyond: participation and gender – more of the same? *Transnational Legal Theory, 13*(2–3), 191–217. https://doi.org/10.1080/20414005.2023.2171347

Nelson, V., Meadows, K., Cannon, T., Morton, J., & Martin, A. (2002). Uncertain predictions, invisible impacts, and the need to mainstream gender in climate change adaptations. *Gender & Development, 10*(2), 51–59.

Ricketts, K. (2023). *On The Global Stage, Jacinda Ardern Was a Climate Champion, But Victories Were Hard to Come by at Home* - Inside Climate News.

Ripple, W. J., Wolf, C., Gregg, J.W., Rockström, J., Mann, M.E., Oreskes, N., Lenton, T.M., Rahmstorf, S., Newsome, T.M., Xu, C., Svenning, J.-C., Cardoso Pereira, C., Law, B.E., Crowther, T.W. (2024). The 2024 state of the climate report: Perilous times on planet Earth, *BioScience*, biae087, https://doi.org/10.1093/biosci/biae087

Stoddard, I., Anderson, K., Capstick, S., Carton, W., Depledge, J., Facer, K., & Williams, M. (2021). Three decades of climate mitigation: why haven't we bent the global emissions curve? *Annual Review of Environment and Resources, 46*(1), 653–689.

Supran, G. and Oreskes, N.J. (2017). "Assessing ExxonMobil's climate change communications (1977–2014)", *Environmental Research Letters*, 12, 084019.

The Feminist Leadership Project (2020). *Meet Srilatha Batliwala* https://feministleadership.org/2020/04/30/meet-srilatha-batliwala/ (Accessed 21 January 2025).

UN HRC. (2024). *A/HRC/56/46 Scene-setting report* https://docs.un.org/en/A/HRC/56/46

UNEP. (2024). *No More Hot Air Please.* UNEP Emissions Gap Report United Nations Environment Programme.

Vowles, K. (2024). *Fuelling denial: The climate change reactionary movement and Swedish far-right media*. Chalmers Tekniska Hogskola (Sweden).

We Stay on the Ground. (n.d.). About us - We Stay on the Ground. (Accessed 07 Jan 2025).

WEGo. (n.d.). Wellbeing Economy Governments (WEGO): Wellbeing Economy Alliance (Accessed 7 Jan 2025).

World Metrological Organization (2021). United In Science 2021.

PART I
Global

PART 1

Global

1

TO PRACTICE WHAT YOU PREACH[*]

Sweden's Feminist Foreign Policy in Diplomatic Work

Malena Rosén Sundström and Ole Elgström

Introduction

In 2014, Sweden became the first country in the world to declare that it would henceforth pursue a feminist foreign policy (FFP), which entailed a "systematic gender equality perspective throughout foreign policy" (Ministry for Foreign Affairs 2019:19). FFP has been described as a normative reorientation, challenging traditional gendered power hierarchies (Aggestam and Bergman Rosamond 2016). This novel foreign policy was introduced by a Red-Green government, under the leadership of Minister for Foreign Affairs, Margot Wallström, who has described how the policy was initially "met by considerable derision" (Rupert 2015). Being both praised and critiqued over the years, FFP has certainly contributed to bringing attention to Sweden's foreign policy in many parts of the world (Rosén Sundström and Elgström 2020, Rosén Sundström et al. 2021). It has also inspired a number of other countries to follow suit and adopt FFPs of their own.

In October 2022, the new Conservative-Liberal government announced it would not continue with FFP. Despite an increasing number of followers and the dismantling of the policy, Sweden's FFP between 2014 and 2022 remains the most comprehensive to date (Papagioti 2023). However, we still have limited knowledge about how, and how much, Sweden's FFP influenced the work "on the ground" by officials and diplomats, during the eight years it existed. The main research contribution to this area is a study by Ann Towns et al. (2023), which primarily focuses on Sweden's bilateral development cooperation. Hence, there are still gaps in our understanding of to what extent FFP was implemented in Swedish Embassies around the globe. Did

[*] This research is funded by Riksbankens jubileumsfond, Grant P19-0712:1.

DOI: 10.4324/9781003461005-3

FFP, as hoped for by the political leadership, come to permeate the everyday activities of the Embassies? Or did it become merely a label or a symbol while activities continued as usual on the ground? Relating to the aims of this book, what lessons could be learnt from Sweden's FFP about the practice of a (state-led) feminist policy?

As a relatively new phenomenon, Feminist Foreign Policy has attracted considerable scholarly attention. Research has been conducted on both the phenomenon in general and on FFP of individual states, not least Sweden's due to its pioneer status. Some of this research has also related Sweden's FFP to diplomacy, mainly digital diplomacy and public diplomacy. Isabella Karlsson investigates FFP and public diplomacy, with a practitioner focus. More specifically, she analyses how officials at the Ministry for Foreign Affairs (MFA), Embassies and the Swedish Institute, an agency connected to the MFA, "translate" FFP into their communication practices (2022). She finds that the officials tend to use the concept of gender equality in their communication, rather than the potentially more provocative term feminism. Katarzyna Jezierska (2022) makes a similar observation in her study of how FFP is communicated on social media by the Swedish Embassies in Poland and Hungary, two states where feminism is highly contested. Sweden has remained relatively silent on FFP and gender equality in its relations with these states, which Jezierska suggests "might be a sign of self-silencing, an attempt to avoid explicit confrontation" (2022: 97–98). In another study, Jezierska and Towns (2018) find that feminism was omitted in documents and public diplomacy on social media by the Swedish Institute, even after the adoption of FFP, in its projections of the "Progressive Sweden" brand. Karin Aggestam et al. (2022) explore how FFP was integrated into the digital diplomacy of the MFA. They argue that the MFA attempted to "project feminist foreign policy through the employment of a 'Feminist Power Sweden' narrative" (2022: 315).

In their assessment of the implementation of Swedish FFP, based primarily on document analysis and a survey, distributed to 1,098 Embassy staff (drawn from all Swedish bilateral foreign missions) with a 50% response rate, Towns et al. conclude that there was a marked increase in gender equality work among foreign policy agencies and Embassies after the FFP was launched. "FFP-implementation consisted of much more than simply re-labeling existing gender equality mechanisms as 'feminist'" (Towns et al. 2023: 12). Still, implementation was uneven and incomplete. While feminist terminology was more frequently used by the political leadership and ambassadors than by other parts of the foreign service, "FFP was never *merely* a label" (Towns et al. 2023: 17): it signalled the government's ambitions and it helped strengthen Sweden's international leadership on gender equality. We will refer to other findings from this study in our text.

Existing theories present different predictions on the extent to which the FFP should be expected to be implemented. Strands of implementation theory argue that bureaucratic actors with a long history and strong identity, such as a Foreign Ministry, tend to resist novel policy ideas and stick to their existing opinions and practices. Others claim that well-organised and carefully executed policy initiatives with strong political backing will be implemented if officials are motivated and feel a shared ownership of the process. Many institutionalists argue that existing institutions tend to be stable and "sticky" and that it is therefore difficult to introduce novel policy initiatives. Other institutionalists on the contrary predict a more positive response if key policy elements of new initiatives already form part of the diplomatic agenda ("path-dependency") (see e.g. Lowndes 1996 for a critical discussion on different forms of institutionalism). In this chapter, we confront this theoretical indecisiveness or puzzle by relating key elements of the theories to the case at hand.

The present study contributes to the knowledge about how, and to what degree, FFP was translated into practice by diplomats. The lessons learnt could provide useful input to the development of a Feminist Climate Policy. The study also investigates factors that seem to have promoted, or prevented, implementation within the foreign policy bureaucracy. It is open to all kinds of practices enacted by the practitioners, rather than focused on specific, predetermined types of practices. The practitioners we investigate are limited to diplomats and officials at Swedish Embassies. Three overarching research questions guide the analysis: 1) *How was FFP conducted in diplomatic work at Swedish Embassies?*; 2) *To what extent was FFP implemented at Swedish Embassies?*; and 3) *What factors have promoted, or prevented, implementation of FFP within the foreign policy bureaucracy?* The study is based on in-depth interviews with 16 Swedish officials and diplomats,[1] which were conducted in 2021, when FFP was still in use.

The chapter proceeds in the following way: Next, Sweden's FFP is introduced. This is followed by a section that briefly outlines the three theories applied: practice theory, feminist institutionalism and implementation theory. Thereafter, the methodology of the study is presented. The findings and analysis are divided into four sections: how FFP was introduced at the MFA, what the diplomats do when enacting FFP, how they do it and the extent to which FFP is implemented in the diplomatic work. In trying to explain the degree of implementation, we relate our findings to the logic and predictions of the various theories. The chapter ends with a summary of the findings and a concluding discussion, suggesting how FFP can inspire a Feminist Climate Policy.

Sweden's Feminist Foreign Policy

There is no singular definition of what Feminist Foreign Policy actually is and there is controversy over what it should entail, both in practice and theory (e.g. Weldon and Alwan 2021). Based on definitions of "feminism" and "foreign policy", Laurel Weldon and Christine Alwan suggest the following definition of

FFP: "a course of action towards those outside national boundaries that is guided by feminist principles and that seeks to solve problems defined in feminist terms, that is, problems of male dominance, gender inequality and the devaluation and denigration of those who do not conform to traditional gender stereotypes" (2021: 7). However, every state that claims to have Feminist Foreign Policy has a somewhat different take on what that course of action should be, as well as on the norms underpinning it (Zhukova, et al 2022).

The initiator of FFP, former Minister for Foreign Affairs Margot Wallström, made an early effort to clarify what Sweden's FFP was about, by presenting it as consisting of three "R:s": rights, resources and representation (Ministry of Foreign Affairs 2015). This meant that women's rights are human rights, women should be represented where decisions are made, including in peace negotiations, and women should have the same resources as men. A fourth "R" was subsequently added, reality, which stated that the policy should be based "on facts and statistics about girls' and women's everyday lives, and shall produce results in people's lives" (Ministry of Foreign Affairs 2019: 3). It was also stated that the feminist perspective should apply to *all areas* of foreign policy: "Throughout our foreign policy, including in peace and security efforts, we will apply *a systematic gender perspective*" (ibid. p. 3, emphasis added). Six specific focus areas were presented, to provide further concretion to what FFP would focus on. These consisted of: women's rights as human rights; freedom from all forms of violence, including sexual violence; participation in preventing and resolving conflicts and post-conflict peace building; political participation in all areas of society; economic rights; and sexual and reproductive health and rights (SRHR) (ibid. p. 4).

FFP provided Sweden with a very specific profile to its foreign policy. It contributed to raise attention for Sweden's foreign policy and on Sweden as a foreign policy actor – at least in the Western world (Rosén Sundström et al. 2021, Zhukova 2023). Sweden's FFP has been met with praise as well as critique in the international media (Rosén Sundström et al. 2021, Rosén Sundström 2022), and by officials and diplomats from other states (Rosén Sundström and Elgström 2020). Scholars have also raised points of criticism towards FFP. Academics and numerous journalists have brought up the ethical dilemma Sweden faced by pursuing FFP at the same time as being one of the leading arms exporters in the world, including supplying authoritarian states violating women's rights (e.g. Aggestam and Bergman Rosamond 2016, Nordberg 2015, Robinson 2021, Taylor 2015).

Practice Theory, Feminist Institutionalism and Implementation Theory

The present study draws upon practice theory, feminist institutionalism and implementation theory. As stated by Emanuel Adler and Vincent Pouliot, "there is no such thing as *the* theory of practice but a variety of theories focused on practices" (2011: 4). Adler and Pouliot define practices as "competent performances" and as "socially meaningful patterns of action, which, in being performed more or less competently, simultaneously embody, act out, and possibly reify background knowledge and discourse in and on the material world" (2011). Rebecca Adler-Nissen argues that a key assumption of practice theory is that "everyday actions are consequential in producing social life" (2016: 7). An important difference between practice theory and many other approaches is that the former "tries to overcome the dualism between interests and ideas by insisting that agents are not (necessarily) socialised into adopting certain norms; instead, norms are often performed rather than internalised" (2016). The focus is therefore on what people are doing and how they are doing it, rather than why (Bremberg and Danielson 2021: 45, cf. Adler and Pouliot 2011). Insights from practice theory will be used in our mapping and analysis of Embassy FFP activities.

Feminist institutionalism (FI), like historical institutionalism (HI), is interested in how institutions react to changes and how they produce their own rules, norms and culture. FI also applies some of the central concepts developed by HI, such as path-dependency and institutional stickiness. The former means that once institutions are created, over time it becomes increasingly difficult to reverse a trend, or "path", in policymaking. This leads to the latter concept: the institutions and their decisions become "sticky", meaning that they lock in certain rules and norms of behaviour, which also make them resistant to change (e.g. Krook and Mackay 2011, Miller 2021, Lowndes 2014, Waylen 2014).

Feminist institutionalism aims to unmask the gendered aspects of institutions. Vivien Lowndes (2014) suggests that Elinor Ostrom's concept of "rules-in-use" is a helpful way to deal with institutions, without having to make a distinction between formal and informal institutions: "Rules-in-use are best described as the distinctive ensemble of dos and don'ts that one learns on the ground" (Lowndes 2014: 688). They "may or may not have a formal manifestation. The researcher's task is to find out "how things are done around here" or "why is X done, but not Y" (2014). Combining practice theory and feminist institutionalism is useful when studying the practices – what is done? how is it done? – and the institutions' relative openness or resistance to change and their gendered aspects. Together they put focus on, for example, how gender is played out in the enactment of certain performances.

The study of implementation processes started in the 1970s when social scientists "discovered" that implementation of major policy decisions could not be taken for granted. Implementation, scholars argued, must be considered "an integral part of the policy process rather than an administrative 'follow up' from policy making" (Barrett and Hill 1984: 220). Attention was paid, inter alia, to the complexity of implementation processes and to the number, role and discretion of implementing agencies (Elmore 1978,

Nakamura and Smallwood 1980). The focus was on obstacles to successful implementation. In the 1990s, scholars started to focus on synthesising the existing, rather fragmented, empirical findings. Researchers produced lists of factors that were supposed to be associated with successful or failed implementation.

In this chapter, we have chosen to rely on the contribution of Paul Sabatier (1986), who distinguished six conditions for "effective implementation": clear and consistent policy goals; an adequate causal theory guiding the policy; a well-planned and structured implementation process; dedicated and competent officials; support from policymakers and interest groups; and the absence of socioeconomic changes that could undermine the policy. We believe that Sabatier's six factors serve to highlight important characteristics of implementation processes, not least regarding the interplay between political and bureaucratic actors.

Methodology

The study is conducted as a qualitative case study. A common objective of practice approaches, as described by Vincent Pouliot and Jérémie Cornut, is to strike a "balance between thick contextual understanding and conceptual abstraction" (2015: 301). In this regard, the study applies a bottom-up approach, to reconstruct the micro-processes of "the everyday, the mundane, the anecdotal", out of which practices are created (Pouliot and Cornut 2015: 302). Feminist institutionalism and implementation theory are concerned with both micro-processes and broader aspects, such as institutional norms.

The main method used to collect empirical material is semi-structured, in-depth interviews. All interviews were carried out in 2021 over Zoom. Two of the interviewees worked, consecutively, as Ambassador of Gender Equality and Feminist Foreign Policy at the MFA. The rest of the interviewees worked at Swedish Embassies, seven of them as ambassadors and seven in other positions. Ten interviewees were female and six were male. The Embassies where the interviewees worked were of diverse types in terms of the size of the Embassy, its geographical location (from all continents except Australia), and – to use a crude distinction – in both developed and developing states (which determines whether the Embassy works with development aid or not). Utilising self-reported interview data means that we may expect some bias towards reporting successful implementation. However, our promise of interviewee anonymity should partially diminish this tendency. We also conducted an interview with Margot Wallström, former Minister of Foreign Affairs, as well as with Eric Sundström, political adviser to Wallström who worked closely with her on developing the FFP.

The interviews were transcribed and the content coded into themes. Richard Boyatzis defines a theme as "a pattern found in the information that at a minimum describes and organises the possible observations and at a maximum interprets aspects of the phenomenon" (1998: 4). Furthermore, a "theme may be identified at the manifest level (directly observable in the information) or at the latent level (underlying the phenomenon)" (1998). The latent level can be related to feminist methodologies for studying what might not be visible or missing in the material, sometimes referred to as "studying silences". This means "'reading' what is not written, or what is 'between the lines'" (Kronsell 2006: 109). Hence, in the coding process, which involves several rounds of close reading and coding, potential "silences" will be taken into account.

The coding process started from noting all activities the officials bring up in relation to FFP, that is, what they are doing when enacting FFP and how they are doing it. Based on previous research on FFP and public diplomacy, particular attention was also paid to whether, how and when they used (or did not use) the word "feminist" in their practices.

Findings and Analysis

Introducing FFP in the Ministry for Foreign Affairs

The concept of FFP was introduced by Margot Wallström immediately after becoming new Minister for Foreign Affairs in October 2014. According to Eric Sundström (interview), political adviser to Wallström, there were two important aspects in the early days of their work with FFP: to learn from academic knowledge about a feminist perspective on foreign affairs and to embed the policy in the organisation. Wallström soon started to use the three "R:s" in speeches but the policy content was not fully outlined from the out-set. Several interviewees state that "many raised eyebrows" when the policy was first announced, "what was this about, we already paid a lot of attention to gender equality". Sweden had already been in the vanguard of women's rights and gender equality in international politics prior to its FFP, especially with regard to its development aid. However, almost every interviewee argues that FFP was definitely a levelling up in the foreign policy work with gender equality. As described by one Ambassador: "Gender equality was not any-thing new with FFP, it has been an important part of our work prior to FFP. But what was different was [...] that it came from the highest level. Margot Wallström was so personally engaged, it was a levelling up".

In early 2015, all employees at the MFA received a letter, inviting them to suggest contributions to FFP. This led to "a big engagement in all parts of the organisation", as described by one interviewee, and which was corroborated by most interviewees. "People sat down and wrote, and when you do that, abstract things become concrete. [...] More than 100 reports were sent in", says one of the ambassadors for Gender Equality and Feminist Foreign Policy (henceforth FFP ambassadors). She explains the impact of this process: "The first effect this had, was that a rather abstract concept became interpreted. [...] The second effect was that everyone was elevated to contributor, it became everyone's task and possibility".

In addition to these efforts to imbue cultural changes, in terms of increased engagement, some organisational changes were also introduced. The early months of 2015 saw the establishment of a small office in the MFA, led by the FFP Ambassador, whose task it was to help coordinate the process and to provide education for employees at all levels. One of these ambassadors describes the preference for a small support team with the idea that FFP was not to be a concern only for a certain group, but for everyone at the MFA. Every Embassy was, however, supposed to appoint one Gender Focal Point, who was to coordinate gender equality work and FFP (Towns et al. 2023: 68). The bottom-up approach, delegating the more specific aspects of FFP work to each Embassy, was combined with leadership and steering from the political level, hence more of a top-down process. In 2015, the MFA launched its Action Plan for FFP, outlining six focus areas (see above). Wallström also decided that the highest priority should be given to sexual violence in conflicts and women in peace processes, according to one of the FFP Ambassadors.

Things That Diplomats Do

According to Iver B. Neumann, three key tasks for a diplomat are "information gathering, negotiation and communication of one's position" (Neumann 2013: 3). Often, representation is added as another major theme (Jönsson 2022). Based on the interview material, the most frequently occurring types of practices can be described as belonging to the overarching tasks of representation and communication. In the material, two practices relating to communication are identified: communicating with the public (public diplomacy) and communicating with representatives of the host state, or political dialogues (communication with government, authorities and Ministry of Foreign Affairs). Other frequently occurring types of practices, mirroring both representation and communication, are the arranging of events by Swedish Embassies and participation of its diplomats in (externally) arranged events in the host state.

Communicating with the Public

"When I started working at the MFA [in the early 2000s], communication was something peripheral. Now it is really central", says one interviewee. As witnessed by previous research on FFP and public and digital diplomacy, communication through social media is important to Swedish Embassies, who have been present on social media since 2013 (Aggestam et al. 2022). As an example of a recent post made on social media by the Embassy, one interviewee mentions an article about Swedish female police officers, published in cooperation with the Swedish Institute. "We definitely have a strategy with social media. We want to convey the Swedish view on gender equality", says another interviewee who considers the host state to be "five to ten years behind Sweden" when it comes to gender equality, while the host state considers itself to be at the forefront; it is therefore important to tread carefully and not "lecture" how it should be done. The importance of avoiding "preaching" is brought up by several interviewees, in contexts ranging from very different to relatively similar to Sweden in terms of gender equality.

There is a wide variety of elements that Embassies prioritise in their communication with the public, which not only depends on the context of the host state but also on international developments and their effects on host states. One interviewee mentions how sexual and reproductive health and rights (SRHR) became more important after the re-introduction of the so-called Mexico City policy by the Trump administration. The policy denies US federal funding to development organisations which provide access to or information about abortion. The interviewee says, "this is not only about abortions, it also actualises family planning and other family-related matters". With regard to family policies, Sweden's parental leave arrangements have been present in the communication with the public at almost all Embassies, often in relation to the photo exhibition "Swedish Dads", which is presented below. Several interviewees bring up how the *Handbook – Feminist Foreign Policy*, or aspects of it, has served as a basis for communication with the public. *The Handbook*, first launched in 2018, is a comprehensive publication of 111 pages, outlining the goals of Sweden's FFP, but also facts, statistics and examples of concrete action taken. *The Handbook* has been translated to English, French, Portuguese and Spanish. It is an example of how the political leadership has guided and supported the implementation process with the help of structured information material.

Communicating with the Host State

The Swedish Ambassador in one state, which now has a Feminist Foreign Policy, has been consulted or invited to speak about FFP by the host state government and authorities at least 40 times. Another Ambassador brings up his/her experience from a previous posting in a developing state with a low degree of gender equality: "In a way it is easier when you also work with development aid to demonstrate what gender equality is. You *do* things, not only talk about them". The communication with the host state is, of course, to a large extent related to what kind of state it is, where it is in its work for gender equality and what focus areas of FFP that are the most important to Sweden exist in the host state. All six focus areas are represented in the material.

Gender-based violence has been a concern in political dialogue with some host states, not least due to increased domestic violence during the pandemic. Some interviewees argue that the current context is challenging for a country pursuing a Feminist Foreign Policy, but the consistent enactment of FFP in political dialogues "at least might serve to dampen negative trends". Another interviewee reasons along the same lines: "We have to fight, also in the EU. If we had not had this very clear policy, it might have been different".

Arranging Events

The interviewees bring up many different types of events arranged by their respective Embassies. Some of the events originated in the Ministry for Foreign Affairs in Stockholm, others are local initiatives. Two centrally originating events brought up are the Wikigap edit-a-thons and the photo exhibition "Swedish Dads". Wikigap edit-a-thons have been arranged by Swedish Embassies, in cooperation with Wikimedia, and local and/or international organisations, such as UN Women. The purpose is to make the internet more gender balanced, by adding articles on women. When it started in 2018, "[n] inety per cent of the content [on Wikipedia] has been created by men, and there are four times more articles about men than about women" (Ministry for Foreign Affairs 2019: 54). The *Handbook – Feminist Foreign Policy* describes Wikigap edit-a-thons as follows:

> The initiative began on International Women's Day, on 8 March 2018, and was carried out in the form of parallel edit-a-thons in almost 50 countries, from Sweden to Indonesia, Egypt and Colombia. More than 1 600 people took part, writing articles in over 30 languages. During the first three months of the campaign alone, participants wrote almost 4 000 new articles.
> *(Ministry for Foreign Affairs 2019: 55)*

The event has been repeated every year since 2018. The photo exhibition "Swedish Dads" has also been one of the more visible events arranged by Swedish Embassies. (It could be categorised as "communicating with the public" as much as "arranging events".) Based on photographs by the Swedish photographer Johan Bävman, "Swedish Dads is based on a series of portraits of fathers who belong to that small percentage of fathers who choose to stay at home with their children for six months or longer" (Embassy of Sweden in Japan 2022). It has been exhibited in 65 countries as of the end of 2022 (Embassy of Sweden in Japan 2022). Sweden is well known for its generous conditions for parental leave and the exhibition has become a way of promoting "gender equal Sweden". The photo exhibition has also been used locally in order to stimulate discussions on parental leave and many Embassies have organised local photo exhibitions and contests, with a focus on fathers in the host state (e.g. Embassy of Sweden in Armenia 2017).

The interviewees state that FFP has clearly affected events organised at or by their Embassies, like receptions, dinners and seminars. FFP influences how the functions are performed, "you make sure that there are not only men represented in the panel", as one interviewee says. It is also reflected in the topics of different events. Due to the varied contexts of the host states, seminars – and during the pandemic, webinars – have covered different FFP topics, reflecting this diversity. One interviewee gives an example of a seminar on women in the mining industry, another of a seminar on sexual and gender-based violence. A number of interviewees talk about seminars relating to the cultural sector. One Embassy has invited mainly female writers, musicians and designers and has brought Swedish culture workers to the host state for inspiration, since the culture sector in the host states is heavily dominated by men. Another Embassy has explicitly focused on gender equality in the film industry.

Participating in (External) Events

"It can be difficult to find female speakers, often you turn to Sweden to find that", says one interviewee. This is corroborated by several interviewees – it is common for local organisers of events to turn to the Swedish Embassy for female speakers or in order to incorporate a feminist or gender equality perspective on a topic. On this topic, one interviewee says: "On Monday, I will give a speech for women about to get involved in politics, about political empowering. [...] I have been the commencement speaker at several events of that type". Another interviewee mentions an invitation to talk about FFP with diplomats in training in the host state. One interviewee argues that there is a huge interest in Swedish gender equality policies, but sometimes Sweden is regarded as something of a utopia, with the implication that it is difficult for other states to reach its status: "Then it is important to talk about how Sweden became the relatively gender equal country that it is today", s/he says.

Sometimes Swedish representatives participate in events arranged by other states' Embassies, or in cooperation with them. One interviewee, in a non-EU host state, describes how the Embassies of EU states have cooperated around "16 days of Activism against Gender-Based Violence", an annual international campaign. One interviewee mentions the Spotlight Initiative, which is a cooperation between UN Women and the EU, also working towards eliminating violence against women and girls.

To summarise this section, we have demonstrated that Swedish Embassies, through their communication and representation, have carried out a large number of activities to spread information about gender equality and the FFP, and Sweden's efforts in this area. In this sense, the FFP has become implemented – practices associated with FFP have become regular parts of what Embassies do. The political leadership has provided instruments (information material and suggested events) to facilitate a structured implementation process, but we have also evidence of manifold local initiatives, showing that officials abroad have incorporated ideas from the feminist agenda in their patterns of activity.

How Diplomats Do What They Do: Focus on Gender Equality Rather Than Feminism

The interviewees work in varied contexts. Some of them describe the host state as strongly dominated by men. Several interviewees notice a backlash when it comes to women's rights and increased violence against women, due to the pandemic. A few interviewees do not consider the word feminist to be very controversial in their host states, but most of them talk about the challenges associated with feminism – and even with the concept of gender equality. One interviewee, who works in the EU context, says: "Hungary and Poland's problem with the concept of gender equality is noticeable every day. [...] I thought it would be uncontroversial to talk about gender equality in Europe, but it is not". Another interviewee says that the discourse in the host state is that "men can't be feminists". One interviewee argues that s/he felt more hopeful when working with FFP in a previous host state, a developing state, than at the present posting in Europe, "but it might be because you start from a lower level". Another interviewee, currently working in a developing country says: "For the people in general, Feminist Foreign Policy is a very abstract concept. We try to explain what Sweden is and why gender equality is important to us".

All interviewees agree that the content of the policy is more important than using the word feminist at any price. One interviewee describes it in the following way:

> "We don't use the concept feminism very much. Sometimes there is reason to emphasise that we have a Feminist Foreign Policy, even if we are not the only ones anymore. It is easier to talk about gender equality. [...] The important thing is what it contains. You can say it [feminism] a few times. But you have to get down to basics, to talk about equal pay for women and men, to include women in peace processes. There is a risk that people will not listen to you as much if you use the word "feminism", but if one uses it often it might become normalised".

Another interviewee describes the use (or not) of feminism in similar terms:

> "We are aware that it can be perceived as provocative. To me, it foremost means gender equality. What the word "feminism" can contribute with is to raise attention and questions, you get an opportunity to explain. For us in the foreign policy administration, it helps to structure and organise our work".

Several interviewees bring up that "the F-word has raised attention". One interviewee says that the fact that foreign policy was given a name beyond gender equality, to use Feminist Foreign Policy, was "more mobilising, it is clearer, easier to communicate". Some interviewees mention the three "R:s" – rights, representation and resources – as important for their work: "I use the three R:s a lot, to remember what aspects to focus on and to explain to others [what FFP is]". Another interviewee also points to the usefulness of the three R:s:

> I think it is an easy concept to communicate. I start all my presentations by saying that 50% of the population are women, but they don't have the same rights, representations and resources as the other 50% – therefore a Feminist Foreign Policy is needed, in order to work for a gender equal society. FFP is needed to create parity. It's easy to understand why it is important. We work for a 50–50-society. [...] No one goes against that, at least not openly.

One interviewee explains how s/he uses "feminist": "You refer to FFP when you give a speech, for example", but then "you talk about gender equality and female representation. You want to see change on the ground". However, sometimes the word "feminism" is avoided altogether. Another interviewee, who describes social media in the host state as containing a lot of hate speech against women, says that s/he rarely uses the word feminism: "If you say feminism to some, there is a backlash".

Our findings in this respect tally with the results from Towns et al.'s study of FFP-implementation. They conclude that the FFP label in part "was used to put old wine in new bottles – 'feminism' was used as a new way to present what had hitherto been referred to as 'gender equality'" (Towns et al. 2023: 36). Notably, officials in countries that receive Swedish foreign assistance seldom use the concept of "feminism" (ibid. pp. 93–94). This could arguably be interpreted as a type of resistance to FFP. However, we contend that it was more of a way to respond to scepticism against the concept of feminism in the recipient country; not to use the word itself made it easier to reach results. In the words of Margot Wallström herself:

> I have always thought of this [FFP] as a practical tool-box... My view is that real-world changes are needed for women. I don't care if they want to label this 'equality' instead – go right ahead as long as the right things are being done. If we can truly see that women are represented, and that they enjoy the same rights, then feel free to call it what you want.
>
> *(interview, authors' translation)*

The bottom-up approach has left ample room for the diplomats to enact FFP in ways they consider most useful in the host state. "It has been much of a bottom-up-approach. The contexts are different, the preconditions are different. [...] It is we at the Embassies around the world who decide what to do here". Almost every interviewee emphasises the importance of relating FFP to the local context: "The focus areas are steered by the action plans from Stockholm to some extent, but the Embassies work with the local context and local preconditions". One interviewee describes how the Embassy is very keen to check with local gender networks, activists, the media and representatives of different sectors in the host state, so as to "get it right" when working with different aspects of FFP, but that FFP is always central in the political dialogue. Locally employed staff at the Embassies are mentioned by some as valuable in the process of identifying what might work or not in the local context.

To summarise, the FFP is implemented in a way that takes the local context into account. This means that in many places and situations, more focus is put on concrete examples of gender equality action than on "feminism" as such. Feminism forms a background, while emphasis is put on how gender equality can be translated into the local context. The decision by the political leadership to have officials on the ground involved in both the shaping of the FFP as such and in the way it is designed locally seems to have been appreciated and to have resulted in local engagement in the implementation of the policy.

Explaining a High Degree of Implementation

In this chapter, we have demonstrated that FFP has been a relatively successful case of implementation. Almost all interviewees state that FFP permeates the daily work of their Embassies. FFP is not something "added" here and there, but reflected in all of their activities: "*Everyone* at the Embassy works with it [FFP] from their angle, in one way or another", says one interviewee. Another interviewee states that: "FFP is definitely integrated in all our work". The general view of our interviewees is that FFP has influenced all kinds of projects and activities. Furthermore, we have already established that activities associated with FFP have become regular elements of Embassies' communication and representation patterns, demonstrating implementation. How can this high degree of implementation be explained?

Practice theory is not engaged in attempts to explain degrees of implementation. Instead, we have looked for answers to our third research question in implementation theory and feminist institutionalism. Consistent with ideas in implementation theory of what factors explain "success", the political leadership provided strong guidance, clear goals and instruments for a well-structured implementation process. They offered a well-known ideological frame in terms of gender equality, a framework that was already widely used and acknowledged with the foreign policy bureaucracy, especially in countries where Sweden had a strong commitment to development assistance (cf. Towns et al. 2023). Politicians also involved officials in Stockholm and in

Embassies, in generating ideas on how to fill the FFP with concrete content. Local officials were also given a relatively free hand in initiating projects in their countries of employment. These traits seem to have developed a sense of ownership and created engagement among officials. According to Towns et al. (2023: 102), the leeway given to implementing officials to interpret FFP also resulted in a "meadow of practices where a thousand flowers bloomed" and to a variety in degrees of implementation across Embassies. The new policy was also supported by many influential NGOs in the area of gender equality. Thus, most of the "helpful factors for successful implementation" provided by Paul Sabatier are to be found in the case of FFP.

Feminist institutionalism puts the spotlight on how institutions once created tend to "stick" and how path-dependency might make them difficult to change. According to one interpretation, this indicates that bureaucracies are resistant to change; they prefer to continue with well-established policies and refrain from introducing novel initiatives that could create uncertainty and risk. Feminist institutionalism, however, also contends that existing policy constructs which work well tend to be continued and even developed. Therefore, the fact that gender equality was a well-established element of Swedish development aid and in its foreign policy in general, could indicate that FFP was considered a continuation and deepening of this policy and that this facilitated a smooth implementation of FFP. In brief, the predictions of Feminist institutionalism depend on what interpretation of the theory you choose (cf. Thomson 2018). Furthermore, FI emphasises the gendered aspects of the "dos and don'ts" that officials learn on the ground. Sweden's previous emphasis on gender equality in foreign policy likely made it easier for officials to develop or adopt their existing rules-in-use to the more ambitious, feminist agenda of FFP. Thus, the everyday activities of diplomats did not require learning something new but rather paying more attention to their gendered aspects.

Conclusion

This chapter has argued that Sweden's FFP demonstrated a relatively high degree of successful implementation, with the main findings being the following:

- FFP permeated everyday work of the diplomats to a large extent. Diplomats were encouraged to pay attention to the feminist agenda through concrete actions.
- At an early stage, the leadership invited all parts of the organisation to help to fill the policy with content. This led to a sense of ownership among officials and diplomats and enabled them to adjust/link the policy to their local contexts.

- The implementation of the policy was guided by a pragmatic approach: the word "feminism" was not always used, rather, the content of the policy mattered more than the packaging. Emphasis was on making a difference on the ground.
- The practice of FFP was very sensitive to context, with the effect that different aspects of FFP were in focus in different locations and at different times. This also links to the finding above – in some contexts the word feminism was used, in others it was not if it was regarded as an obstacle to the actual enactment of FFP.
- The policy benefitted from Sweden's previous work on gender equality in foreign policy and the officials' knowledge in this area. This helped officials to adjust their practices to the more ambitious framework of FFP.

Sweden's FFP (2014–2022) is an example of a state-led feminist policy. The relatively successful implementation of FFP demonstrates that a feminist agenda, such as a Feminist Climate Policy, can be formulated and also put in practice, if the political will exists, if clear goals and instruments for a well-structured implementation process are provided, and if those "doing" the policy on the ground are closely involved in the process. However, it is important to stress that Swedish foreign policy officials had a long tradition of work on gender equality to draw upon when the new policy was introduced. Also, in addition to the factors contributing to its success as mentioned above, FFP was backed up by a larger structure, i.e. through a feminist government, making its standing stronger than if simply pursued on its own.

Notes

1 A discussion of the selection process follows in our methodology section.

References

Adler, Emanuel and Pouliot, Vincent (2011). "International practices", *International Theory*, 3 (1): 1–36.

Adler-Nissen, Rebecca (2016). "Towards a Practice Turn in EU Studies. The Everyday of European Integration", *Journal of Common Market Studies*, 54 (1): 87–103.

Aggestam, Karin and Bergman Rosamond, Annika (2016). Swedish feminist foreign policy in the making: Ethics, politics and gender. *Ethics and International Affairs*, 30 (3): 323–334.

Aggestam, Karin; Bergman Rosamond, Annika and Hedling, Elsa (2022). Feminist digital diplomacy and foreign policy change in Sweden, *Place Branding and Public Diplomacy*, 18: 314–324.

Barrett, Susan and Hill, Michael (1984). Policy, bargaining and structure in implementation theory, *Policy and Politics*, 12: 219–240.

Boyatzis, Richard E. (1998). *Transforming qualitative information: Thematic analysis and code development*. London: Sage.

Bremberg, Niklas and Danielson, August (2021). Communities of practice and the everyday making of EU foreign and security policy in Bremberg, N.; Danielson, A.; Hedling, E, and Michailski, A. (Eds.) *The everyday making of EU foreign and security policy*. Cheltenham: Edgar Elgar Publishing. (pp. 37–55).

Elmore, Richard F. (1978). "Organizational Models of Social Program Implementation", *Public Policy*, 26: 185–228.

Embassy of Sweden in Armenia (2017). "'Swedish Dads' and 'Armenian Dads' photo exhibitions on display at Republic Square Metro Station". https://www.swedenabroad. se/es/embajada/armenia-yerevan/current/news/swedish-dads-and-armenian-dads-photo-exhibitions/ (retrieved 3 March 2023).

Embassy of Sweden in Japan (2022). Travelling Photo Exhibition 'Swedish Dads'. https://www.swedenabroad.se/fr/ambassade/japan-tokyo/current/calendar/ exhibition-travelling-photo-exhibition-swedish-dads/# (retrieved 3 March 2023)

Jezierska, Katarzyna (2022). Incredibly loud and extremely silent: Feminist foreign policy on Twitter, *Conflict and Cooperation*, 57 (1): 84–107.

Jezierska, Katarzyna and Towns, Ann (2018). "Taming feminism? The place of gender equality in the 'Progressive Sweden' brand", *Place Branding and Public Diplomacy* 14 (1): 55–63.

Jönsson, Christer (2022). "Theorising Diplomacy", In B.J.C. McKersher, *The Routledge Handbook of Diplomacy and Statecraft*. 2nd edition. London: Routledge.

Karlsson, Isabelle (2022). "'We try to be nuanced everywhere all the time': Sweden's feminist foreign policy and discursive closure in public diplomacy", *Place Branding and Public Diplomacy*, 18: 325–334.

Kronsell, Annica (2006). Methods for studying silences: gender analysis in institutions of hegemonic masculinity in Ackerly, B. A.; Stern, M. and True, J. (Eds.) *Feminist methodologies for international relations*. Cambridge: Cambridge University Press.

Krook, Mona Lena and Mackay, Fiona (2011). *Gender, politics and institutions: Towards a feminist institutionalism*. Basingstoke: Palgrave Macmillan.

Lowndes, Vivien (1996). "Varieties of new institutionalism: A critical appraisal", *Public Administration* 74 (2): 181–197.

Lowndes, Vivien (2014). How are things done around here? Uncovering institutional rules and their gendered effects, critical perspectives. *Politics & Gender* 10 (4): 685–691.

Miller, Cherry (2021). "Parliamentary ethnography and feminist institutionalism: gendering institutions – but how?", *European Journal of Politics and Gender*, 4 (3): 361–380.

Ministry for Foreign Affairs (2015). *Swedish Foreign Service action plan for feminist foreign policy 2015–2018 including focus areas for 2016*.

Ministry for Foreign Affairs (2019). *Handbook Sweden's Feminist Foreign Policy*.

Nakamura, Robert and Smallwood, Frank (1980). *The Politics of Policy Implementation*. New York: S.t Martin's.

Neumann, Iver B. (2013). *Diplomatic Sites: A Critical Enquiry*. Oxford: Oxford University Press.

Nordberg, Jenny (2015). Who's Afraid of a Feminist Foreign Policy?, *The New Yorker*, 15 April 2015.

Papagioti, Foteini (2023). *Feminist Foreign Policy Index: A Qualitative Evaluation of Feminist Commitments*. Washington, DC: International Center for Research on Women.

Pouliot, Vincent and Cornut, Jérémie (2015). Practice theory and the study of diplomacy: A research agenda, *Cooperation and Conflict*, 50 (3): 297–315.

Robinson, Fiona (2021). Feminist foreign policy as ethical foreign policy? A care ethics perspective, *Journal of International Political Theory*, 17 (1): 20–37.

Rosén Sundström, Malena (2022). Inspiration or provocation? Sweden's Feminist Foreign Policy in national newspapers in EU Member States, *European Foreign Affairs Review*, 27 (2): 283–306.

Rosén Sundström, Malena and Elgström, Ole (2020). Praise or critique? Sweden's Feminist Foreign Policy in the eyes of its fellow EU members, *European Politics and Society* 21 (4): 418–433.

Rosén Sundström, Malena; Zhukova, Ekatherina and Elgström, Ole (2021). Spreading a norm-based policy? Sweden's Feminist Foreign Policy in international media, *Contemporary Politics* 27 (4): 439–460.

Rupert, James (2015). "Sweden's Foreign Minister Explains Feminist Foreign Policy. Margot Wallström and Colleagues Face 'The Giggling Factor'", *United States Institute of Peace*, 2015-02-09. https://www.usip.org/publications/2015/02/swedens-foreign-minister-explains-feminist-foreign-policy (retrieved 11 December 2022).

Sabatier, Paul (1986). "Top-down and Bottom-up Approaches to Implementation", *Journal of Public Policy*, 6: 21–48.

Taylor, Adam (2015, October 7). Sweden's subtly radical 'feminist' foreign policy is causing a stir, *The Washington Post*, 7 October 2015.

Thomson, Jennifer (2018). "Resisting gendered change: Feminist institutionalism and critical actors", *International Political Science Review* 39 (2): 178–191.

Towns, Ann, Bjarnegård, Elin and Jezierska, Kasia (2023). *More Than a Label, Less Than a Revolution: Sweden's Feminist Foreign Policy*, EBA Report 2023:02, The Expert Group for Aid Studies (EBA), Sweden.

Waylen, Georgina (2014). Informal institutions, institutional change, and gender equality, *Political Research Quarterly*, 67 (1): 212–223.

Weldon, S. Laurel and Alwan, Christine (2021). "What is Feminist Foreign Policy? An exploratory evaluation of foreign policy in OECD countries". Paper presented at *International Studies Association (ISA)*, April 2021.

Zhukova, Ekatherina (2023). "Postcolonial logic and silences in strategic narratives: Sweden's feminist foreign policy in conflict-affected states", *Global Society*, 37 (1): 1–22.

Zhukova, Ekatherina; Rosén Sundström, Malena and Elgström, Ole (2022). "Feminist foreign policies (FFPs) as strategic narratives: Norm translation in Sweden, Canada, France, and Mexico", *Review of International Studies*, 48 (1): 195–216.

Interview 1

CATHERINE MCKENNA, FORMER MINISTER OF ENVIRONMENT AND CLIMATE CHANGE, CANADA: INTERNATIONAL AND NATIONAL ROLE IN CLIMATE POLICY

Interviewed by Dory Reeves and Julie MacArthur,
12 October 2023

Introduction: The Canadian context

Climate leadership in Canada is highly politicised due to the significant role that fossil-fuel dominated energy industries play in the federation's regional political economies – totalling nearly 12% of GDP at CAD309 billion in 2022 (Natural Resources Canada (NRCan), 2024, p. 7; Carter, 2020). Canada is the second largest country, by area, in the world with the third highest GHG intensity per capita in the OECD (after Australia and the United States), and fifth highest total emissions in the OECD (OECD, 2023). The country's geography, climate and culture of high energy use have led to significant challenges related to decarbonising transport, buildings and the energy sector, and this is without even scratching the surface of addressing the history and practice of colonisation and its effects on Indigenous (First Nations, Metis and Inuit) peoples. Canada's rapidly expanding population consists of an ethnically diverse 40 million people in 2023, of whom the fastest growing and youngest segment is Indigenous peoples.

Subnational variation in resource endowments and the federal nature of environmental and resource governance leads to deep political and economic cleavages between the 'petro-provinces' of Alberta and Saskatchewan and those with more diversified economies or large hydropower or mixed energy systems such as Québec, British Columbia and Ontario. Despite this, a series of federal level plans have made commitments to climate action through, for example, the Pan-Canadian Framework on Clean Growth and Climate Change (2016), and the 2030 Emissions Reduction Plan which sets a target of 40–45% emissions reductions below 2005 levels by 2030.

DOI: 10.4324/9781003461005-4

Catherine McKenna served as Canada's Minister of Environment and Climate Change from 2015–2019 where she led the Pan-Canadian Framework on Clean Growth and Climate Change (2016), and Minister of Infrastructure and Communities (2019–2021). She was appointed to the ministerial role shortly after being elected for the first time becoming the first woman to represent her electoral district (Ottawa-Centre). She is currently the founder and principal of Climate and Nature Solutions, an advisory firm that works with various institutions to scale practical climate- and nature-based solutions. McKenna has also been in international climate leadership roles launching "Women Leading on Climate" at COP26 and was the Chair of the UN Secretary-General's High-Level Expert Group on Commitments of Non-State Entities, which delivered its key recommendations report to COP27 in Sharm-el Sheikh.

During her tenure as Minister of Environment and Climate Change, McKenna was the target of a range of abuse and misogyny for her leadership role in climate policy. She was assigned a Royal Canadian Mounted Police (RCMP) security detail due to the persistent and systemic verbal abuse and threats to her and her family both online and in person. Despite this deeply personalised opposition she has and continues to play a strong role at national and international levels drawing together women-centred climate policy action.

What motivated your work in climate policy, leadership with the Canadian Federal Government?

I've always cared about nature and the planet. I'm a swimmer and someone who does a lot outdoors. So it was something that personally I care about. I started a charitable organization, Canadian Lawyers Abroad, that worked on environmental issues but not climate issues full time.

I came to [climate policy] because I was appointed Minister of Environment and Climate Change. I was the first Canadian Minister with Climate Change in the title, which was a clear departure from our previous government. It really became something I was immersed in very quickly, because I was a minister just a few days before I went to the Paris climate talks. When I came back [from Paris] we had to do the work and get a climate plan in Canada. I felt that there was an opportunity for me to play a role because I had worked in negotiations before, and Canadian people were excited to have Canada there.

The real knowledge about Climate policy came when I was minister. It was a massive learning curve. It actually helped sometimes that I wasn't a climate expert, because I feel like climate environmentalists often talk in a way that no one understands, including me. When I started, they were using all these acronyms and words like mitigation and adaptation and now we use net zero; they are really not understandable to folks and they aren't motivational. I also learned that it is really important to translate climate into a way that motivates people. As much as I was a policy expert, understanding the policy, the only way you can motivate people is actually reaching people. So you have to ... get to people and motivate them to take action, including supporting governments on difficult things.

Could you tell us about the importance of your feminism and intersectionality to the work on climate?

It's actually very practical. I'm feminist, and I've worked on gender issues, including human rights and social justice. But when I went to the climate negotiations, I felt like the people doing the most were all women. These included Christiana Figueres, the UN Framework Convention on Climate Change (UNFCCC) head, Laurence Tubiana, the French climate Ambassador, Mary Robinson (2018), who was bringing in the social justice angle, and Jennifer Morgan, then at Greenpeace. You also had Sharan Burrow who was bringing the labor movement angle, young environmentalists that were really pushing hard for change, and you had Indigenous women. So there's this whole network of women who are 'the high ambition coalition' (see Profiles of Paris PP). All these women were pushing really hard. And I thought, Okay, this is a network we need to activate in a more strategic way. It all came together organically.

But we need to really think about how we bring together women to support each other to raise ambition, and also to help mentor women who are doing great work in their communities and may not have access to support, or in some cases might even be under threat. It was a very practical thing. It wasn't a theoretical thing. All these women are amazing, and actually, if they all ran things, we'd be way further ahead.

Now I am coming back to that because transformational change happens in a variety of different ways. So yes, it is getting off fossil fuels. But sometimes you need preconditions, and you need courageous leadership, and I often think it is the women that are the most courageous. They understand better than anyone what is at stake when it comes to their children and their communities. And they're the most practical.

We also did 'women kicking it on climate' when I was Minister (MRF 2018). We brought all these women together to talk about [climate action] 'I think even my public servants at the time thought this is a waste of money'. I said that's not to celebrate ourselves, it's to say, 'what are you doing, and how do we support you? And what's the next agenda?'

The UN Secretary General asked me to take on a very large volunteer job on the net zero commitments, so it's been harder to focus on this. So in the Paris Agreement, we really worked on recognizing the disproportionate impact of climate change on women. There's a gender action plan. I've supported the UNFCCC as part of that plan to get half the negotiators to be women, including in senior roles and they found it's actually worth it. They do analysis of what happens at COPs. One of the funniest things is that male negotiators were more likely to talk twice as long (McKenna and Jaffe, 2022). Everyone was failing in making the gender quota; now in Canada our negotiating team was mainly women, so that was awesome.

We also have a feminist foreign policy, and we do a gender-based analysis which I required, when looking at our policies. And then we trained women negotiators, so we were really focused on this and really supporting this agenda because there are forces against a gender agenda within the COP. I mean it. It has to all be mainstreamed and needs 'buy in' because there's some countries who do not believe in gender equality.

Which of the policies you've been instrumental in developing regarding climate that were the most successful?

You need to do real things and policy should lead to real things. I think just using a gender-based analysis is a really [important initiative], and that is more a testament to our government. But a gender-based analysis in particular, I think, is important, because you don't really necessarily understand how policies that you think are gender neutral often disadvantaged women. Let's take an international example. So there's some massive flooding, and there's an evacuation: who gets on the bus first? It's the men. And who are the most vulnerable? It's often single mothers with children.

But in the Canadian context, there were so many things we did. In the job space women entrepreneurs were not being supported in the same way as men. They had good ideas. Or Indigenous women who are medicine keepers. I mean, obviously, they're disproportionately affected because they're in these communities and the impacts of climate change. It's very hard to find traditional medicines, and those communities often are flooded. So there are multiple impacts. So I thought that that policy was really important.

But, as I say actually bringing together women and celebrating women and recognising women, and then saying, okay, we're now calling the shots a bit more. And I found that women were really excited because it's not that people didn't notice, but they don't always link it. And then you often feel in your own country that you're isolated. I remember the Japanese minister who was the only woman in the cabinet. Or you know you're in your community fighting for clean air and clean water, and that is not an easy place to often be. I think that [creating] networks of women is also important, it's not just work at the policy level.

What personal qualities do you think are needed for developing and implementing the successful policies that you've just been talking about to address the climate emergency?

I don't know if there's a gender angle. I'm just very clear about what needs to be achieved, because I think in climate you can get very distracted. I generally work in threes, although with our [UN Credibility Matters] report we had ten recommendations. But I think you just need to be very clear on what are the most important things you need to do, and every single day you need to wake up and require people to work on that. Whether it is your political team, the public servants, or even yourself, you need to be focused on landing outcomes, real outcomes. Otherwise, you can have every day disrupted by millions of things.

On inclusivity, who do you need to be talking to? And who should you be bringing around the table? I was thoughtful about pointing to women, but it wasn't actually that hard. The opposite is that we need to find a binder [full of women to pick from]. Someone's gonna give me a binder because I don't know where the women are. There's millions of women that are amazing. I guess I have trouble finding men in this context, but maybe I'm hyper aware of it. I did some pieces on women getting into politics called 'Run like a girl' (McKenna, 2018).

When I was at COP22 there was this really amazing Inuit woman, young woman, Maatelii, (Maatalii Aneraq Okalik, National Inuit Youth Council President at COP21) and she was part of our delegation. I'm [a] pretty usual suspect in the sense that I'm the minister, and you know, Ministers always wanna say loads of things. But who's gonna be compelling enough to motivate people. And so I told my team I wanted her to speak, and they said, 'We'll ask the UN'. Of course, the UN said no, so I was like, 'Okay, that's great'. So I just brought her up with me because, what are they gonna do? Are they gonna pull us off stage? So I just said my opening words and then, 'actually, I'm gonna hand it over to Maatelii, cause she's gonna talk about her homeland'. So it was over to her. We wrote about this in a Scientific American article about giving more women the microphone at COPs (McKenna and Jaffe, 2022).

And then the last thing is you have to fight entrenched interests and people often don't wanna do that. I'm not saying women don't wanna do it. I just think people don't wanna do it. And often it's men because they are part of the entrenched interest. But you have to fight entrenchments and that is not the most fun part of my job. I was constantly under attack, but it was really important to call things out. There were personal attacks I called out, but also oil and gas as they need to be held accountable. I worry about complicity or doing things that are easy because we don't wanna do [the hard things]. The things that are hard that everyone talks about are tripling renewables. Oh, okay, great. We gotta phase out fossil fuels because tripling of renewables is not gonna solve the problem in the timeline we have. To be honest [some of the things we did] actually worked quite well, because I felt like we were able to achieve things, and it was very clear to folks where I stood. Like it or not, and I would try to be a reasonable person, and I feel like, you know, I could go for a beer with anyone. But I also said 'this is what we're getting done here, folks' If we don't do that in climate, I'm not at all sure how we'll do it, because there's always reasons [not to act]. Hard things are hard. So there's always reasons not to do things.

What are the main obstacles to developing socially and economically, just climate policies in Canada?

Entrenched interests are there, so that's a real problem. Because fossil fuel company goals and interests are not the same interests as those of workers. Sometimes they frame it like that though, so it gets confusing. Their interest is actually to make as much money as possible, to extend the life of fossil fuels for as long as possible, and they make other people pay to clean up their pollution. So that is a problem, and you have to fight that. Often our financial interests are tied to that. That's not even me trying to make some real, deep economic statements, it is the reality.

But the flip side to that is, there are workers and communities that rely on fossil fuels for their well-being right now, and if you don't recognise that you are also gonna run into massive problems because people need to put food on the table and they want a job. Generally, they care about climate change, even if they don't call it climate change. They care about clean air and clean water and good jobs and the economy growing. And you have to be careful that you don't ignore that; you listen to workers and work with them in their communities. Where there are opportunities is really important to know.

With the coal phase-out we had fights internally with our department of finance because I said, 'Okay, well, we're gonna have to make significant investments, we have to think of policies', I said, 'No, no, we're actually getting rid of their jobs. It's called phase out of coal folks'. And so we actually sent people, including labor leaders, to these communities. And 700 people would come to an auditorium, and they'd be upset and angry and worried. And I think they were shocked that the Federal Government would actually come and have people that would actually listen to them, including people that represented workers. That didn't make everything happy necessarily, but at least they said, 'Okay, people are taking it seriously'.

People aren't dumb. So I think they recognise things have to change. But what does that mean for them in their community? And they ask for things like, of course, early retirement and retraining. But they also wanted investments in infrastructure in their communities that would allow them to have other opportunities. And that's totally fair. So I think we have to be realistic. We have to not be like, what did my dad say? Smoked salmon socialists who sit around and like having your latte in downtown Ottawa or Toronto and be like, oh, we're all gonna transition. And you know, we don't care about people and jobs. You can't get policy done, you actually can't get re-elected [with that attitude]. That's a tension in our country. Because I think there are a lot of people who want to portray these [climate policies] as anti jobs. I actually think that they're pro jobs because there are more jobs in the clean economy. And we're gonna have to compete with the US. So all the jobs are gonna go there, or to China or Europe. I think there's also a lack of imagination in government. I actually found it very hard to get things done, so you would say, 'this is what needs to be done [based on] the science'. Then this is what fossil fuel interests want and, oh, that's probably terrible. Environmentalists generally don't even go as far as the science, because they know no one will do that. Then [t]here are the bureaucrats. I love the public service, but they are risk averse in a very risky way when it comes to the planet because they're like, okay, we'll do a policy over here. For me, that's gonna do nothing.

Take transit, you're not investing any real money, so no one's gonna believe that this is real and it's not gonna be real cause you haven't invested any money to do this. And I felt this on every policy: everyone's like, 'Oh, my gosh, McKenna is so difficult'. No, actually, I'm a realist. So I know where there's limits [to] what we gotta do, but it's transformational change we need. I really think there's a lack of understanding about how risky it is to do incremental change in a place where we need transformational change because it creates this idea that you're doing things, even if you're not really doing things that are having real impacts.

In Canada, we have a climate plan, but we're still missing some of the big planks, because they're really hard. It's like a cap on emissions from fossil fuels. We're not meeting our targets and everyone knows that. [Fossil fuel companies] have had lots of time and notice, and they're making loads of profit. So sending it back to their shareholders and doing executive conversations just call them out like they're bad guys. I think that the way we make decisions is not aligned with what is needed. And people think that's okay because they don't want to be too rash. But doing that is so risky.

As an example in Canada when you do memorandum to Cabinet, the ministers, of course, direct them, but it's really drafted by public servants, and they do three options. How is this even possible? It's like, we're in Goldilocks, land too hot, too cold, just right? Okay, [usually it is] too expensive, not enough, just right. But just right is not just right.

The last thing I'll say, like climate is THE thing. Climate is a national security issue. Climate is a health issue. Climate is an economic issue. Climate is a jobs issue. It's an environmental issue. So it's an everything issue. And if you just treat it as something like gender, and you don't mainstream it you're just not gonna be able to get the change you need, but you also will not get the benefits you need, and you will get all the downside.

What networks and relationships that we find most useful in developing and implementing the policies that you've been talking about?

I have a network of unusual suspects. That's what I call it. This week, there's going to be a Supreme Court decision on our new Impact Assessment Act that I brought in (Dryden, 2023). We had one on our carbon pricing, and I'm actually looking at the decision. And what happened was actually quite funny, there was 'the resistance'. Our national magazine [*Maclean's*] (2018) put a cover, and 'the resistance' was all these dudes: like Provincial Premiers, and the leader of the opposition. And it was like, great, if this is a resistance, obviously it doesn't look like a great group of people.

But I created my own [resistance], who are the unusual suspects that I need to bring to the table. There are two categories. So first of all, I did the dudes, cause I knew I had to. I got George Schultz because he was a Conservative Republican, but he talked to me about carbon pricing, and he said: "give it back". I promoted that I talked to a Republican that knew how to do hard things. Then I got Arnold Schwarzenegger to do a video for me because I thought, well, "who can turn down Arnie?." We worked in a bipartisan way. California is now the fastest growing economy and their emissions are down. So those were good. I got conservative premiers in Canada. We've lost progressive Conservatives, but these are the folks that did things on climate. So I actually did it in a way where I was protecting myself, cause I didn't want people to say, oh, of course you can't talk to real people.

I worked in the corporate world, so I'm not unused to how this works. So I was gonna turn it against them by taking more credible people on [climate]. And so I brought in the former premier of Quebec, who brought in carbon pricing in Quebec, the former small C conservative premier of BC, who also brought in carbon pricing, and then the former [now late] prime minister, Brian Mulroney, who cared about the environment and had done hard things, including on acid rain.

I created a carbon pricing leadership coalition, when we had no businesses supporting me. We had environmentalists, but that was kind of obvious environmental support for us and so I needed to be able to push back. So I went and actually knocked on doors. I got two banks in Canada, and then all three big banks had to come on board, even though I'm not sure they were all on board. But they joined, and then we had consumer goods companies, telecoms, manufacturing companies, one oil and gas company. So when I was slammed in the House of Commons for a 'job killing carbon tax' by the current leader of the opposition, I'd say 'you're the one really killing jobs. These people [signed on] represent X number jobs' (I had the number). And then I would just give quotes, or I'd say the names, and it was good because I could answer a lot of questions that way.

But then I also had my real unusual suspects. I also got health professionals. So the Canadian Association of Physicians for the environment, they actually became an intervenor in the litigation. Indigenous leaders. Young people. So there was a group of young people who are interveners that represented Generation Squeeze (GS, 2024), and they were talking about how they were like losers on everything. They just inherited all the costs, and all the impacts, and they weren't decision makers. I did have environmentalists and economists, and a lot of work with religious groups. I said, everyone's gonna think I just have environmentalists. If you wanna look at that person and say, I identify with Conservatives. I identify with young people. So I just went and found whomever so that I could get the support I needed to be able to stand up and say, this is ridiculous, it's very easy to lose carbon pricing, because people just say it's a tax, and people don't like taxes. So that was my strategy.

How do you see some of the policies that you worked on, or the work that you're doing now, contributing to gender equality?

I've resurrected women 'kicking on climate', now, we have 'women leading on climate' and I'm going to be working with Columbia University on that front. So we're actually still building that network. And looking at how do we use that? Including with the UNFCCC, it's just the geopolitics are very hard. Building this network of women is really important. And I'm gonna spend more time on this going forward because I need to. I think we're spending so much time on these COPs. I am called often to support young women or women initiatives, whether it's in politics or on climate. The good news is I run my own business so I feel like that's the gender empowerment piece, because I decide everything now. When I was in politics, there were limits right? You're gonna have the Prime Minister's office telling you what you gotta do and people telling you what you gotta do. That doesn't affect me anymore.

When I was minister I had to do something very well. I had to live through something very hard. We announced the climate emergency, and we had a whole climate plan. And then the next day we announced that we were buying a pipeline. So I don't have to do that anymore. I only work with people I like on issues that I think are important. And I'm very selective. That is consistent with what I want to do with my life. But it also leads me to working with a lot of women and folks that actually understand gender. It's much less strategic than in government.

Who do you draw your inspiration and strength from?

Well, my Granny, number one. She's obviously passed away a long time ago, but she was Irish, and she had six boys and she was lovely. But you didn't fool around with her and she kept them all in line. They didn't have a lot of money. Her husband fought for Irish Independence; he was in the army and then they bought a pub, and that's what she did. And she kept them all in line and was a teacher before that. So I always think. Oh, is it hard? No. So like what I'm worried about even how I sound. Do I sound like I'm too much? Granny didn't care about that. Who cares? So she's very good.

But then there are other women like Mary Robinson. She's really inspirational, because she's been able to link climate to social justice issues. Because I think sometimes we focus on climate as an environmental issue and it becomes very sciencey. Temperature is important, of course. But it's also really about people ultimately. And so, I think she's really focused on the inequality aspect. And I think that's really important to remember, like Canada can have a great climate plan, and our emissions can be going down. But if we're still exporting oil and gas, and that's actually having massive impacts, disproportionate impacts on vulnerable people, why are we not thinking about that?

You talked about the importance of discipline and clarity. How do you do it? How do you maintain that discipline, especially in a ministerial office?

I was a competitive swimmer. People always ask me, 'You work really hard. How do you do things?' I have a goal. With a goal like tackling climate change, it's not like a little goal. So one has to be disciplined like you have to actually have a long-term stretch goal. This is why I don't have time when people say we can't make 1.5 degrees]. I was like, no one is going to the Olympics by saying they're gonna do the same time as when they were 12? No, it's not the thing. So you have a stretch goal. But then you have to actually be very clear on what you need to achieve, and you have to be held accountable. You have to get up on the blocks, and you have to do the work. So I mean, that's just the way I think it's just ingrained.

It's not even like I spend tons of time. I would say, 'Okay, I want to talk about how we advanced our three things. I know we've got crisis after crisis. But what are we doing on that?' And I think it's really helped, because it makes it easier, right, because there's so much stuff in life. And in a way, maybe it's too simplistic. But it's a way I find I'm able to do things because I'm just very clear in what I want to do and what's important. And then the other things one has to deal with, and hopefully prevent, [is] crises. But if you don't do that it is very hard to imagine how we could ever tackle climate change because it's so complex. And it's such a multi-faceted issue. It's not something you can just like one day be like, 'Okay, I'm gonna do climate change today', and then, like 364 days later, I'll come back to it. I'm just very focused because I say, I came in swimming. I had to run for three years, and we knocked [on] more than 100,000 doors like that was just what we needed to do. So, I knew, that's what we need to do. So that's what we did.

Often people want to [discuss] the online hate, which you just can Google, and how it was actually organised. It went up after we brought in carbon pricing. But there were particular incidents; someone would come to my community office, and I got called 'climate Barbie'. I don't think that really defines me though. In fact, I somewhat use that to my advantage sometimes because I just call it out. And Canadians are like, 'yeah, that's bad. We don't like that'. So I'd be like great. I've reminded Canadians that they need to know when I'm fighting, like literally fighting body blows.

Maybe other things should have been prioritised. Sometimes I don't know. That's the only way I can get through this, because there's too much. But I also set limits. I say I'm going home and I'm gonna be with my kids. And I'm turning off my phone between 6 and 8 pm. So that was a priority for me, too. So I don't know, somehow, maybe because I got old and then I was like, 'this is what I care about'. I'm not doing things I don't really care about, and people can't tell me that I have to do that. So [I'm] just gonna focus on that which doesn't always make you like me. I'm not like the easiest, I think, in government, because people are like, actually, that's something we don't want to do or have to do. I'm a minister. I got elected. I can. I'll do this for a while, and then I'm gonna do other things. I'm not here to be here forever. I'm here to get things done.

References

Carter, A. (2020) *Fossilized: Environmental Policy in Canada's Petro Provinces*. Vancouver: UBC Press.

Dryden, J. (2023) *Supreme Court rules environmental impact legislation largely unconstiutional* [Online] Available: https://ottawacitizen.com/opinion/columnists/mckenna-how-to-run-like-a-girl-for-political-office (accessed 8 May 2024).

Generation Squeeze (2024) [Online] Available: https://www.gensqueeze.ca/ (accessed 8 May 2024)

Macleans Magazine (2018) *The Resistance - Front Cover*, November 7. [Online] Available: https://macleans.ca/news/canada/a-carbon-tax-just-try-them (accessed 8 May 2024).

Mary Robinson Foundation MRF (2018) *Women kicking it on climate change*. [Online] Available: https://www.mrfcj.org/resources/women-kicking-it-on-climate/ (accessed 8 May 2024).

McKenna, C. (2018) *How to run like a girl for political office*. [Online] Available: https://ottawacitizen.com/opinion/columnists/mckenna-how-to-run-like-a-girl-for-political-office (accessed 8 May 2024).

McKenna, C. and Jaffe, A.M. (2022) *Give more women the microphone at COP27*. [Online] Available: https://www.scientificamerican.com/article/give-more-women-the-microphone-at-cop-27/ (accessed 8 May 2024).

Natural Resources Canada 2024, 'Energy Factbook 2023-2024', https://energy-information.canada.ca/sites/default/files/2023-10/energy-factbook-2023-2024.pdf. Accessed May 26.

OECD (2023) *Air and climate: Greenhouse gas emissions by source*, OECD Environment Statistics (database). [Online] Available: https://doi.org/10.1787/data-00594-en. Accessed 8 May 2024).

2

A GLOBE OF ONE'S OWN

The Inverse Effect of Women's Political Representation on GHG Emissions

Laura Winther Engelsbak

Introduction

> In every country, in every society, it's a struggle to ensure that we have women in leadership positions when it comes to climate, and we'll be doing everything we can to support it.
>
> (UN Secretary-General, António Guterres 2022)

Climate change and gender inequality are among the most pressing challenges of our time. The year 2023 was the warmest ever with a temperature increase of 1.43°C, alarmingly close to the Paris Agreement's 1.5°C target (UNFCCC 2023). As extreme weather events become more frequent, and temperatures and sea levels rise, world leaders are discussing how to accelerate global climate action. This effort requires transformative action to reduce greenhouse gas emissions, scale up adaptation, and build resilience (World Bank 2022). Parallel to climate change, gender inequality is worsening, and it is now expected to take 135 years to achieve gender equality globally (World Economic Forum 2021, OECD 2022a, 2022b).

While climate change and gender inequality are each complex challenges, together they pose a severe threat to people, climate, and nature worldwide. A growing body of research establishes that climate change exacerbates existing gender inequalities in society, as the consequences of climate change disproportionately affect girls, women, and other structurally excluded groups (World Bank 2022). For example, women are 14 times more likely to lose their lives compared to men when natural disasters hit (UNDP 2022). Women's greater vulnerability to climate change is often attributed to their

DOI: 10.4324/9781003461005-5

limited access to and control over decision-making, environmental and socio-economic resources, and limited rights (Antwi-Agyei et al. 2021).

Gender has been a recurrent theme in the context of sustainable development since the 1992 Rio Declaration on Environment and Development. The nexus between climate change and gender equality has, however, gained greater political attention in recent years. At the Conference of the Parties (COP) there are gender days, dedicated to discussing the relationship between gender equality and climate change (UNFCCC 2023), and in 2023 the UN adopted a new resolution on gender equality and the role of women and girls in sustainable development (UN Press 2023). On a national level, Pakistan, one of the world's most climate vulnerable countries and one of the countries with most gender inequality in the world, has implemented their first action plan for climate change and gender equality (IUCN 2022), while Norway's Minister of Development and Denmark's former Minister for Global Climate Politics and Development Cooperation have stated that gender equality should be at the core of climate action (Newsweek 2023).

A wealth of studies indicate that women generally exhibit greater awareness of and concern for climate change, have lower climate impacts, and are more positive towards climate and environmental policies compared to men (McCright 2010, Alam & Rahman 2014, Magnusdottir & Kronsell 2015, Pearse 2017, Knight & Givens 2021, Kanyama et al. 2021). Moreover, research shows that women's participation and leadership in climate action are associated with better resource management, increased environmental protection, and more ambitious climate policies (Nugent & Shandra 2009, Kronsell 2011, Mavisakalyan & Tarverdi 2019, MacGregor 2021). Yet, women are significantly underrepresented in the top decision-making climate policy bodies across the world, including at domestic level (UNFCCC 2019, EIGE 2021, World Bank 2022).

Despite increased focus on the intersection between gender and climate change, the connection between women's political representation and countries' climate footprint remains underexplored (Ergas & York 2012, Mavisakalyan & Tarverdi 2019, Rainard et al. 2023).

Motivated by this gap, this chapter raises a central question: *What role can women's political representation play in reducing the global climate footprint?* To address this issue, I have compiled a longitudinal panel dataset. Through fixed effects regression, the effect of women's political representation on countries' GHG emissions is investigated.

With a robust methodological design, this chapter not only serves to fill an empirical gap, but also to inform and guide political decision-making processes worldwide by highlighting how solutions to the climate crisis and gender inequality can function synergistically.

The chapter is structured as follows: the second section briefly introduces previous research in the area of gender and climate change. The third section outlines the research design, including the data and methodology used. The data analysis and results are presented in section four. The primary analysis of the effect of women's political representation on countries' climate footprint is presented, followed by additional analyses with alternative variables to qualify and nuance the results of the primary analysis. The final section presents the concluding discussion, including contributions, limitations, and political implications.

Gender and Climate Change: What Do We Know?

In order to link women's political representation and countries' climate footprint, this section presents previous research in the area of gender equality and climate change. The purpose is to account for existing studies in the area, aiming to shed light on the knowledge gap this chapter addresses. The section begins with previous research that identifies gender differences in attitudes, behaviours, impacts, and vulnerabilities concerning climate change. It then presents research indicating that women's political representation influences countries' climate and environmental policies. Subsequently, the few studies examining the effect of women's political representation and gender equality on countries' climate and environmental footprints are presented. Finally, the identified knowledge gaps and limitations of existing studies in the field, as well as the contributions of this chapter, are presented.

Gendered Attitudes, Behaviours, Impacts, and Vulnerabilities

Gender differences in relation to climate change are well documented in the literature (see Pearse 2017 for a comprehensive review). Previous Danish and Swedish studies from 2009 and 2013, respectively, show that women and men, on average, have different climate impacts, perceive the need for climate action differently, and are affected differently by both climate change and climate policies (Oldrup et al. 2009, Kronsell 2013). A Swedish study finds that men emit 16% more greenhouse gases than women through their consumption (Kanyama et al. 2021), while another Swedish study finds that women exhibit more sustainable behaviour and are more likely to change their behaviour for environmental reasons (Sand for Nordisk Ministerråd 2022).

A significant amount of literature supports that women express greater concerns about and awareness of the environment and the climate (McCright 2010, McCright & Dunlap 2011, Lewis et al. 2019, Knight & Givens 2021), perceive climate and environmental risks as a greater threat (Buckingham 2010, McCright 2010), and tend to be more positively inclined towards progressive climate and environmental policies compared to men (Wängnerud 2009, Magnusdottir & Kronsell 2015). Moreover, several studies show that it is particularly older, white, conservative men in Western countries that report lower levels of concern about climate change and greater resistance towards environmental regulation (McCright & Dunlap 2011, 2013; Davidson & Haan 2012; Tranter & Booth 2015; Salehi et al. 2015; Knight & Givens 2021).

Gender differences in attitudes, behaviours, and impacts on climate change are thus well established. In addition, a larger number of studies find that the consequences of climate change are gender differentiated, with women bearing the brunt to a greater extent (Cannon 2002, Alam & Rahman 2014, Seager et al. 2016, Mavisakalyan & Tarverdi 2019). These studies also conclude that women's greater vulnerability to the consequences of climate change is not an expression of inherent or natural characteristics of women but rather an expression of existing gender inequalities and power relations in society (Cannon 2002, Alam & Rahman 2014, Seager et al. 2016, Mavisakalyan & Tarverdi 2019). These studies conclude that gender inequalities are predominantly created by gendered division of labour and caregiving responsibilities, which limit women's opportunities to cope with the negative consequences of climate change (Cannon 2002, Alam & Rahman 2014, Seager et al. 2016, Mavisakalyan & Tarverdi 2019).

Climate Policy and Political Gender Representation

As described, women express greater concern about climate change, are more likely to change behaviour due to climate change, are more affected by the consequences of climate change, and are more positively inclined towards progressive climate policies. In addition, several researchers argue that women's political representation influences both the substance and weighting of climate and environmental policies (Zalewski 1993, Kronsell 2011, Lovenduski 2005, Wängnerud 2009, Nugent & Shandra 2009, Mavisakalyan & Tarverdi 2019, MacGregor 2021).

Lv and Deng (2018) find that women's political participation plays a central role in emission reduction strategies and sustainable development policies. Norgaard and York's study (2005) shows that nations with a higher proportion of women in parliament ratify a greater number of environmental treaties. A study across 25 countries shows that countries with higher female parliamentary representation are more likely to allocate land areas for protection (UNDP 2016). Nugent and Shandra's quantitative study (2009) demonstrates a strong correlation between women's political representation and increased environmental protection. A cross-national, quantitative study shows that female representation leads countries to adopt stricter climate policies (Mavisakalyan & Tarverdi 2019).

In studies of the relationship between political gender representation and climate policy, a distinction is made between descriptive and substantive representation (Phillips 1995, Wängnerud 2009, Magnusdottir & Kronsell 2015). Descriptive representation refers to the demographic composition of political bodies, including the relative number of women/men. Substantive representation refers to the actual effect of women's political representation on climate policy decisions and outcomes (Wängnerud 2009, Magnusdottir & Kronsell 2015). Several studies find that descriptive representation of women does not necessarily result in substantive effects on actual climate policies. Sundström and McCright (2014) find no robust evidence of gender differences in environmental considerations among Swedish parliamentarians, although such gender differences are observed in the broader Swedish public. Magnusdottir and Kronsell (2015) similarly conclude that women's descriptive representation does not automatically result in substantive representation due to the presence of masculine norms in political institutions.

Climate Footprint and Political Gender Representation: Knowledge Gaps and Contributions

Several existing studies thus report that women's political representation leads to more progressive climate and environmental policies. The question is whether women's political representation also affects countries' climate footprints. This intersection is precisely what is examined in this chapter. There are very few existing studies that have addressed the relationship between women's political representation and climate footprint (Ergas & York 2012, Mavisakalyan & Tarverdi 2019, Rainard et al. 2023).

Both Ergas and York (2012) and Mavisakalyan & Tarverdi (2019) find that CO_2 emissions are significantly lower in countries where women have higher political status. Rainard et al. (2023) find that environmental performance is significantly better in countries with greater gender equality. While these studies have made progress in unfolding the relationships between gender equality and countries' climate and environmental footprints, each of these studies has limitations in terms of data and methodological design which this chapter addresses.

The existing studies are based on limited data. Ergas and York's study primarily relies on data from 2004, Mavisakalyan and Tarverdi's study relies on an average of values from 2005 to 2010, and Rainard et al.'s study relies on data from 2010 and 2020, respectively. In contrast, this chapter includes data collected for each year between 2009 and 2018. The large number of countries and observations makes it the most comprehensive dataset identified, examining the effect of women's political representation on countries' climate footprint. This has various advantages.

First, there has been a massive change in our climate and environmental awareness, and in our understanding of gender equality since 2004 and 2005–2010. The use of a panel dataset with annual data from 2009 to 2018 thus creates a more up-to-date and accurate picture of the relationship between women's political representation and countries' climate footprints. Second, the large and balanced number of countries in this chapter provides a more complete picture of the relationship between women's political representation and countries' climate footprints. Third, the panel structure of the data enables the use of fixed effects (FE) regression models, allowing for more accurate estimation of causal relationships, by controlling for time-invariant heterogeneity. Advantages and disadvantages of FE models are further elaborated in the research design section.

Furthermore, the relationship between women's political representation and countries' climate footprints is qualified and nuanced by testing the primary model with an alternative dependent variable (environmental performance) and an alternative primary, independent variable (gender equality).

In summary, this chapter contributes with an extensive and more up-to-date panel dataset, and the application of FE estimation, helping to fill an empirical knowledge gap in an area where limited knowledge currently exists.

Research Design

In this chapter, quantitative methods are used to examine the effect of women's political representation on climate footprint. Specifically, FE regression models are estimated with data from 137 countries over the period 2009–2018. This is explained in the sub-sections below.

Data and Sample

For this chapter, I have compiled a longitudinal panel dataset with 24 variables and a total of 2,145 observations across 193 countries over 11 years (2009–2019).[1] The dataset is compiled from various data sources: World Bank Open Data, Polity IV, Database of Political Institutions from the Inter-American Development Bank, UNDP, and Yale University. These sources are considered to be reliable data sources used by researchers, politicians, and media sources. The large number of countries and observations makes it the most comprehensive dataset identified so far that investigates the effect of women's political representation on countries' climate footprint.

The sample used in this chapter to examine the effect of women's political representation on countries' climate footprint, is based on 14 variables, including 993 observations across 137 countries over ten years (2009–2018). The reduction of the population to the sample is due to the elimination of variables and missing values in several variables that reduce the number of observations. The empirical measures of climate footprint (dependent variable) and women's political representation (primary, independent variable) are described in detail below, with the remaining 12 variables summarised in Table 2.1.

TABLE 2.1 Description of included variables

Variable	Description	Data source
GHG emissions	GHG emissions per capita	World Bank
Women's political representation	Women's share of parliamentary seats in percentage	World Bank
GDP	GDP per capita	World Bank
Urbanisation	Percentage share of the total population living in urban areas	World Bank
Industrialisation	Value added created by industry and construction as a percentage of GDP	World Bank
Agriculture	Value added created by agriculture, forestry, and fishing as a percentage of GDP	World Bank
Export	Value of all goods exported to the rest of the world, measured as a percentage of GDP	World Bank
Import	Value of all goods imported from the rest of the world, measured as a percentage of GDP	World Bank
Population growth	Annual growth of people in per cent	World Bank
Fertility	Fertility rate	World Bank
Female labour force	Women's participation in the labour force as a percentage of the total workforce	World Bank
Girls' education	Completion rate in primary school for girls	World Bank
Political regime	Political regime on a democracy-autocracy scale from -10 (strong autocracy) to 10 (strong democracy)	Polity IV
Political system	Measures whether a country's political system is parliamentary, assembly-elected presidential, or presidential	DPI

Dependent Variable: Climate Footprint

Climate footprint can be challenging to operationalise as it can include many different aspects. This is also reflected in the fact that measures of climate footprint vary across different data sources, including Climate Watch, PIK PRIMAP, UNFCCC, and GCP. For example, there is a difference in whether emissions are attributed to the countries that produce or the countries that consume goods. The empirical measure of climate footprint used in this chapter is "greenhouse gas emissions per capita", measured in kilotons. The measure is from World Bank Open Data that collects data from Climate Watch, managed by World Resources Institute. In this chapter, climate footprint is thus based on production-based emissions data and not consumption-based. This means that GHG emissions are attributed to the country where the product is produced, not the country where the end products are consumed. This can lead to countries with high consumption but limited production appearing to have lower emissions, while the opposite can be true for countries with high production and low consumption. This can create misleading greenhouse gas accounts that under- or over-estimate countries' actual climate footprints. However, the measures for climate footprint under the Paris Agreement and the UN Climate Convention (UNFCCC) are also production-based. In addition, World Bank Open Data and Climate Watch are considered reliable sources used by decision-makers worldwide and researchers, including in previous studies in the field (Ergas & York 2012, Mavisakalyan & Tarverdi 2019, Knight & Givens 2021). Likewise, their carbon footprint measure is one of the most comprehensive available that brings together dozens of datasets (World Resources Institute 2023).

Primary, Independent Variable: Women's Political Representation

The empirical measure of women's political representation in this chapter is "women's share of parliamentary seats in percentage", from World Bank Open Data that collects data from the Inter- Parliamentary Union (IPU). IPU is an international organisation of parliaments, and provides data directly from national parliaments on their structure, composition, working methods, and activities. Parliaments vary significantly across countries in their internal workflows and procedures, but generally, they legislate, control (to varying degrees) the government, and represent the voters. It can also be challenging to operationalise women's political representation and assess whether it measures only women's formal, descriptive representation, or women's real, substantive contributions to political decision-making. This is significant because social norms and beliefs can lead to barriers for women to perform their parliamentary mandate fully and effectively, despite their formal representation. The choice of this measure of women's political representation is motivated by the fact that it is used in other empirical studies where it is considered the best measure (Ergas & York 2012, Mavisakalyan & Tarverdi 2019, Rainard et al. 2023).

Methodology

The panel structure of the dataset enables the use of FE regression estimation. FE models are particularly suitable for investigating the effect of women's political representation on climate footprint across countries and time. FE eliminates the effect of variables that disturb this effect (such as the date of women's suffrage, which would vary between countries but does not vary over time within each country, and is expected to have a more or less homogeneous effect across countries), without having to measure them or knowing exactly what they are, as long as they are stable over time. This is called 'the advantage of fixed effects'. By including fixed effects for each country, the model controls for time-invariant heterogeneity, such as cultural, institutional, or economic differences between countries.

In social science research, it is virtually impossible to identify, measure, and include all relevant variables. This is called "omitted variable bias", and refers to a distortion in the coefficient estimates when important variables are omitted, which can create inconsistent coefficients. It is possible to mitigate "omitted variable bias" by reducing the correlation between the omitted variables and the included variables. By using panel data, FE models can control for time-invariant, unobserved variables, reducing the risk of omitted variable bias, and enabling a more accurate estimation of causal relationships between women's political representation and countries' climate footprint. However, FE models focus on the variation over time within the units, making it impossible to estimate the effect of variables that do not vary over time. Likewise, it is important to emphasise that FE models do not control for time-varying, omitted variables. This can result in biased coefficient estimates. The fact that FE focuses on the variation within countries also reduces its statistical power, as FE removes all variation between countries. The models are thus estimated with a 10% significance level in the analysis.

In summary, FE regression is a suitable estimation technique for testing the effect of women's political representation on climate footprint, as FE models control for time-invariant, omitted variables. This reduces the risk of omitted variable bias and results in more accurate estimates of the effect of women's political representation on climate footprint. With the application of FE, this chapter stands out as a methodological advance compared to the few existing studies in the field. Based on the robust methodological design, the possibilities for making causal interpretations are much better than in previous studies. However, it is important to emphasise that it is always complicated to speak of causality in social science research, and it is unrealistic to assume that I control for all relevant explanatory variables when conducting social science research.

Analysis and Results

The aim of this section is to unfold the relationship between women's political representation and climate footprint. Based on existing studies, I have formulated the following hypotheses (Table 2.2).

First, the results of the primary model – the effect of women's political representation on countries' climate footprint – are presented. Second, three new models are presented, where the primary model is tested with an alternative dependent variable (environmental performance) and an alternative primary, independent variable (gender equality). As outlined in the methodological design, the models are tested using fixed effects regression, and all analyses are performed in Stata.

TABLE 2.2 Main and sub-hypotheses

Hypothesis	Expected relationship
Main hypothesis	
(1)	In countries where women's political representation is higher, the climate footprint will be lower than in countries where women's political representation is lower.
Sub-hypotheses	
(1.1)	In countries where women's political representation is higher, the country's environmental performance will be better than in countries where women's political representation is lower.
(1.2)	In countries with greater gender equality, the climate footprint will be lower than in countries with less gender equality.
(1.3)	In countries with greater gender equality, the country's environmental performance will be better than in countries with less gender equality.

Primary Model: Women's Political Representation and Climate Footprint

This section presents the primary analysis, aiming to confirm or reject the main hypothesis:

> In countries where women's political representation is higher, the climate footprint will be lower than in countries where women's political representation is lower.

To test the hypothesis, the effect of women's political representation on countries' climate footprint is estimated using FE regression.

Table 2.3 summarises the results of the FE regression of the effect of women's political representation on countries' climate footprint.

The results of the FE regression show a significant, negative effect of women's political representation on the climate footprint, thus confirming the main hypothesis: *In countries where women's political representation is higher, the climate footprint will be lower than in countries where women's political representation is lower.* When women's political representation increases by 1%, the country's GHG emissions per capita decrease by 0.014 kilotons, holding all other variables constant. The average for GHG emissions per capita across the included countries is 6.4 kilotons. Thus, 0.014 kilotons are a relatively large decrease. The results are significant and remain robust when accounting for country characteristics using clustered standard errors. The relationship is

TABLE 2.3 Primary model: Women's political representation and GHG emissions

GHG emissions	(2)
Women's political representation	−0,014*
	(0,008)
GDP	0,000***
	(0,000)
Industrialisation	0,019*
	(0,010)
Agriculture	−0,003
	(0,010)
Urbanisation	0,000
	(0,023)
Female labour force	−0,023
	(0,047)
Girls' education	0,007*
	(0,004)
Population growth	0,175***
	(0,053)
Fertility	0,160
	(0,252)
Political regime	−0,017
	(0,014)
Political system	−0,278
	(0,222)
Export	−0,011
	(0,010)
Import	0,004
	(0,005)
R^2 (within)	0,15
N	993

Note: The table shows the results of the FE regression between women's political representation and GHG emissions. The bottom row contains information on the adjusted within-R^2 (coefficient of determination) and N (sample size). Standard errors are clustered at country level. Significance level: * $p < .1$, ** $p < .05$, *** $p < .01$

illustrated in Figure 2.1, showing the marginal effects of women's political representation on the predicted values of the GHG emissions.

Among the control variables in the model, GDP, industrialisation, girls' education, and population growth have a significant, positive effect on climate footprint. In summary, this means that the higher a country's GDP, industrialisation, and population growth, and the more girls who complete primary school, the higher the country's climate footprint will be. No other control variables in the model have significant effects.

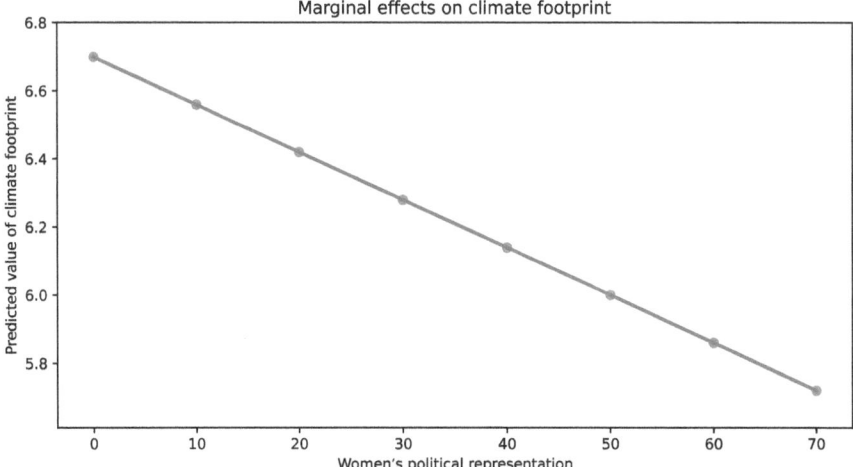

FIGURE 2.1   Marginal effects of women's political representation on climate footprint.

Note: The figure shows the marginal effects of women's political representation on the predicted values of climate footprint, controlling for measures of modernisation, development, and democracy. The figure illustrates that when women's political representation increases by 1%, the country's greenhouse gas emissions per capita decrease by 0.014 kilotons.

In summary, the analysis of the primary model shows that women's political representation has a significant, negative effect on countries' climate footprint, and the main hypothesis can thus be confirmed. Even when controlling for various measures of modernisation, economic development, democracy, and for time-invariant, unobserved effects, the climate footprint is significantly lower in countries where more women are politically represented than in countries where fewer women are politically represented. Before drawing the final conclusions from the results of the analysis, I will conduct an analysis of the primary model with an alternative dependent variable and alternative primary, independent variable, to nuance the relationship between women's political representation and countries' climate footprint.

Alternative Models: Gender Equality and Environmental Performance

The analysis of the primary model shows that in countries where women's political representation is higher, the climate footprint will be significantly lower. To qualify this result, three additional models are tested with the same control variables as in the primary model, but with an alternative dependent variable (environmental performance) and an alternative primary, independent variable (gender equality).

TABLE 2.4 Models with alternative dependent and primary independent variables

	Dependent variable	*Primary independent variable*	*Expected relationship*
Model 1	Climate footprint	Women's political representation	–
Model 2	Environmental performance	Women's political representation	+
Model 3	Climate footprint	Gender inequality	+
Model 4	Environmental performance	Gender inequality	–

Note: The table summarises the expected relationship for four models with different dependent and independent variables, where "–" indicates an expected negative relationship, and "+" indicates an expected positive relationship.

Environmental performance is measured by Yale University's Environmental Performance Index (EPI). The index consists of 40 indicators across 11 sub-categories within greenhouse gas emissions, environmental health, and eco-system vitality. This variable thus nuances the measure of climate footprint with a more comprehensive measure of countries' sustainability.

Gender equality is measured by the UNDP's Gender Inequality Index (GII), which is an index of inequality between genders. The index is created from three categories: reproductive health, empowerment, and the labour market. This variable nuances the measure of women's political representation by contributing with a more comprehensive measure of a country's gender equality.

Table 2.4 summarises the models tested and the expected relationships. To allow for comparisons, Model 1 is included, which is the primary model already tested in the previous section (Table 2.3). For Models 1 and 4, a negative relationship is expected, while a positive relationship is expected for Models 2 and 3.

Table 2.5 summarises the results for Models 1, 2, 3, and 4. The primary independent variable in Models 1 and 2 is women's political representation, while the primary independent variable in models 3 and 4 is gender equality. The models are estimated with the same control variables as in the primary model but are removed in the table for clarity.

As Table 2.5 illustrates, all models live up to the expected relationships presented in Table 2.4. Model 1 is the primary model and shows, as previously described, a significant, negative relationship between women's political representation and climate footprint. Model 2, measuring the effect of women's political representation on countries' environmental performance, confirms a positive relationship (0.133), although the coefficient is insignificant. This may be due to the reduced sample size on 468 observations. Model 3, measuring the effect of a country's general gender inequality on climate footprint, confirms a significant, positive relationship (0.002). That is, in countries with greater gender equality, the climate footprint will be smaller than in countries where there is less gender equality. Model 4, measuring the

TABLE 2.5 Fixed effect models with alternative dependent and primary independent variables

	Model 1	Model 2	Model 3	Model 4
	Climate footprint	Environmental performance	Climate footprint	Environmental performance
Women's political representation	−0,014*	0,133		
	(0,008)	(0,125)		
Gender inequality			0,002*	−0,040*
			(0,001)	(0,024)
Adjusted R^2	0,15	0,06	0,15	0,07
Observations	993	468	984	463

Note: This table shows the results of the FE regression for Models 1, 2, 3, and 4 from Table 2.4. The models are estimated with the same control variables as the primary model but are removed in this table for clarity. The bottom row contains information on N (sample size) and adjusted R^2 (coefficient of determination). Note that N is not constant across the different models. Clustered standard errors in parentheses. Refer to section 4.3 for definitions of the variables. Significance level: * p < .1, ** p < .05, *** p < .01

effect of a country's general gender inequality on environmental performance, confirms a significant, negative relationship (−0.040). That is, the environmental performance will be better in countries with greater gender equality than in countries where there is less gender equality.

Summary

This chapter has investigated the relationship between women's political representation and countries' climate footprint. To address this relationship, I have estimated fixed effects regression models with data for 137 countries over the period 2009–2018, providing insights into the following key findings.

I find that women's political representation has a significant negative effect on countries' climate footprint. The results show that as women's political representation increases by 1%, countries' GHG emissions per capita are reduced by 0.014 kilotons, controlling for modernisation, economic development, democracy, and time-invariant unobserved effects. The results are significant and remain robust when accounting for country characteristics using clustered standard errors. Additionally, the large and balanced number of countries provides good opportunities for generalisation across all countries. The main findings are qualified and nuanced by analyses with alternative primary variables. The results of the models with an alternative dependent and independent variable show that countries' environmental performance is significantly better, and the climate footprint significantly lower, in countries with greater gender equality. The consistency across alternative model specifications suggests that the observed effects are robust to changes in variable selection, thus reflecting the underlying associations. The results also nuance the primary model and show that the relationship has broader generalisability to environmental performance and general gender equality.

Concluding Discussion

This chapter ends with a concluding discussion. First, this chapter's findings, limitations, and further research is discussed. Second, the results' political implications are discussed.

Findings, Limitations, and Further Research

The results of this chapter demonstrate a significant, negative relationship between women's political representation and climate footprint across 137 countries over ten years (2009–2018). As women's political representation increases, countries' greenhouse gas emissions decrease. This is a significant contribution to an area characterised by limited research. Looking into differences between the included countries, interaction model analysis from the research on which this chapter is based (Engelsbak 2023), shows stronger effects of women's political representation on greenhouse gas emissions in low-income countries compared to upper-middle, and high-income countries, and a stronger effect of women's political representation in countries with high, upper-middle, and lower-middle climate vulnerability compared to countries with low climate vulnerability (Engelsbak 2023). This chapter contributes with an extensive panel dataset and the use of fixed effects estimation, which controls for time-invariant heterogeneity and reduces the risk of omitted variable bias. Thus, this chapter provides a robust framework to examine the effect of women's political representation on countries' climate footprint. Although this chapter provides valuable methodological insights into the effect of women's political representation on countries' climate footprint, some limitations should be acknowledged. These limitations offer opportunities for future research to build upon and extend the knowledge base in this area.

Firstly, this quantitative macro-analysis cannot directly identify the underlying mechanisms moderating the effect of women's political representation on countries' climate footprint. Identifying and isolating these mechanisms is an important task for future research. This could involve country-based case studies of whether collective behaviour in political bodies changes when women achieve greater representation, or whether women face barriers in decision-making processes when they are politically represented. Further, this study could be supplemented by in-depth qualitative interviews with politicians about decision-making in climate and environmental policy. Such studies can contribute to examining the underlying norms and values that underpin women's political representation. In this way, qualitative studies can support and nuance these quantitative findings.

Secondly, while this chapter suggests a significant negative effect of women's political representation on countries' climate footprint, it is crucial to critically examine the assumption that descriptive representation – simply having more women in political roles – automatically leads to substantive policy changes and outcomes. Descriptive representation does not inherently ensure that women's political participation translates into lower emissions. Several factors may mediate or moderate this relationship, including the political context, the degree of women's influence within political institutions, and the presence of supportive structures and policies. The mere presence of women in political positions does not guarantee that they have the power or influence to shape significant climate policies. Institutional barriers, gender norms, and political dynamics can limit the substantive impact of women's representation (Magnusdottir & Kronsell 2021, 2024). Further, the alignment between women's representation and climate and environmental outcomes may be influenced by the overall commitment of the political system to gender equality and sustainability. In summary, the effectiveness of women's political representation in addressing climate change depends on a complex interplay of institutional, political, and societal factors that enable or constrain their influence. Further research should explore these dynamics to better understand the relationship between descriptive and substantive representation. This chapter does, however, show a significant effect of women's political representation on countries' climate footprint, even when controlling for many different factors.

Thirdly, the quantitative macro-approach in this chapter may create challenges regarding the understanding of gender and women. First, there is a risk of essentialising women as inherently closer, and with a special relationship, to nature. Within research on gender equality and climate change, there is often criticism of how women are referred to as inherently vulnerable and marginalised victims of climate change, reinforcing the idea of women as defenceless. To avoid this essentialisation, I have emphasised that women's relationship to nature is not based on natural characteristics but is rooted in individuals' and groups' socialisation into society-created gender stereotypes. Likewise, the findings of this chapter create a basis for viewing women not as vulnerable and defenceless, but as important actors in the solutions to the climate crisis. Additionally, there is a risk of viewing all women as one homogenous group with the same experiences, interests, and knowledge. The purpose of this chapter is to examine the effect of women's political representation on countries' climate footprint across the entire world. This generalisation involves a certain homogenisation of women. It is, however, important to acknowledge that gender is linked to other social categories such as class, ethnicity, race, and sexuality. This means there are significant differences between women, both within and across countries. Thus, the explanations for the significant effect of women's political representation on countries' climate footprint may be divergent. Despite these acknowledgements, there is still a risk that this chapter reduces the diversity of women, as well as risks ignoring other structural inequalities. An important task for future research is to improve the collection, analysis, and reporting of aggregated data across gender and other identities. On this basis, future research will benefit from incorporating an intersectional perspective that considers the connections between different identity dimensions and climate change.

While this chapter's findings can serve as inspiration for future research, they also have important political implications. These are discussed in the next section.

Political Implications

The findings of this chapter show a significant, negative effect of women's political representation on countries' climate footprint. This indicates that increased political representation of women can work synergistically in reducing the global climate footprint. These results have important political implications that can contribute to informing political decisions and actions related to climate change and gender equality.

While the findings of this chapter emphasise the importance of women's political representation in reducing the global climate footprint, women's political equality is also a necessity in itself. In 2022, UN Secretary-General António Guterres emphasised that centuries of discrimination and stereotypes have created a vast gender gap across societies and cultures. To change the male-dominated culture and create balance, we need gender equality in terms of decision-making and participation at all levels, according to Guterres (UN News 2022).

Promoting political gender equality can include implementing various targeted initiatives, interventions, and campaigns. This involves promoting female role models who can serve as inspiration for other women and girls, and who can help break down and challenge traditional stereotypes about women's competencies in decision-making positions. Additionally, various forms of affirmative action have increasingly been introduced by countries in recent years to increase women's representation. This can involve strengthening rules for gender balance, as well as a significant increase in the implementation of gender quotas that require the inclusion of women on an equal footing with men (Latura & Weeks 2022). However, women's formal political representation is often not enough to ensure that women's voices are included on an equal footing with men's. It is important to create inclusive political institutions that are not characterised by gendered hierarchies and work cultures (Kronsell 2011; Magnusdottir & Kronsell 2021, 2024). Inclusive political processes that involve different perspectives can lead to more comprehensive and effective climate strategies that consider a wider range of issues (Kronsell 2011, 2013).

The findings of this chapter motivate how the general promotion of gender equality and solutions to climate changes can go hand in hand. These findings also have important implications in international and development policy cooperation. This involves knowledge-sharing of best practices and experiences that can contribute to a global effort to address equality and climate change more effectively. As analysed in the previous chapter, experiences from Sweden show a significant increase in the integration of gender perspectives in development policy with the introduction of a Swedish feminist foreign policy in 2014 (DanChurchAid 2021). With the change of government in 2022, Sweden no longer has a declared feminist foreign policy, and there is already a noticeable decline in ambitions for gender equality in its development cooperation (ACT Alliance 2023). A feminist foreign policy can help emphasise the importance of gender equality and women's representation in addressing global challenges such as climate change.

Additionally, many argue that climate change and gender equality should be two sides of the same coin in development cooperation and aid, including the Danish Minister for Global Climate Policy and Development Cooperation, Dan Jørgensen (Itad for the Danish Ministry of Foreign Affairs 2023, Globalnyt 2023, Newsweek 2023). A new evaluation of support for gender equality in Danish development cooperation shows that considerations of gender equality are very rarely incorporated into Danish efforts for climate change and green transition (Globalnyt 2023). The evaluation describes this situation as "a missed opportunity" and concludes that improving gender equality and solutions to climate change can and should go hand in hand, and that climate aid holds the greatest potential for ensuring a more solid impact on gender equality (Globalnyt 2023). The results of the analyses are an important contribution to the argument for integrating climate and gender equality in development policy.

The findings of this chapter also indicate the importance of integrating gender perspectives in climate policy, recognising and incorporating gender differences in relation to climate change. The Beijing Declaration from 1995 is considered the most progressive plan for promoting women's rights globally. The text established the necessity of "gender mainstreaming", which involves the integration of gender equality in the preparation, design, and implementation of all policies and legislation. While European countries have committed to this since the Amsterdam Treaty of 1997 (KVINFO 2023), many emphasise the marginal effect of gender mainstreaming in practice (Prügl 2010, Wittman 2010, KVINFO 2023). A study on gender equality assessments across Danish legislative proposals shows that more than half of the ministries have not conducted one single gender equality assessment since the parliamentary year 2016/2017 (KVINFO 2023). The Ministry of Climate, Energy and Utilities have assessed 2% of their legislative proposals for gender equality since the parliamentary year 2016/2017, while the Ministry of Environment and the Ministry of Foreign Affairs have assessed none of their legislative proposals (KVINFO 2023). Gender mainstreaming can be useful for identifying gendered differences of the consequences of climate change but can also function as a "tick box exercise", thereby diluting the opportunities for real change (Rainard et al. 2023).

The results of this chapter suggest that an approach is needed that not only recognises the relationship between gender equality and climate change but also actively incorporates this relationship into political climate solutions. Many countries and key actors, including Denmark, often advocate for a gender-transformative approach rather than a gender-responsive or gender-sensitive approach in foreign and development policy (Global Gender and Climate Alliance 2016, Resurrección et al. 2019, Stern and Dietz 2015, Urry 2015, Anderson & Bows 2012, O'Brien 2016, Eizenberg & Jabareen 2017). While gender-sensitive and gender-responsive approaches recognise gender inequalities in relation to climate impacts and address the fundamental causes behind the inequalities, gender-transformative approaches systematically address the fundamental causes of gendered inequalities in relation to climate change. With a gender-transformative approach, it is possible to address other social norms and factors, and the intersections between them, that are relevant to the relationships between gender equality and climate change, rather than simply concluding that political gender representation is the solution to the climate crisis. This involves changing and transforming social norms and values that contribute to gender inequality, including breaking with traditional, stereotypical perceptions of gender, gender roles, and gendered divisions of labour; stereotypes that are also limiting for men and boys if they act outside their traditional gender roles. Based on this, the gender-transformative approach also emphasises the importance of engaging and including men and boys in the green transition.

The findings of this chapter indicate that gender equality and the reduction of the global climate footprint can work synergistically. However, it is not certain that a "one size fits all" approach exists for integrating gender equality and climate change across all countries. Additionally, the chapter's findings do not suggest that the responsibility for solving the climate crisis should be placed on women but instead that improving gender equality across societies can address challenges of social injustice and global challenges like the climate crisis. This chapter thus establishes that there is a clear trend that the climate footprint is lower in countries where more women are politically represented.

Note

1 The full dataset consists of data for 11 years (2009–2019), but the sample used in the analysis in this chapter consists of data for ten years (2009–2018).

References

ACT Alliance (2023). Climate finance and gender: lessons from Nordic efforts to integrate gender equality in climate-related development finance. https://actalliance.org/wp-content/uploads/2023/12/Gender-and-climate-finance-ACT-INKA-report-04.12.23.pdf

Alam, K., Rahman, M. H. (2014). Women in natural disasters: a case study from southern coastal region of Bangladesh. *International Journal of Disaster Risk Reduction*, 8:68–82.

Anderson, K., & Bows, A. (2012). A new paradigm for climate change. *Nature Climate Change* 2, 639–640. doi: 10.1038/nclimate1646

Antwi-Agyei, P., Abu, M., & Okyere-Nyako, A. (2021). *Climate Change Gender Action Plan*. Acra: Ministry of Environment, Science, Technology and Innovation and Ministry of Gender, Children and Social Protection, Government of Ghana. https://www.undp.org/sites/g/files/zskgke326/files/migration/gh/UNDP_Gh-Ghana_-Gender_-Action_Plan.pdf

Buckingham, S. (2010). Call in the women. *Nature* 468, 502.

Cannon, T. (2002). Gender and climate hazards in Bangladesh. *Gender Development*, 10:45–50.

DanChurchAid (2021). From Words to Action. Lessons from Nordic Efforts to Integrate Gender Equality In Climate Finance. https://actalliance.org/wp-content/uploads/2021/11/Climate_FromWordToAction-final.pdf

Davidson, D. J. & Haan, M. (2012). Gender, political ideology, and climate change beliefs in an extractive industry community. *Population and Environment*, 34:217–234.

EIGE (2021). Decision-making in environment and climate change: women woefully under-represented in the EU Member States. https://eige.europa.eu/gender-statistics/dgs/data-talks/decision-making-environment-and-climate-change-women-woefully-under-represented-eu-member-states

Eizenberg, E., & Jabareen, Y. (2017). Social sustainability: a new conceptual framework. *Sustainability* 9, 68. doi: 10.3390/su9010068

Engelsbak, L. W. (2023). KØN OG KLIMA Et kvantitativt, paneldatastudie af sammenhængen mellem kvinders politiske repræsentation og klimaaftryk. University of Copenhagen. https://curis.ku.dk/ws/files/380401361/Speciale_Laura_Winther_Engelsbak.pdf

Ergas, C., & York, R. (2012). Women's status and carbon dioxide emissions: a quantitative cross-national analysis. *Social Science Research* 41, 965–976. doi:10.1016/j.ssresearch.2012.03.008

Global Gender and Climate Alliance (2016). *Gender and Climate Change: A Closer Look at Existing Evidence.* New York: Global Gender and Climate Alliance.

Globalnyt (2023). Ny evaluering roser dansk indsats for ligestilling, men peger på klimabistanden som en øm tå. https://globalnyt.dk/ny-evaluering-roser-dansk-indsats-for-ligestilling-men-peger-paa-klimabistanden-som-en-oem-taa/

Itad for the Danish Ministry of Foreign Affairs (2023). Evaluation of support to gender equality in Danish development cooperation 2014–2021.

IUCN (2022). Climate Change Gender Action Plan of the Government and People of Pakistan. https://iucnhq-my.sharepoint.com/personal/saeedh_iucn_org/_layouts/15/onedrive.aspx?id=%2Fpersonal%2Fsaeedh%5Fiucn%5Forg%2FDocuments%2FccGAP%2FClimate%20Change%20Gender%20Action%20Plan%2Epdf&parent=%2Fpersonal%2Fsaeedh%5Fiucn%5Forg%2FDocuments%2FccGAP&ga=1

Kanyama, A. C., Nässén, J. & Benders, R. (2021). Shifting expenditure on food, holidays, and furnishings could lower greenhouse gas emissions by almost 40%. *Journal of Industrial Ecology,* 25:6. https://doi.org/10.1111/jiec.13176

Knight, K. W. & Givens, J. E. (2021). Gender and climate change views in context: a cross-national multilevel analysis, *The Social Science Journal,* DOI:10.1080/03623319.2021.1913041

Kronsell, A. (2011). Gendered Practices in Institutions of Hegemonic Masculinity. *International Feminist Journal of Politics,* 7(2), 280–298, DOI: 10.1080/14616740500065170

Kronsell, A. (2013). Gender transition in climate governance. *Environmental Innovation and Societal Transitions,* 7, 1–15.

KVINFO (2023). *Ligestillingsvurdering Af Lovforslag.* ANALYSENOTAT.

Latura, A., & Weeks, A. C. (2022). Corporate Board Quotas and Gender Equality Policies in the Workplace. *American Journal of Political Science.* Vol. 67, Issue 3.

Lewis, G. B., Palm, R., & Feng, B. (2019). Cross-national variation in determinants of climate change concern. *Environmental Politics,* 28(5), 793–821. https://doi.org/10.1080/09644016.2018.1512261

Lovenduski, J. (2005). *State Feminism and Political Representation.* Cambridge University Press.

Lv, Z. & Deng, C. (2018). Does women's political empowerment matter for improving the environment? A heterogeneous dynamic panel analysis. *Sustainable Development* 27(4), 603–612.

MacGregor, S. (2021). Making matter great again? Ecofeminism, new materialism and the everyday turn in environmental politics. *Environmental Politics* 2021, 1–20. doi:10.1080/09644016.2020.1846954

Magnusdottir, G. L., & Kronsell, A. (2015). The (in)visibility of gender in Scandinavian climate policy-making. *International Feminist Journal of Politics,* 17, 308–326.

Magnusdottir, G. L., & Kronsell, A. (eds) (2021). *Gender, Intersectionality and Climate Institutions in Industrialized States*. Routledge: Earthscan.

Magnusdottir, G. L., & Kronsell, A. (2024). Climate institutions matter: The challenges of making gender-sensitive and inclusive climate policies. *Cooperation and Conflict*, 0(0). https://doi.org/10.1177/00108367241230011

Mavisakalyan, A., & Tarverdi, Y. (2019). Gender and climate change: do female parliamentarians make difference? *European Journal of Political Economy 56*, 151–164. doi:10.1016/j.ejpoleco.2018.08.001

McCright, A. M. (2010). The effects of gender on climate change knowledge and concern in the American public. *Population and Environment 32*, 66–87.

McCright, A. M., & Dunlap, R. E. (2011). Cool dudes: The denial of climate change among conservative white males in the United States. *Global Environmental Change*, 21(4), 1163–1172. https://doi.org/10.1016/j.gloenvcha.2011.06.003

McCright, A. M. & Dunlap, R. E. (2013). Bringing ideology in the conservative white male effect on worry about environmental problems in the USA. *Journal of Risk Research*, 16, 211–226.

Newsweek (2023). Gender Equality Must Be at the Heart of Climate Action | Opinion. https://www.newsweek.com/gender-equality-must-heart-climate-action-opinion-1821430

Norgaard, K. & York, R. (2005). Gender Equality and State Environmentalism. *Gender & Society*, 19(4), 506–522. DOI:10.1177/0891243204273612

Nugent, C. & Shandra, J. M. (2009). State Environmental Protection Efforts, Women's Status, and World Polity A Cross-National Analysis. *Organization & Environment*, 22(2), 208–229. DOI:10.1177/1086026609338166

O'Brien, K. L. (2016). Climate change and social transformations: is it time for a quantum leap? *Wiley Interdisciplinary Reviews: Climate Change 7*, 618–626. doi: 10.1002/wcc.413

OECD (2022a). *Development finance for gender-repsonsive climate action*. OECD Development Co-operation Directorate, OECD Publishing: Paris.

OECD (2022b). Social Expenditure Database (SOCX). https://www.oecd.org/social/expenditure.htm

Oldrup, H. & Breengaard, M. H. for Nordisk Ministerråd (2009). *Gender and Climate Change*. København: Nordic Council of Ministers.

Pearse, R. (2017). Gender and climate change. *WIREs Climate Change*, 8, 451. doi: 10.1002/wcc.451

Phillips, A. (1995). *The Politics of Presence*. Oxford: Clarendon Press. Print.

Prügl, E. (2010). Feminism and the postmodern state: gender mainstreaming in European rural development. *Journal of Women in Culture and Society 35*, 447–475. doi: 10.1086/605484

Rainard, M., Smith, J. C., & Pachauri, S. (2023). Gender equality and climate change mitigation: Are women a secret weapon?. *Frontiers in Climate*. doi:10.3389/fclim.2023.946712

Resurrección, B. P., Bee, B. A., Dankelman, I., Park, C. M. Y., Haldar, M. & McMullen, C. P. (2019). *Gender-Transformative Climate Change Adaptation: Advancing Social Equity*. Paper Commissioned by the Global Commission on Adaptation.

Salehi, S., Pazuki, N. Z., Mahmoudi, H., & Knierim, A. (2015). Gender, responsible citizenship and global climate change. *Women Studies International Forum*, 50, 30–36.

Sand, J. for Nordisk Ministerråd. (2022). Climate, Gender and Consumption: A research overview of gender perspectives on sustainable lifestyles. https://norden.diva-portal.org/smash/record.jsf?pid=diva2%3A1700810&dswid=4131

Seager, J., Bechtel, J., Bock, S., Dankelman, I., Fordham, M., Gabizon, S., Thuy Trang, N., Perch, L., Qayum, S., & Roehr, U. (2016). *Global Gender and Environment Outlook*. United Nations Environment Programme (UNEP).

Stern, P. C., & Dietz, T. (2015). IPCC: social scientists are ready. *Nature* 521, 161–161. doi: 10.1038/521161a

Sundström, A. & McCright, A. M. (2014). Gender differences in environmental concern among Swedish citizens and politicians, *Environmental Politics*, 23:6, 1082–1095, https://doi.org/10.1080/09644016.2014.921462

Tranter, B., & Booth, K. (2015). Skepticism in a changing climate: A cross-national study. *Global Environmental Change*, 33, 154–164. https://doi.org/10.1016/j.gloenvcha.2015.05.003

UN News (2022). Gender parity, the only path to gender equality: Guterres. https://news.un.org/en/story/2022/12/1131642

UN Press (2023). Second Committee Approves 16 Resolutions, including on Achieving Gender Equality, Eliminating Unauthorized Unilateral Trade Measures. https://press.un.org/en/2023/gaef3596.doc.htm

UNDP (2016). Gender and Climate Change. Overview of linkages between gender and climate change. https://www.undp.org/sites/g/files/zskgke326/files/publications/UNDP%20Linkages%20Gender%20and%20CC%20Policy%20Brief%201-WEB.pdf

UNDP (2022). Women are hit hardest in disasters, so why are responses too often gender-blind? https://www.undp.org/blog/women-are-hit-hardest-disasters-so-why-are-responses-too-often-gender-blind

UNFCCC (2019). Women Still Underrepresented in Decision-Making on Climate Issues under the UN. https://unfccc.int/news/women-still-underrepresented-in-decision-making-on-climate-issues-under-the-un

UNFCCC (2023). New Analysis of National Climate Plans: Insufficient Progress Made, COP28 Must Set Stage for Immediate Action. https://unfccc.int/news/new-analysis-of-national-climate-plans-insufficient-progress-made-cop28-must-set-stage-for-immediate

Urry, J. (2015). Climate change and society. In *Why the social sciences matter* (pp. 45–59). Palgrave Macmillan, London. doi: 10.1057/9781137269 928_4

Wängnerud, L. (2009). Women in Parliaments: Descriptive and Substantive Representation. *Annual Review of Political Science*, 12, 51–69. https://doi.org/10.1146/annurev.polisci.11.053106.123839

Wittman, A. B. (2010). *What Happens to the Radical Potential of Gender Mainstreaming? Problems of Implementation and Institutionalisation in Gendered Organisations*. Edinburgh: Edimburgh Research Archive (ERA).

World Bank (2022). People and planet together: Why women and girls are at the heart of climate action. https://blogs.worldbank.org/climatechange/people-and-planet-together-why-women-and-girls-are-heart-climate-action

World Economic Forum (2021). Global Gender Gap Report. https://www.weforum.org/reports/ab6795a1-960c-42b2-b3d5-587eccda6023

World Resources Institute (2023). Climate Watch. https://www.wri.org/initiatives/climate-watch

Zalewski, M. (1993). "Feminist Standpoint Theory Meets International Relations Theory: A Feminist Version of David and Goliath?" *Fletcher Forum of World Affairs*, 17(2), 13–32.

3

TO WHAT EXTENT CAN THE EUROPEAN UNION CONTRIBUTE TO A FEMINIST CLIMATE POLICY?

Gill Allwood

Introduction

The European Union (EU) presents itself as a global climate actor and a global gender actor. On the opening day of the UN Climate Change Conference COP29 in Baku, Azerbaijan, the EU and its Member States issued a joint statement reaffirming their commitment to strengthening the integration of gender in global climate action. By the end of the first week, many countries had agreed to support this statement (European Commission 2024). Speaking at the launch event, Jan Dusík, Deputy Director General for Climate Action at the European Commission, said, 'The EU is committed to integrating gender considerations in all our climate strategies and actions to guarantee an effective transition to climate neutrality that leaves no one behind. By making gender equality a cross-cutting priority in our climate efforts, we are not only building resilience – we are creating a fairer, more inclusive world for generations to come'. The EU's global leadership on gender and climate is particularly important in the context of rising opposition from the Vatican, Saudi Arabia, Russia, Egypt and Iran to the inclusion in international climate agreements of the term 'gender' (Stallard, 2024). However, as will be demonstrated below, the EU's own internal climate policy does not consistently mainstream gender, leaving a gap between external declarations and internal practice.

Following the adoption by a number of countries of what they refer to as a 'Feminist Foreign Policy' (FFP), there has been a growth in references to 'Feminist Climate Policy' (FCP), including within some EU institutions. Tracing the relation between FFP and FCP, this chapter assesses the extent to which the EU can contribute to the latter, addressing the following research questions: to what extent is gender mainstreamed into EU

DOI: 10.4324/9781003461005-6

climate policy and what obstructs or facilitates a transformative EU FCP? It answers these through a combination of content analysis of key EU policy documents and a feminist institutionalist analysis of EU climate policy. Civil society documents, drawn from organisations that advocate for an FCP in the EU, are used to inform the analysis and to expose the differences between the meanings attributed to the term 'feminist' by the various actors. The analysis finds that EU climate policy is not gender mainstreamed; there are differences in the meanings attributed to gender, gender equality, and gender mainstreaming; references to gender equality appear more frequently in external than in internal policy; and rhetorical commitments to gender equality disappear before policy is implemented.

The chapter concludes that the EU has the potential to contribute to an FCP, but this potential has not yet been realised. It argues that, to be meaningful, an EU FCP has to be more than a climate-focused version of FFP. Although foreign policy has to integrate climate change, climate policy is internal as well as external, so an FCP modelled only on FFP is inadequate. Moreover, the 'feminist' in FFP is rarely transformative, whereas transformation is required if climate disaster is to be averted. As the UN Climate Change Executive Secretary, Simon Stiell (2024), stressed at COP29, 'Business-as-usual won't get us there. We need urgent action. We need transformation'.

In the following sections, I first introduce FFP, then explore the meaning of 'feminist' in FFP. I then trace the evolution from FFP to FCP, before introducing the EU as a case study. This is followed by the methodology, findings, and conclusions.

What is FFP?

FFP originated in Sweden, which announced in 2014 that its foreign policy would henceforth be 'feminist'. The then Minister for Foreign Affairs, Margot Wallström, presented her initiative as consisting of 'three Rs': rights (women's rights are human rights), resources (women should have the same resources as men), and representation (women should be represented in decision-making). Further clarifications accompanied developments in Sweden's FFP during its eight years of existence (Sundström and Elgström, this volume), until it was abandoned in 2022 by the new right-wing government.

Other countries rapidly followed Sweden's example, declaring their foreign policies also 'feminist'. In the EU, this includes France (2019), Germany (2021), Luxembourg (2018), Netherlands (2022), Slovenia (2023) and Spain (2021). There have also been calls, primarily from within the European Parliament and from some civil society organisations (CSOs), for the EU to follow suit.

However, what FFP means in each case varies, depending on the social, economic, political and cultural context (Aggestam and Bergman-Rosamond, 2016; Sundström and Elgström, this volume). Sweden's FFP was introduced by a government that had already declared itself feminist and was built on a solid base of gender equality activity. Nevertheless, it has not been without its feminist critics, who have challenged the binary conceptualisations of gender on which it rests; the absence of the rights and needs of LGBTQI+ individuals; the continuing export of arms, including to authoritarian regimes; its migration and asylum policies; and the treatment of indigenous people (Bernarding and Lunz, 2020, p. 23).

What is the 'feminist' in 'Feminist Foreign Policy' and why does this matter?

The term 'feminist' appeared in Sweden's FFP to describe the integration of a gender equality perspective throughout all areas of foreign policy. In this context, it equates to 'gender mainstreaming', the idea that gender equality should be integrated into all areas of policy and at all stages of policy-making. 'Gender mainstreaming', like 'gender equality' and even 'gender', has been used by different actors to mean very different things (Cornwall, 2007; Zalewski, 2010). It was introduced at the fourth UN conference on women in Beijing in 1995 and widely adopted by governments around the world as a means of meeting the objectives set out in the Beijing Platform for Action. Gender mainstreaming countered the idea that gender equality could be achieved by separate ministries or departments for women's rights, and instead needed to be integrated in all sectors from initial policy proposals, to implementation, monitoring and evaluation (Daly, 2005; Stratigaki, 2005). Gender mainstreaming's popularity has been accompanied by a watering-down of its initial transformative potential, to the extent that many see it as a tick box exercise or a means of adding 'gender', or even just 'women', to unchanged structures and processes (Arora-Jonsson and Sijapati, 2018; Rao et al., 2016). This may explain the decision by some to use the less neutral term 'feminist' instead of 'gender mainstreaming', drawing attention to their transformative intent. 'Feminism', for many at least, carries with it the mission to undo gendered power relations, so its use in this context seems important, but also somewhat surprising. When and why did it become acceptable to use the term 'feminist' in such a mainstream context as foreign policy? What work is the term 'feminist' doing?

There are many types of feminism, from economically liberal feminism that seeks equality between (some) women and men within existing structures and institutions to intersectional, anti-racist, anti-capitalist feminism that seeks transformative change. The 'feminist' in FFP could contain any of a wide range of meanings. For example, the Centre for Feminist Foreign Policy (CFFP) (2021), a membership-based research organisation promoting a feminist approach to foreign policy, and a highly visible participant in debates around FFP in the EU, defines feminism as:

[...] a set of philosophies which drive political organising. Feminism acts as a tool to both analyse and question existing power hierarchies as well as present new and alternative visions for equal and just societies for all. Its primary goal is to end all types of oppression, injustices, and power hierarchies, including sexism, racism, classism, colonialism, and imperialism, among others. The everyday lived experiences of marginalised people with diverse backgrounds, experiences, and identities are reflected, and so, intersectionality is a core tenet to feminism.

If this meaning of feminism were to be applied to FFP, it would radically transform foreign policy and the social, political, and economic landscape in which it is formulated and implemented. So, is the spread of FFP an indicator of the success of feminism? Or is it evidence that the term 'feminism' has been co-opted and is used to serve neoliberal ends, as Nancy Fraser (2009) warned?

The debate around the institutionalisation or co-option of feminism is not new (Kantola, 2023). While some feminists have rejected work with the state on the grounds that it dilutes their demands and removes their control over the meaning of key terms such as gender, equality, and empowerment, others have seen engagement with the state as a way of advancing their long-term objectives. Some scholars have shown that it is the combination of feminists within state structures and outside them which has led to key feminist victories (Weldon and Htun, 2013; Woodward, 2003). Feminists have worked within and outside the state and international organisations to bring about changes in public policy and international law which advance feminist goals (Morrow, 2021).

It could also be that the ambiguity of the term 'FFP' itself serves a purpose, as Andrea Cornwall (2007) argues in relation to international development's 'buzzwords and fuzzwords', such as 'poverty reduction', 'empowerment' and 'development' itself. She writes: 'Development's buzzwords gain their purchase and power through their vague and euphemistic qualities, their capacity to embrace a multitude of possible meanings, and their normative resonance. [...] Buzzwords do not just cloud meanings: they combine performative qualities with an absence of real definition and a strong belief in what the notion is supposed to bring about'. Following this argument, it could be that a depoliticised version of the term 'feminist' in FFP enables the endorsement of a wide range of political actors and provides opportunities for them to fill this concept with meanings they find acceptable and useful. It could also be that the recent backlash against the term 'gender' has repoliticised this previously neutralised term, causing some actors to avoid it (Butler, 2024).

The relatively sudden appearance in the mainstream of the term FFP has not been greeted uncritically by feminist scholars. Guerrina, Haastrup and Wright (2023, p. 489), for example, state:

The question for feminist and gender scholars is whether FFP represents a shift in the way gender perspectives have been integrated into global politics, or whether it is a mere branding exercise by states seeking to affirm their position in the international system. Ultimately, it is probably a bit of both, insofar as it has elevated discussions about gender and other feminist principles to the area of "high politics", but it has done so in limited and limiting ways.

From FFP to FCP

Once the term FFP had begun to spread, the term FCP started to appear, often in the context of aspects of foreign policy that are interwoven with the effects of climate change and concerns about ensuing threats. In EU policy discourse, these connections are frequently labelled nexuses, for example, the 'climate-migration-security nexus' (Allwood, 2020b). FCP in this sense has therefore emerged from FFP and remains closely connected with its focus on foreign and security policy. Within foreign policy, climate change features heavily in climate diplomacy and international development, as well as migration control. We therefore see FCP developing in these same areas.

The disadvantage of building on FFP to design FCP is that climate policy is not solely external. Excluding internal climate mitigation and adaptation policy from FCP is problematic. It suggests a failure to practise at home what you preach abroad, and wrongly suggests that the problem of gender inequality only exists elsewhere, having been resolved in Europe (see Montoya and Agustin, 2013).

We can trace the connections between FFP and FCP and see the latter as a logical successor to the former. However, we can also trace alternative genealogies of FCP, seeing its origins in the rich heritage of eco-feminisms and feminist work in other sectors (MacGregor, 2014; Buckingham and Le Masson, 2017). This appears to be a more fertile ground for the growth of an FCP that challenges gender and climate injustice. FFP as a label has caught the imagination of some policymakers, but feminist policy has a longer history and is not restricted to foreign policy (Abels et al., 2021; Mazur, 2017). We need to draw on this wider literature in order to inform our understanding of FCP, rather than restricting ourselves solely to FFP, which is partly driven by the co-opted versions put forward by governments, rather than the radical transformative versions put forward by scholars and activists.

The EU as case study

The EU is an interesting case study, because it presents itself both as a global gender actor and as a global climate actor. The latest iteration of the EU's Gender Equality Strategy (European Commission, 2020) calls for the EU to be a 'global frontrunner in promoting gender equality' and for it to contribute to a gender-equal world. The EU has grown as a foreign policy actor, and climate change has gained an increasingly prominent place within EU foreign policy. When it created its new foreign office, the European External Action Service (EEAS), in 2010, the EU expanded its foreign policy, subsequently including external climate policy, to comprise climate diplomacy (engagement in international negotiations and engagement with partner countries about climate action); climate action in international development; climate change in its nexuses with migration and security; and climate change as a cross-cutting issue which needs to be mainstreamed throughout all external (and internal) actions. Although the EU has not adopted an FFP or an FCP, there are calls from within and outside EU institutions for it to do so.

Climate action and gender equality have been explicitly identified as two of the EU's political priorities (von der Leyen, 2019), but they have been pursued in parallel, rather than in an integrated fashion (Allwood, 2020b, 2021). EU climate policy continues to be largely impervious to a gender equality perspective, despite the pursuit of a Union of Equality and the EU's long-standing commitment to mainstreaming gender throughout all of its internal and external actions (Allwood, 2023; Allwood and Kronsell, 2025). The latest iteration of the EU's Gender Action Plan for its external action (GAP III) (European Commission and High Representative of the Union for Foreign Affairs and Security Policy, 2020) claims to aim for gender transformation, 'tackling the root causes of gender inequality', 'examining, questioning and changing rigid gender norms and imbalances of power', and 'address[ing] intersectionality'. However, a gender equality approach, to which the EU regularly expresses commitment, is not necessarily the same as a feminist approach, although it could be, as both terms are open to framing and re-framing.

The EU is a multi-level governance actor. Its Member States influence policy direction, normative trends, and play a key role in implementation. Sweden and like-minded Member States have helped shape gender equality policy and have raised awareness of FFP (Sundström and Elgström, 2020). Some EU countries have adopted an FFP and there have been some calls for an EU FFP (Greens/EFA, 2021). Civil society organisations, such as the European Environment Bureau and Women Engage for a Common Future (2021), are also advocating for an EU FFP. However, resistance from Member States, such as Poland and Hungary, has grown in recent years, in a context of both an anti-green and an anti-gender backlash (Tocci, 2022).

Within the EU, the various institutions play different roles in establishing policy priorities, formulating and adopting policy, and monitoring and evaluating its implementation. Relations between the institutions influence policy outcomes, as does the broader socio-political context. For example, a growing backlash against gender equality and climate action is having a marked impact on policy priorities and content. The European Commission proposes new policies, which are then discussed, shaped, and adopted by the Council and the European Parliament. Within the European Commission, the Directorate General for Climate Action (DG CLIMA) plays a leading role, although the principle of climate mainstreaming means that EU action in all policy sectors should align with the objectives of the European Green Deal, the framework announced in 2019 for achieving climate neutrality (European Commission, 2019). This is important, since DG CLIMA cannot address the problem of climate change if its initiatives are simultaneously undermined by the actions of the parts of the Commission responsible for energy, industry, transport, agriculture, and others. The Executive Vice-President and Commissioner for climate action is responsible for the European Green Deal, including delivering on the EU's domestic climate targets, and negotiates internationally on behalf of the EU. Departments within the Commission are responsible for climate action within the EU and globally, climate main-streaming and policy coordination, international climate negotiations, energy cooperation with neighbouring countries, and contributing to negotiations on Free Trade Agreements with third countries (Kahlen et al., 2023, pp. 32–33).

The Council of the EU adopts conclusions on climate diplomacy (since 2021, climate and energy diplomacy) every year, setting out the EU's position and priorities on climate change in its external relations. The Council and the European Parliament together adopt legislation on the basis of European Commission proposals. The Council is made up of ministers from the Member States and reflects the range of national interests and positions. The European Parliament has traditionally been greener and more gender equal-ity-friendly than the Council, but can be squeezed out of decision-making, particularly in areas where Member States seek to retain control. The Council has retained exceptional control of climate policy-making, meaning that the Member States have more influence and the European Parliament has less influence than is set out in the Treaties (Dupont, 2019). The European Parliament has a Green group and an environmental committee, which scru-tinise policy proposals, initiate studies, and resolutions. Members of the European Parliament from other political groups include opponents to Green and feminist policies (Kantola and Lombardo, 2021).

Methodology

The objective of this chapter is to assess the extent to which the EU can con-tribute to an FCP. In order to meet this objective, it addresses the following research questions: To what extent is gender mainstreamed into EU climate policy and what obstructs or facilitates a transformative EU FCP? The research questions are answered using a combination of content analysis of key policy documents (listed in the references) and a feminist institutionalist analysis of climate policy. The documents selected for analysis are those which set out EU strategic priorities in foreign and climate policy, and those which pay particular attention to gender equality in these policy sectors. They were selected to cover a range of institutions, from the Council, European Commission and European Parliament to the Committee of the Regions, and include internal and external climate policy. The content analysis firstly identified all references to gender/gender equality/gender mainstreaming/men/women, then asked whether gender was confined solely to a statement of commitment, contained in discrete sections, or mainstreamed throughout. To this end, questions concerned: whether concrete actions are proposed; are the goals transformative; does it challenge existing power relations and institutional structures? A second phase exposed similarities and differences within and between institutions in the meanings attached to these terms and the role they placed in the documents as a whole, drawing out integrationist and transformative approaches to gender mainstreaming, and the presence or absence of gendered perspectives in internal and external climate policy.

A feminist institutionalist approach (Chappell and Waylen, 2013; Krook and MacKay, 2011) is applied to show which institutions facilitate or obstruct moves towards an EU FCP. Feminist institutionalism tells us that institutions matter, that they are riddled with gendered power relations, and that they are resistant to change. Feminist institutionalism can help explain the gap between, on the one hand, formal commitments to gender mainstreaming and gender equality in all policy areas and at all stages of policy-making and, on the other hand, persistently gender-blind policy in particular areas, in this case, climate change (Magnusdottir and Kronsell, 2021). It enables us to examine the institutional constraints, opportunities, and resistances that affect gender mainstreaming within climate change policy-making. Feminist discursive institutionalism in particular enables us to focus on the construction and contestation of meaning in the interaction between gender mainstreaming and climate change policies (Schmidt, 2012). Individual and collective actors engage in struggles to impose their understandings of gender mainstreaming, and this is affected by the broader context of institutional power imbalances that push issues such as gender equality to the centre or the margins of particular policy debates. Feminist institutionalism can reveal ways in which gender mainstreaming is imbued with new meanings in day-to-day policy-making practices and can highlight the ways in which issues are constructed as certain types of problem requiring certain types of solution.

Civil society organisations' documents (listed in the references) were used to inform the analysis. They were selected to represent the views of key feminist environmental and climate advocacy organisations which attempt to influence EU policy and are frequently cited by policymakers in the European Parliament and the European Commission. The analysis of CSO documents, alongside EU policy documents, exposes the struggles around meaning and how the various actors use and redefine the terms FFP and FCP, as well as gender, gender equality, and gender mainstreaming.

The analysis finds that EU climate policy is not gender mainstreamed; there are differences in the meanings attributed to gender, gender equality, and gender mainstreaming; references to gender equality appear more frequently in external than in internal policy; and rhetorical commitments to gender equality disappear before policy is implemented. These findings are presented in the following sections.

EU climate policy is not gender mainstreamed

The analysis of EU policy documents shows that a gender equality perspective is not integrated into climate policy. This is despite the EU's longstanding commitment to gender equality and gender mainstreaming as a means to achieve it, and despite the promising appearance of both climate policy and a Union of Equality amongst the political priorities of the incoming European Commission announced by Ursula von der Leyen in 2019. As a political priority, EU climate policy was articulated in the European Green Deal (COM(2019)640 final), the overarching framework for a transition to a climate-neutral economy. However, the European Green Deal does not contain a single mention of gender, women, or equality. A series of laws have been adopted fleshing out the overall aims of the European Green Deal and adding binding targets and actions for Member States. Many of these also ignore gender entirely, as demonstrated by Allwood and Kronsell (2025).

EU institutions frame gender differently

Where there are references to gender equality in the selected policy documents, there are clear differences between the institutions in the way in which it is framed, and this affects the potential for an emerging FCP. EU institutions differ in their support for, or opposition to, the integration of a gender equality approach, or gender mainstreaming. As the different institutions exert more or less influence in different policy sectors, this also affects the extent to which gender equality is integrated. The European Parliament, and in particular its various committees on the environment, development, and gender equality, have been increasingly active in advocating the mainstreaming of these issues throughout all European Parliament decision-making, but the European Parliament can be excluded from forms of decision-making dominated by intergovernmentalism, and this applies to most of the Union's climate change policy. The Council plays a more important role in intergovernmental decision-making, and is the key decision-maker in foreign and security policy, much of which requires unanimity on the part of the Member

States. Each EU institution plays a specific role in the formulation, adoption, and implementation of EU climate policy, and external agencies and civil society organisations engage differently with each of them. Calls for an FCP will therefore resonate differently with the different EU institutions and will gain more traction in some than in others.

The analysis shows that gender equality is more frequently mentioned in the documents produced by the European Parliament than those produced by the European Commission or Council. For example, the European Parliament (2020, 2022) states that women and girls are disproportionately and negatively impacted by climate change; that gender inequality is exacerbated by climate change; and that gender equality is essential to achieving climate objectives and sustainable development. It emphasises the importance of gender mainstreaming as well as women's participation in climate decision-making; and women's organisations' access to international climate funds. It also advocates the empowerment of women and girls in the design and implementation of effective approaches to mitigation and adaptation in partner countries (European Parliament, 2022). This goes beyond participation in decision-making, demonstrating the influence of gender and development approaches, which have a strong tradition in EU development policy and in the work of the European Parliament's development committee (Allwood, 2020a). The Resolution on GAPIII also refers to effective gender-transformative climate action (European Parliament, 2022), in line with the CFFP report produced for the Greens/EFA group in the European Parliament (Greens/EFA, 2021).

In contrast to the European Parliament's efforts to address climate policy from a gender equality perspective, the Council tends to frame the problem of climate change as a complex security challenge or threat (Council of the European Union, 2022, 2023a, 2024; von der Leyen, 2019), and 'as a core component of EU foreign and security policy' (Council of the European Union, 2024). This has implications for the way in which it is perceived to relate to gender equality. For example, the Council Conclusions on Women, Peace and Security (Council of the European Union, 2022) include climate change in a long list of 'complex security challenges'. Towards the end of the Council Conclusions on Green Diplomacy (Council of the European Union, 2024), there is a strong reassertion that: 'climate change and environmental degradation lead to increased instability and conflicts and vice versa, as well as to human suffering, resource scarcity, including water and food insecurity, internal displacement and forced migration [...] The Council therefore calls for further engagement on these issues in relevant multilateral and international fora while paying specific attention to the disproportionate effects on vulnerable people, as well as women and children, including children in armed conflict'. Climate change is consistently represented as disproportionately harmful for women and girls (Council of the European Union, 2022,

2024). While the relevance of gender to climate change and climate action is acknowledged in these statements, it is done in a way that emphasises the vulnerability of women and girls and the securitisation of solutions proposed. Council declarations often lack detail, having been produced by the Member States in a quest for unanimity, and assertions that the 'gender dimension' will be taken into account do not specify what this means or how it might be addressed.

External versus internal climate policy

Gender equality is more frequently mentioned in EU external climate policy than internal. This is despite the fact that the EU is bound to mainstream gender throughout all internal and external actions. While gender equality appears in policy documents focused on aspects of external climate policy, such as climate diplomacy and climate change in international development cooperation, instances of internal climate policy engaging with gender in any way are scarce. One rare example is the Commission's Recommendation on Energy Poverty (European Commission, 2023), which recognises the gendered nature of energy poverty, stating that, 'Women, and in particular those who are single parents and older women, are also particularly affected by energy poverty due to structural inequalities in income distribution, socio-economic status and the gender care gap'. To address this, it states its preference for 'Structural measures [...] that address energy poverty at its root causes through investments into energy efficiency or renewable energy sources', rather than income support measures, which, although necessary during the energy crisis, 'are not an alternative to measures such as building renovations'. In an approach which links structural inequalities of gender and income, and with potential to contribute to an FCP, the Commission states that 'Giving priority to the renovation of the buildings with the worst energy performance allows to directly address energy poverty, since people affected by energy poverty and vulnerable people tend to live in such buildings'. However, this recognition of the link between gender and energy poverty is exceptional. Moreover, even in this document, the link is not articulated further, and there is no evidence that it informs broader energy policy formulation, which continues to ignore gender.

Commitments to gender disappear before implementation

While policy formulation is initiated by the European Commission and revised and adopted by the Council and the European Parliament, most climate policy is implemented at the regional and local level. According to a report for the European Committee of the Regions (Martinos et al., 2022): 'Local and Regional Authorities (LRAs) implement 70% of climate mitigation measures, [and] 90% of climate adaptation policies'. This makes it particularly significant that the European Committee of the Regions' (2022) Handbook on Implementing the European Green Deal is very scant on references to gender and is certainly not gender mainstreamed. One of the few mentions of gender notes that women are more likely to be affected by transport poverty; that car-reliant infrastructure may exacerbate gender gaps and reduce overall accessibility, as well as quality of life; and that, *as transport users, women, children and the elderly have specific needs* (my emphasis). However, this is itself a distortion. According to Eurostat (2023), 15% of the EU population is under 15 and 21.2% of the EU population is over 65. This means that 36.2% of the population are children and the elderly (both of these categories are, of course, also gendered). Just over half of the remaining 63.8% are women. So, more than 68% of the EU population are described in this handbook as *transport users with specific needs (women, children and the elderly)*, whereas the adult men (32% of the population) are understood to constitute the norm. Women are consistently represented as exceptions to the norm, with specific needs. Mainstream policy, including climate, energy, and transport, is assumed to be gender-neutral with a few exceptions. Gender mainstreaming, introduced in the 1990s to illuminate and correct this problem, has clearly not succeeded in EU climate, energy, and transport policy.

Along with implementation, monitoring and evaluation are also essential to ensure that grand statements on gender equality and gender mainstreaming are translated into concrete action and impact. The 8th Environmental Action Programme (European Parliament and Council, 2022) states that 'a gender perspective on actions and goals related to the attainment of the priority objectives of the 8th Environmental Action Programme is necessary in order to help ensure that gender inequalities are not perpetuated'. This is to be achieved by 'gender mainstreaming throughout climate and environmental policies, including by incorporating a gender perspective at all stages of the policy-making process'. However, there are no references to gender in the accompanying monitoring framework (European Commission, 2022). Without monitoring, we will not know whether and how a gender perspective is being – or could be – implemented at the local, regional, and EU levels. This is an example of the commonly noted phenomenon of gender evaporating between rhetorical commitments and effective implementation (Longwe, 1997).

How have civil society organisations contributed to gendering policy?

Since 2020, civil society organisations and thinktanks have been producing reports, studies, and recommendations on an EU FFP and, to a lesser extent, an EU FCP. The feminism in these reports tends to be more intersectional than that of the EU institutions, and their proposals more transformative and cross-cutting. The importance of care features in these analyses, and the conceptualisation of an FFP evolves in line with the work of the Centre for Feminist Foreign Policy (CFFP) and others. European CSOs have advocated for gender-responsive climate policy, including a feminist European Green Deal (European Environment Bureau and Women Engage for a Common Future 2021). The European Environment Bureau and Women Engage for a Common Future have used the term 'feminist' to describe 'a systemic transformation of our economic, social and political system away from the fixation on GDP growth towards values of inclusion, care and wellbeing for people and planet, where nature and its resources are regarded as the essential life support on which we all depend' (Heffernan et al., 2021, pp. 6–7).

The CFFP's report (Bernarding and Lunz, 2020) 'A Feminist Foreign Policy for the European Union', commissioned by the Greens/EFA group in the European Parliament, argues that the EU repeatedly asserts its commitment to gender mainstreaming, but faces considerable challenges when it comes to the implementation of an FFP. These challenges include the persistent understanding of gender as synonymous with (white, heterosexual) women and difficulty acknowledging intersecting discriminations and gender as structural power relations. The authors claim that the EU treats gender equality as an afterthought that is added onto gender-blind policies. They argue that an EU FFP would need to adopt and institutionalise an intersectional definition of gender that recognises it as a system of power, intersecting with, and reinforcing, other systems of power, including colonialism, slavery, class, race, and caste.

CSOs continue to play an important role in advocacy for more gender-just approaches to addressing the climate crisis. The findings of some of their reports are debated in the European Parliament and are occasionally referred to in Council Conclusions and European Commission Staff Working Papers, and some positive impact can be identified in EU policy documents. For example, CAN Europe (2023) finds that the climate and energy diplomacy conclusions from March 2023 are at least a step in a more ambitious and feminist direction than the Climate Diplomacy Conclusions of 2019, 2020, 2021, and 2022, which all contained the same vague statement that the 'EU will continue to uphold, promote, and protect gender equality and women's empowerment'.

Conclusion: to what extent can the EU contribute to an FCP?

The policy documents analysed reveal elements of potential contributions to an FCP. On paper, some of the EU's stated positions align with FCP objectives. However the depth of engagement is ambiguous, for example, the 2023 updated Nationally Determined Contribution of the EU and its Member States (Spain and the European Commission, 2023) mentions gender only once, but it does so in an ambitious and overarching statement: 'The EU is committed to promoting a human rights-based and gender-responsive approach to climate action, promoting social justice, fairness and inclusiveness in the global transition towards climate neutrality, full, equal and meaningful participation and engagement of women in climate related decision-making and fully meeting our human rights obligations when taking action to address climate change' (Article 27). Similar statements appear in, for example, the Council Conclusions on Climate and Energy Diplomacy (Council of the European Union, 2023b) and the Council Conclusions on Green Diplomacy (Council of the European Union, 2024).

Such statements are not meaningless. They establish or reiterate norms to which EU actors can then be held accountable. A growing interest in FFP, prompted by some Member States, starting with Sweden, and pursued by the Green group and the women's rights and gender equality committee in the European Parliament, has the potential to evolve into calls for an FCP, and some evidence of this is already present. However, the route to an FCP would need to include both internal and external climate policy and would also need to extend across all policy sectors. Greater coherence between the European Green Deal and a Union of Equality; and between climate mainstreaming and gender mainstreaming would enhance the EU's contribution to an FCP. The persistence of siloed policy-making, despite the rhetoric of nexuses and mainstreaming (including climate and gender mainstreaming) hinders the move towards an FCP, and the context of the 2024 European Parliament elections, in which the number of right-wing, populist, anti-gender, and anti-Green MEPs increased, combined with a global context of [international gender equality] 'norm spoiling' (Holmes, 2024) makes this work more challenging.

The EU has the potential to contribute to an FCP, but this potential has not yet been realised. Although an FCP does not stop at gender mainstreaming, ensuring that gender is mainstreamed into all policy sectors and at all stages of decision-making is one of its essential components. Despite the EU's longstanding rhetorical commitment to gender equality and to gender mainstreaming along with targeted measures as the means to achieve it, it has not successfully mainstreamed gender into climate policy. In most of the EU climate policy documents analysed, gender is absent, is used to mean 'women and men', or is a synonym for 'women'. When gender equality is present, it is added on, not mainstreamed. Policies, structures, and institutions remain unchanged.

This chapter has shown that the meaning of the term 'feminist' in FCP (and in FFP) matters. If the feminist in FCP is neoliberal, the EU can continue to pursue growth through the European Green Deal and call it feminist as long as more women are added to decision-making and something is done about access to energy and to transport that suits mothers/carers. If the feminist in FCP is decolonial, anti-capitalist and intersectional, in line with definitions proposed by the CFFP, including in its report for the Greens/EFA group in the European Parliament, then the EU will need to pursue a transformation of existing power relations and build a Green and socially just Europe. Addressing the environmental crises, including climate change, requires transformation of structures, institutions and power relations and tackling root causes, including inequalities.

An understanding of climate justice which is socially just and gender just must necessarily challenge the EU's framing of the European Green Deal as a growth strategy. Unrelenting economic growth is fundamentally incompatible with the efforts to live within planetary boundaries, to halt climate heating, pollution and the biodiversity crisis (Khalfan et al., 2023). As Cohn and Duncanson (2023) observe, endless growth 'depends on endless extraction, expansion of production and consumption, and endless resource-use and exploitation – a recipe for climate- and eco-disaster'.

FCP cannot be a gentle call to add 'the gender perspective' to EU climate policy; it is instead a fundamental challenge to existing climate policy, and it has the potential to be transformative. In order to achieve this, it will be necessary to maintain pressure to ensure that the term feminist is not co-opted and depoliticised, but instead retains its critical and political meanings.

References

Abels, G. et al. (2021) *Routledge Handbook on Gender and EU Politics*. New York: Routledge.

Aggestam, K. and Bergman-Rosamond, A. (2016) 'Swedish Feminist Foreign Policy in the Making: Ethics, Politics, and Gender', *Ethics & International Affairs*, 30(3), pp. 323–334.

Allwood, G. (2020a) 'Gender Equality in European Union Development Policy in Times of Crisis', *Political Studies Review*, 18(3), pp. 329–345.

Allwood, G. (2020b) 'Mainstreaming Gender and Climate Change to Achieve a Just Transition to a Climate-Neutral Europe', *Journal of Common Market Studies*, 58, pp. 173–186.

Allwood, G. (2021) 'EU External Climate Policy', in G. Magnusdottir and A. Kronsell (eds) *Gender, Intersectionality and Climate Institutions in Industrialised States*. Routledge, pp. 36–51.

Allwood, G. (2023) *Climate Mainstreaming: Climate and Gender Policy*. Brussels: Foundation for European Progressive Studies.

Allwood, G. and Kronsell, A. (2025) 'The European Green Deal through an Equalities Lens: A Missed Opportunity?', in G. Abels et al. (eds) *The European Commission under President Ursula von der Leyen: Gender, Leadership, Policies and Crises*. Oxford: Oxford University Press.

Arora-Jonsson, S. and Sijapati, B.B. (2018) 'Disciplining Gender in Environmental Organisations: The Texts and Practices of Gender Mainstreaming', *Gender, Work and Organisation*, 25(3), pp. 309–325.

Bernarding, N. and Lunz, K. (2020) *A Feminist Foreign Policy for the European Union*. Brussels: The Greens/EFA in the European Parliament.

Buckingham, S. and Le Masson, V. (2017) 'Introduction', in S. Buckingham and V. Le Masson (eds) *Understanding Climate Change Through Gender Relations*. London and New York: Routledge, pp. 1–12.

Butler, Judith (2024) *Who's Afraid of Gender?*, Penguin.

Centre for Feminist Foreign Policy (CFFP) (2021) 'The Centre for Feminist Foreign Policy Glossary', https://centreforfeministforeignpolicy.org/wordpress/wp-content/uploads/2023/06/CFFPGlossaryfinal.pdf

Chappell, Louise, and Waylen, Georgina. 2013. "Gender and the Hidden Life of Institutions." *Public Administration* 91(3): 599–615.

Climate Action Network (CAN) Europe (2023) *Towards a Feminist Foreign Climate Policy: Considerations for the EU*, Brussels: Climate Action Network.

Cohn, C. and Duncanson, C. (2023) 'Critical Feminist Engagements with Green New Deals', *Feminist Economics*, 29(3), pp. 15–39.

Cornwall, A. (2007) 'Buzzwords and Fuzzwords: Deconstructing Development Discourse', *Development in Practice*, 17(4/5), pp. 471–484.

Council of the European Union (2022) '*Council Conclusions of 14 November 2022 on Women, Peace and Security*'.

Council of the European Union (2023a) '*Council Conclusions of 17 October 2023 on Preparations for COP28*'.

Council of the European Union (2023b) '*Council Conclusions of 9 March 2023 on Climate and Energy Diplomacy*'.

Council of the European Union (2024) '*Council Conclusions of 18 March 2024 on Green Diplomacy*'.

Daly, M. (2005) 'Gender Mainstreaming in Theory and Practice', *Social Politics*, 12(3), pp. 433–434.

Dupont, Claire. 2019. "The EU's Collective Securitisation of Climate Change." *West European Politics* 42(2): 369–390. https://doi.org/10.1080/01402382.2018.1510199

European Commission (2019) 'The European Green Deal, COM(2019)640 final, 11 December 2019'.

European Commission (2020) 'A Union of Equality: Gender Equality Strategy 2020-2025 COM(2020)152 final'.

European Commission (2022) '8th Environmental Action Programme Monitoring Framework COM(2022)357final'.

European Commission (2023) 'Recommendation of 20 October 2023 on Energy Poverty, 2023/2407'.

European Commission (2024) *COP29: The EU commits to ambitious climate action on gender and climate change*, Brussels: Directorate-General for Climate Action.

European Commission and High Representative of the Union for Foreign Affairs and Security Policy (2020) 'EU Gender Action Plan (GAP) III - An Ambitious Agenda for Gender Equality and Women's Empowerment in EU External Action. Joint Communication. JOIN(2020)17 final'.

European Committee of the Regions (2022) *Implementing the European Green Deal, Handbook for Local and Regional Governments*, Brussels: Publications Office of the European Union.

European Environment Bureau and Women Engage for a Common Future (2021) *Why the European Green Deal Needs Ecofeminism*, Brussels: EEB and WECF.

European Parliament (2020) 'Resolution of 23 October 2020 on Gender Equality in the EU's Foreign and Security Policy (2019/2167(INI))'.

European Parliament (2022) 'Resolution of 10 March 2022 on the EU Gender Action Plan III (2021/2003(INI))'.

European Parliament and Council (2022) 'The 8th Environmental Action Programme 2022 (Decision (EU) 2022/591 of the European Parliament and of the Council of 6 April 2022 on a General Union Environment Action Programme to 2030)'.

Eurostat (2023) *Demography of Europe 2023*, http://ec.europa.eu, accessed 14 December 2023.

Fraser, N. (2009) 'Feminism, Capitalism and the Cunning of History', *New Left Review*, 56, pp. 1–11.

Greens/EFA (2021) *Making the EU Foreign Policy a Feminist One. Practising What We Preach.* Brussels.

Guerrina, R. et al. (2023) 'Contesting Feminist Power Europe: Is Feminist Foreign Policy Possible for the EU?', *European Security*, 32(3), pp. 485–507.

Heffernan, R. et al. (2021) *A Feminist European Green Deal – towards an ecological and gender just transition*, Berlin: Friedrich Ebert Stiftung, Women Engage for a Common Future and European Environment Bureau.

Holmes, R. (2024) 'Feminist responses to "norm-spoiling" at the United Nations', *ODI Briefing Note.*

Kahlen, L. et al. (2023) *Climate Audit of the European Union's Foreign Policy*, Brussels: New Climate Institute.

Kantola, J. (2023) 'Feminist Governance and the State', in *Handbook of Feminist Governance.* Edward Elgar Publishing, pp. 51–62.

Kantola, J. and Lombardo, E. (2021) 'Strategies of Right Populists in Opposing Gender Equality in a Polarised European Parliament', *International Political Science Review*, 42(5), pp. 565–579.

Khalfan, A. et al. (2023) *Climate Equality: A planet for the 99%.* Oxford: Oxfam International.

Krook, Mona Lena, and Mackay, Fiona. 2011. "Introduction: Gender, Politics, and Institutions." In *Gender, Politics and Institutions*, London: Palgrave Macmillan UK, 1–20. doi:10.1057/9780230303911_1

Longwe, S.H. (1997) 'The Evaporation of Gender Policies in the Patriarchal Cooking Pot', *Development in Practice*, 7(2), pp. 148–156.

MacGregor, S. (2014) 'Only Resist: Feminist Ecological Citizenship and the Post-Politics of Climate Change', *Hypatia*, 29(3), pp. 617–633.

Magnusdottir, G.L. and Kronsell, A. (eds) (2021) *Gender, Intersectionality and Climate Institutions in Industrialised States*, Abingdon: Routledge.

Martinos, H. et al., *Equal opportunities and responsibilities in the implementation of the European Green Deal*, European Committee of the Regions, 2022.

Mazur, A.G. (2017) 'Toward the Systematic Study of Feminist Policy in Practice: An Essential First Step', *Journal of Women, Politics & Policy*, 38(1), pp. 64–83.

Montoya, C. and Agustin, L.R. (2013) 'The Othering of Domestic Violence: the EU and Cultural Framings of Violence Against Women', *Social Politics*, 20(4), pp. 543–557.

Morrow, K. (2021) 'Gender in the Global Climate Governance Regime: A Day Late and a Dollar Short?', in G.L. Magnusdottir and A. Kronsell (eds) *Gender, Intersectionality and Climate Institutions in Industrialised States.* Abingdon: Routledge, pp. 17–35.

Rao, A. et al. (2016) *Gender at Work. Theory and Practice for 21st Century Organisations*. Abingdon and New York: Routledge.

Schmidt, Vivien A. 'Discursive Institutionalism: Scope, Dynamics, and Philosophical Underpinnings' in Frank Fischer and John Forester eds. *The Argumentative Turn Revised: Public Policy as Communicative Practice*. Durham, NC: Duke University Press (2012)

Spain and the European Commission (2023) *Update of the Nationally Determined Contribution of the EU and Its Member States*, Brussels: Publications Office of the EU.

Stallard, Esme (2024) 'Vatican in row at climate talks over gender rights', *BBC News*, 20 November.

Stiell, S. (2024) UN Climate Change Executive Secretary Simon Stiell at a High-Level Dialogue on National Adaptation Plans during the UN Climate Change Conference COP29 in Baku, Azerbaijan, on 18 November 2024. https://unfccc.int/news/national-adaptation-plans-key-to-unleashing-the-transformative-power-of-resilience-and-protecting, accessed 20 November 2024.

Stratigaki, M. (2005) 'Gender Mainstreaming vs Positive Action: An Ongoing Conflict in EU Gender Equality Policy', *European Journal of Women's Studies*, 12(2), pp. 165–186.

Sundström, M.R. and Elgström, O. (2020) 'Praise or Critique? Sweden's Feminist Foreign Policy in the Eyes of its Fellow EU Members', *European Politics and Society*, 21(4), pp. 418–433.

Tocci, N. (2022) *A Green and Global Europe*. Cambridge: Polity.

von der Leyen, U. (2019) *My Agenda for Europe, Political Guidelines for the Next European Commission 2019-2024*.

Weldon, S.L. and Htun, M. (2013) 'Feminist Mobilisation and Progressive Policy Change: Why Governments take Action to Combat Violence against Women', *Gender & Development*, 21(2), pp. 231–247.

Woodward, A. (2003) 'Building Velvet Triangles: Gender and Informal Governance', in T. Christiansen and S. Piattoni (eds) *Informal Governance in the European Union*. Cheltenham: Edward Elgar, pp. 76–93.

Zalewski, M. (2010) '"I don't even know what Gender is": A Discussion of the Connections between Gender, Gender Mainstreaming and Feminist Theory', *Review of International Studies*, (36), pp. 3–27.

4

THE OCEAN WE WANT

A Feminist Approach to the Ocean Decade

*Susan Buckingham, Mariamalia Rodríguez-Chaves,
Ellen Johannesen, Renis Auma Ojwala, Zhen Sun,
Momoko Kitada, Francis Neat and Ronán Long*

Introduction

Ocean governance has been slow to adopt gender equality in international environmental policy-making. Though sustainable ocean and seas are identified as a sustainable development goal (SDG14) and therefore subject to articulation with the other 16 goals, including SDG5 'Achieve Gender Equality and Empower all Women and Girls', ocean policy has been very late to address gender equality. The United Nations Decade of Ocean Science for Sustainable Development, officially launched in 2021, committed in principle to systematically identifying and removing barriers to achieve full gender equality in ocean science (IOC UNESCO, 2020). The Agreement under the United Nations Convention on the Law of the Sea on the Conservation and Sustainable Use of Marine Biological Diversity of Areas Beyond National Jurisdiction (BBNJ Agreement), adopted in 2023, became the first international law of the sea instrument that contains specific gender-sensitive provisions (BBNJ Agreement, 2023). This is later than the United Nations Framework Convention on Climate Change (UNFCCC), whose Lima Work Programme on Gender was finally agreed in 2014; and considerably later than the International Union for Conservation of Nature (IUCN) which formally adopted gender equality measures in 1998, Beijing Platform for Action for Women (1995), and Agenda 21 (1992). The first Ocean and Climate Change Dialogue was mandated by UNFCCC COP25 in 2019 to facilitate interactive discussions between the two fields of oceans and climate change in recognition of the need for greenhouse gas emissions reductions to achieve a healthy and sustainable ocean. However, COP28 was the first climate conference to consider the ocean in a meaningful way (UNFCCC, 2023), although

DOI: 10.4324/9781003461005-7

there were few references to gender equality in the documentation. For example, the UNFCCC's 'Ocean Breakthroughs' identified as being needed to secure a healthy and sustainable ocean, reveal only two mentions of women: in the context of 'victims' and 'labour and human rights' (UNFCCC, 2023).

The working environment in the ocean context can be seen as exclusionary or even hostile for women: ocean and marine research often requires long residencies at sea on vessels which are dominated by men. Problems for women range from sexual abuse to inappropriate facilities and kit (Johannesen et al., 2022). Experience of working on board ship can often be a prerequisite for promotion in research or for involvement in networks which advise ocean policy. In any event, ocean policy-making at the international scale is male-dominated, while participation in ocean research and higher education, which provides the next generation of research professionals and policy makers, is also still mostly by men (Ojwala et al., 2022).

Because ocean, and therefore ocean research and governance, is so critical for climate regulation, it is important to be interrogated for its male/masculine biases. Much gender equality work in the environmental field is still focused on bringing equal numbers of women as men into decision-making, and yet it is abundantly clear that quantity alone does not achieve equality, let alone women's empowerment (Buckingham and Kulcur, 2017). Feminist leadership has been increasingly identified as necessary to achieve sustainability goals (social, economic, and environmental). For example, the Wellbeing Economy Governments Partnership, formed by six nations at a time when four of these were led by women: New Zealand, Finland, Scotland and Iceland, and the remaining two by governments led by men who publicly committed their governments to gender equality goals (Wales and Canada), issued a statement to COP26 in Glasgow which demanded more attention to gender equality and its links to sustainability. Their principles highlighted that the economy should serve human and ecological wellbeing, and that community ownership and fair distribution of goods and services needed to be achieved through participatory governance (Wellbeing Economy Alliance, n.d.). As this book's introduction has identified, feminist leadership sets out to be transformative, reflective, caring, responsible, transparent, non-violent, inclusive, courageous and with zero tolerance for discrimination and the abuse of power (Action Aid, n.d.; Fair Share of Women Leaders, n.d.; Centre for Feminist Foreign Policy, 2023). These qualities have a better chance of delivering a sustainable and fair world than those which currently dominate.

The book to which this chapter contributes seeks to establish what feminist climate policy and feminist climate leadership look like, and how this can make a difference to the, to date, failed attempts to deliver gender equality in sustainability (UN, 2023). Given the centrality of ocean governance, and the research which underpins this, to climate sustainability, this is an important though hitherto poorly understood context in which to understand how gender inequality works, and what can be done to address this. The chapter considers three research projects conducted under the auspices of the UN Ocean Decade, examining practices of a range of international intergovernmental and non-governmental organisations with ocean interests; an intergovernmental ocean science network moving towards setting a gender equality action plan; and gender (in)equality at the micro level of university departments specialising in ocean science. It will conclude by considering the extent to which ocean science and governance is achieving gender equality, and the importance of feminist leadership in achieving this.

Climate Change and the Ocean

There is a clear relationship between climate change and the ocean, but it is complex. The ocean covers 71% of the Earth's surface and contains about 97% of the Earth's water. The ocean supports a wide range of habitats and ecosystems and is interconnected with atmospheric and terrestrial components of the climate system through the global exchange of water, energy, and carbon. Climate change is affecting the ocean in many ways. There are three key pathways for climate change impacts on oceans, each of which then leads to a range of secondary impacts. First, climate change results in oceans taking up additional energy in the form of heat, resulting in a rise in ocean temperature. It is estimated that 90% of the excess heat caused by greenhouse gas emissions has been absorbed by the ocean. Second, the increasing levels of carbon dioxide result in increased ocean acidification changing the chemicophysical properties of the ocean. This is especially problematic for calcifying marine organisms, such as corals, molluscs, and crustaceans. Third, climate change results in the melting of sea ice and land-based ice, and the resulting run-off into oceans causes changes in salinity, circulation patterns, and sea-level rise. (IPCC, 2019). Rising ocean temperatures are changing weather patterns, with more severe storms being recorded in recent decades. The ocean can potentially play an important role in mitigating the impacts of climate change and it is increasingly recognised as a major carbon sink. Certain coastal ecosystems including mangrove forests, saltmarshes, and seagrass habitats are highly effective at carbon sequestration. The deep ocean also stores carbon derived from dead organisms for centuries to millennia before it is eventually released back to the surface through upwelling. If the oceans were not as expansive and deep as they are, the effects of climate change to date would be far more severe.

Human communities in close connection with coastal environments and small islands are particularly exposed to the social and economic impacts of ocean changes, such as sea-level rise and extreme weather events (Caron, 1990; Kulp and Strauss, 2019). It has further been acknowledged that among human communities, women and other minority groups are disproportionately affected by impacts of the ocean driven by climate change (UN 2023 Global Sustainable Development Report). While changes are already taking place, the ultimate scale and rate of climate-related changes to the ocean and human communities will depend on the level of effort of the global community to decarbonise, and on choices made about approaches to adaptation. The adoption of a feminist approach to ocean governance will be central to developing transformative climate actions to deliver the ocean we want.

Legal Framework of Ocean Governance

Law has for centuries structured the governance of the ocean, from the *Mare Liberum* influentially envisioned by the 17th-century Dutch legal theorist Hugo Grotius, to the United Nations Convention on the Law of the Sea (UNCLOS or the Convention) first agreed in 1982 well before climate change took centre stage on the international agenda. As a result, there was no direct reference to the impacts of climate change on the ocean. Nevertheless, provisions on States' rights and duties to protect and preserve the marine environment under Part XII are of relevance to addressing the effects of climate change to the ocean (ITLOS, 2024). In addition, considering its status as the 'constitution of the oceans', the Convention must be interpreted and applied with subsequent developments in mind, including emerging new international environmental law and climate change law (Koh, 1982; Redgwell, 2019).

While the interactions between climate change and the ocean are complex, the primary connection between human activities, climate change and the ocean is the emission of greenhouse gases (GHG) that result in increases in ocean temperatures, the melting of grounded ice, changes in the chemical composition of seawaters, and a long list of secondary impacts (IPCC, 2021). As such, GHG emissions fall under the UNCLOS (Article 1(1)(4)) definition of 'pollution', which is in the form of 'energy' and 'substance' introduced by human activities and 'results or is likely to result in' damage to the marine environment. (Doelle, 2006; Burns, 2007; Boyle and Redgwell, 2021; Harrison, 2021; ITLOS, 2024). The general obligation of States to protect and preserve the marine environment therefore requires them to take all necessary measures, including applying international rules and national legislation, to prevent, reduce and control GHG emissions from all sources (ITLOS, 2024).

The general obligation of States to protect the marine environment is supplemented and implemented through other international instruments (ITLOS, 2024). International rules and standards concerning vessel source GHG emissions are developed through the International Maritime Organization (IMO), which include legally binding and non-binding instruments (UNCLOS, Articles 194(3)(b) and 211(1); IMO, LEG/MISC.8, 2014). Annex VI of the International Convention for the Prevention of Pollution from Ships (MARPOL) establishes the legally binding technical standards to reduce

GHG emissions, while IMO GHG Strategy sets out the levels of ambition for its further reduction and requires States to develop and implement mid- and long-term further measures with possible timelines. (MARPOL, 1997; IMO, 2023). The UNFCCC and the Paris Agreement address the prevention and reduction of dangerous human interference with the climate system, including impacts on the ocean (UNFCCC, 1992; UNFCCC, 2015). These ocean-related legal instruments facilitate States to identify and undertake all necessary measures, both collectively and individually, to prevent and reduce GHG emissions from all sources.

A common feature of the legal instruments that support ocean governance is that they are gender-blind. Take UNCLOS for example; except for the expressed reference to various aspects of equality of all peoples of the world in the Preamble, no specific provisions contain considerations of the gender dimension of what happens at sea (Papanicolopulu, 2019). It requires further examination from a gender perspective to identify the legal or factual obstacles that result in the lack of representation or absence of women in specific activities and sectors relating to the governance of the ocean. For example, it can be observed that women are severely underrepresented in the bodies established by UNCLOS (Goettsche-Wanli, 2019). Interviews with representatives of IGOs and NGOs reported in Kitada et al. (2023) revealed the scope of women's representation at senior levels. The first female judge of the International Tribunal for the Law of the Sea, Elsa Kelly, was elected only in 2011, 15 years after the establishment of the Tribunal in 1996. The election of the first woman, Wanda-Lee De Landro Clarke, to the Commission on the Limits of the Continental Shelf was in 2017, 20 years after the Commission was established. The International Seabed Authority (ISA) has never elected a female Secretary-General or Deputy Secretary-General since its establishment in 1996. This lack of representation inevitably adds to gender inequality and marginalises women's contribution to creating the framework where rules will be developed and implemented.

Feminist legal scholars argue that there is a distinctively feminine way of thinking or solving problems, and that recognising and respecting the different voices of women in the law of the sea context can contribute to the progressive development of the law (Charlesworth et al., 1991). The need for change has been reflected in the law of the sea. For example, in March 2021, ITLOS decided to amend its rules to render them gender inclusive in relation to its constitution and functions (ITLOS/Press 314, 2021). The ISA also has launched and implemented various initiatives to promote women's participation and representation in deep ocean scientific research (ISA, 2023).

While the UNCLOS was mostly negotiated by men, recent treaties have seen increasing participation of women diplomats and experts. The BBNJ Agreement, adopted in 2023, is the first law of the sea instrument to include gender-sensitive provisions, which reflects the strong leadership of the

President of the Intergovernmental Conference, Rena Lee, as well as the female facilitators of the associated working groups. This Agreement also embedded gender-neutral language in its general principles and approaches. In addition, the nomination of members for subsidiary bodies established under the Agreement, must take into account gender balance (BBNJ Agreement, Articles 15, 46, 49, and 55). Furthermore, the modalities for capacity-building and the transfer of marine technology are also required to be gender-responsive (BBNJ Agreement, Articles 43).

Another notable development is that the third cycle of the Regular Process for Global Reporting and Assessment of the States of the Marine Environment, including Socioeconomic Aspects, will incorporate a dedicated chapter on gender, with the aim of providing specific content on high-level global aspects of gender in Section 5A, on a sustainable and inclusive ocean economy. This includes a) the role of gender in achieving the Sustainable Development Goals; b) gender issues in ocean science and ocean governance systems and the importance of the collection of gender-disaggregated data in these systems; and c) the consideration of how the gender lens has been incorporated into ocean science and technology to strengthen science-based approaches for addressing marine issues (inter alia: fisheries, aquaculture, climate change, shipping, marine conservation) (United Nations DOALOS, 2023).

Gender Mainstreaming in the Conduct of Ocean Science

Ocean governance systems depend on a network of knowledge produced by various national and international ocean science entities. Such knowledge is, however, disproportionately produced by men, who dominate both ocean science research and governance; a systemic issue that is also observed in other STEM fields and international environmental frameworks such as the UNFCCC. This results in failure to provide the diversity of knowledge and solutions required to solve complex ocean governance challenges related to the conservation and use of marine resources, as well as ocean health. Although recognised as an issue, the actual underrepresentation of women in ocean science is not well understood or documented due to the lack of baseline data about the number and contribution of women ocean scientists globally and nationally. It is important to not only demonstrate the true underrepresentation of women in ocean science but also to understand why it persists. The reasons for this can be found within organisational cultures and workplaces, but specifically for ocean science, they may relate to the challenges associated with sea-going research. Johannesen et al. (2022) suggest that four main factors may act as barriers to women: (1) behavioural/ social norms and gender-biased culture in science and at sea; (2) failure to provide for balancing duties of family care with extended periods away from home; (3) gender-insensitive design of ship facilities, operations, and personal protective equipment (PPE); and (4) the need for a safe working environment at sea, with attention paid to gender-related aspects of health, safety, and personal security at sea.

Three recent case studies have investigated the degree of gender inequality and the challenges women face in embarking on and developing a career in ocean science. The first case considers the role of gender and of women in senior leadership positions in ocean governance bodies; the second examines the case of an intergovernmental marine science organisation covering the North Atlantic Ocean; while the third is a study of ocean science institutions in Kenya with a remit in the Western Indian Ocean. This section reviews key lessons learned from these case studies and discusses how gender can be mainstreamed in the conduct of ocean science to bring positive change in ocean governance worldwide.

Gender Mainstreaming in Ocean Governance Bodies

Ocean governance entails a wide variety of actors and dynamics. At the international level, a number of intergovernmental organisations (IGOs) have been advancing gender equality strategies within their Secretariats and in coordination with member States, which the DFO-WMU Empowering Women for the UN Decade of Ocean Science for Sustainable Development programme has researched. The research focused on the inclusion of: a) a gender policy or action plan; b) a gender focal point; c) capacity-building/ education programmes; d) official commitments on gender and oceans; and e) publications and events.[1] The findings revealed that all IGOs have a gender policy in place and action plans/strategies, however, updated statistics on the implementation of gender targets or indicators need to be reflected in reports which address those instruments in order to identify weaknesses, monitor progress, and establish targeted actions. Likewise, gender focal points exist but in different formats throughout the IGOs. For example, ISA and Division for Ocean Affairs and the Law of the Sea (DOALOS) have designated officers, the Intergovernmental Oceanographic Commission (IOC) is part of the UNESCO and supported by its Gender Equality Division, the Food and Agriculture Organization of the United Nations (FAO) has a network of gender focal points, and the Joint Group of Experts on the Scientific Aspects of Marine Environmental Protection (GESAMP) has a spokesperson on gender. In addition, while all the IGOs have capacity-building/education programmes, some are more specialised than others in the field of marine science (Sun et al., 2021).

In relation to official commitments on gender and the ocean, only ISA has a specific voluntary commitment to enhancing the role of women in marine scientific research; FAO, DOALOS and IOC commitments respond to broader gender equality strategic objectives from United Nations Agencies (for example, the Beijing Declaration and Platform for Action, as well as UN General Assembly resolutions on gender parity). The majority of IGOs have publications on gender equality and organise events to raise awareness on women's empowerment.

Interviews with gender focal points and other officers focusing on gender mainstreaming in ocean governance bodies highlighted some critical challenges to equality. The following areas lacked or had poor performance regarding disaggregated gender data, a key component for informed decision-making processes and building baseline information; investment in capacity development programmes that seek gender balance; strengthening accountability (reporting and audit to monitor gender mainstreaming activities) (Sun et al., 2023). Other challenges identified included the ineffective implementation of gender equality policies, including disconnections between policy and practice regarding effective women's representation in decision-making bodies/senior leadership positions within States' delegations (also a shared responsibility with member States); a need for updated data on progress in implementing gender targets; inadequate recruitment processes to improve women's representation in high-level positions; lack of consideration of care responsibilities when designing fieldwork; and challenges to achieving structural changes to deliver gender equality including through a bias-free, inclusive, and enabling organisational culture.

On the positive side, diverse good practices to improve gender equality included flexible working hours; increased child-care measures; mentoring programmes to encourage younger women to progress professionally; monitoring and reporting requirements to track progress; establishing minimum standards for gender mainstreaming (e.g. statistical databases with gender-disaggregated data; gender balance assessments, performance indicators); having dedicated time and allocated resources, including funds, for gender equality actions within projects/programmes; facilitating support structures/networks for women, including frequent training (Sun et al., 2021).

The development of a Gender Equality Strategy and Action Plan informed by the data analysis and the identification of key areas to provide practical and policy-relevant recommendations was a key output of the Programme. This is designed to achieve gender equality in ocean science and ocean governance systems and identify how women's participation and representation in the conduct of marine scientific research in ocean science/ocean governance structures, and in decision-making processes, can be increased.

The proposed Gender Equality Strategy has six building-blocks: a) women in power, decision-making, and leadership; b) organisational culture; c) gender-responsive resources (including financial resource allocation, gender-responsive budgeting, and operational work allocation); d) monitoring and reporting; e) professional development and capacity-building; and f) communication and advocacy. The Strategy is designed to support organisations in advancing knowledge and understanding of gender-responsive actions and their coherent mainstreaming and implementation. It is also intended to enable IGOs to develop, update and/or improve their own gender equality action plan as an instrument to achieve gender equality and women's empowerment in ocean sciences. The Strategy can also guide member States and enhance synergies and collaboration for transformative actions, as well as address a continuing need for gender mainstreaming through targets and goals as an important contribution to increasing sustainability.

Mainstreaming Gender in an Intergovernmental Organisation

The second case study examined how an intergovernmental organisation concerned with ocean science – International Council for the Exploration of the Sea (ICES) – can mainstream gender through the development of a gender action plan. Further, it examines the specific challenges to achieving women's representation and leadership. The study took a Feminist Participatory Action Research (FPAR) approach, using both quantitative and qualitative analyses. An FPAR approach aims to rectify power imbalances, challenge assumptions of researcher neutrality, and view research as a transformative force, amplifying participant voices and the co-creation of knowledge. The researcher was simultaneously employed by ICES which gave her privileged access to ICES' members and staff as well as data from the day-to-day institutional operations, such as organising conferences and developing a gender equality action plan. This hybrid position of employee and researcher enabled a particularly rich form of participatory action research.

In the absence of self-identified gender-disaggregated data, a list of participants in selected ICES conferences was generated and gender was inferred based on first name, to make gender inequality visible. This quantitative approach also helped to show that while women make up approximately 50% of people in the organisation/network, they are underrepresented in decision-making and leadership roles, as well as recipients of awards (Johannesen et al., 2023). Qualitative approaches were also important for documenting the experiences and perceptions of the ICES community to help tailor specific interventions and help develop a common understanding of the challenges and solutions needed to improve institutional gender equality. The first female General Secretary of ICES, in its more than 100-year history, served from 2012 to 2021. It was under her leadership that issues of gender equality became an organisational priority and were included in the 2019 Strategic Plan (Sun et al., 2021).

Using a participatory and context-specific approach was critical for building community support, not only for the Secretariat but also for the entire organisation and its 20 member States. The researcher's insider access undoubtedly facilitated the work of developing and approving a gender equality plan (Johannesen, 2024). Although the structure of IGOs is complex, and access to the network is controlled by member States, the Secretariat can help move the needle on organisational gender equality. By raising awareness, and providing a platform for member countries to exchange knowledge and identify their own goals and solutions for equitable representation, it can create more inclusive working environments. It remains the case, however, that the organisation is unable to mandate the gender equality of scientists representing individual States. One conclusion of the research is that greater gender equality in IGOs requires "concerted actions, and policy interventions that address both the measurable and immeasurable, formal and informal processes, cultures, values, and norms" (Johannesen, 2024). Another is the need for institutional reforms which make it easier for women, minority ethnic people, care-givers, and others disadvantaged in the workplace, to work in and thrive in the organisation. This is preferable to focusing on individuals who are encouraged to fit into structures which remain masculinised and exclusionary.

Gendered Ocean Science Institutions in Kenya

This study focused on ocean science institutions in Kenya for a number of reasons. The country has a significant coastline along the Indian Ocean providing a rich and diverse marine environment that offers numerous benefits to the national economy, including fisheries, tourism, and maritime transport. Kenya has been actively involved in ocean-related initiatives globally, including hosting the Blue Economy Conference and co-hosting the UN Ocean Conference with the Government of Portugal in 2018 and 2022, respectively. The country is also home to the Nairobi Convention, part of UNEP's Regional Seas Programme, which aims to protect, manage, and develop the Western Indian Ocean.

The government of Kenya has also shown commitment to gender equality through the ratification of various international and legal frameworks and policies geared toward gender equality. It has also developed national and institutional gender policies to promote gender equality in different sectors, including public universities. Despite this, however, discrimination, biases, and marginalisation experienced by Kenyan women at different levels in ocean science are not well documented. This case study evaluated gender equality and women's empowerment in ocean science institutions in Kenya. A combination of evidence was used, including institutional gender policies, gender-disaggregated data of students and staff, and surveys and interviews with students and staff. Using Feminist Political Ecology (FPE) as a theoretical framework to analyse gender-biased access to higher education and job opportunities in the ocean science sector led to adopting an intersectional approach in which multiple factors, such as gender and age, family support, education, ethnicity, and class were examined. This approach provided a better understanding of gendered power relations and how these shaped different forms of inequalities in the institutions studied.

Several key findings emerged (Ojwala et al., 2022, 2024; Ojwala, 2023, 2024). First, strategies written into institutional gender policies were more likely to be gender-neutral than gender-specific. Consequently, the presence of these institutional policies did not translate to gender equality, a point reinforced by the finding that some institutions without gender equality policies performed the same as – or in some cases better than – those that had gender equality policies in place. There was also no relationship between the existence of a gender equality policy and higher numbers and proportions of women scientists and managers. It was also notable that every single gender equality policy reviewed for the research was outdated.

Second, similar to other STEM subjects, the proportion of female students enrolled in undergraduate ocean science programmes (35% of the total) progressively reduced through postgraduate levels (22% of all masters registrations and negligible in PhD programmes). Correspondingly, there were fewer women staff than men in ocean science-related departments in public universities, with an average of 32%. The majority of these women (60%) occupied junior positions, with only 14% of women staff employed as professors. Of non-academic institutions, including government agencies and international NGOs dealing with ocean science, government agencies recorded the lowest percentages of women in senior positions, with only 20% of women directors (Ojwala, 2023). Also important to note is that government-affiliated institutions seemed to be meeting, though never exceeding, the 33% minimum requirement for women (and men) stipulated in Kenya's 2010 Constitution. There was a better representation (45%) of women in NGOs, where women were also well represented in management positions (75%). Sixty-seven per cent of directors in IGOs were women.

Third, several barriers to accessing higher education by women students and to the career progression of women staff were identified. Cultural barriers (identified by 58% of students who responded to a survey), gender discrimination (by 58%) and gender stereotypes (by 53%) were perceived as the major impediments to women's access to higher education and career opportunities in ocean science (Ojwala, 2024). Several other barriers were also mentioned by staff participants, including discriminatory promotion guidelines, work-family conflicts, sexual harassment, lack of support and recognition, and lack of institutional gender policies.

Fourth, there was clear evidence that gender and ethnicity interacted to influence the enrolment and participation of female students in ocean science-related courses. Politically dominant ethnic groups, including Luo, Kikuyu, Luhya and Kalenjin were overrepresented among Kenyan institutions' students and staff groups (Ojwala, 2024). The majority of women, unlike men, felt they were discriminated against based on their gender, ethnicity, age, education, and social class. Ethnicity is one source of power in Kenya that can influence who can access opportunities in an institution, a pattern which reflects the dominant ethnic-related power relations in national politics (Weinreb, 2001; Taaliu, 2017; Kenya Human Rights Commission, 2018). The research established the salience of an intersectional lens which provides a better understanding of the subtle differences in individuals' access to educational resources. Further, the research also confirmed the FPE claim that biased representation ignores knowledge and perspectives that can ensure better environmental (and in this case, ocean) governance and emphasises the need for the involvement of women and diverse ethnic groups in decision-making positions.

Towards a Feminist Approach

The chapter has set out the importance of the ocean in climate change, and how recognition of this has only slowly been incorporated into UNFCCC discussions. This is paralleled by the tardiness of including any reference to, or action towards, gender equality in ocean science and governance. In terms of practical recommendations, systematic and participatory strategies need to be developed and implemented with monitoring and accountability actions.

The pioneering research programme that this chapter reflects on has identified a handful of cases where strong female leadership (even though those involved might not have identified as feminist at the time) appears to have made a positive contribution towards gender equality. These are: the decision to work towards gender equality through a gender action plan at ICES, changes to the BBNJ Agreement to include gender-sensitive provisions and adding an independent chapter on gender equality alongside treating gender as a cross-cutting issue in the World Ocean Assessment process. Elsewhere, some indications of isolated policies which support women have been noted, though other gender policies appear to be in name only.

We can conclude, then, that the journey to address the first part of the aim of the United Nations Decade of Ocean Science for Sustainable Development – to identify barriers to achieve full gender equality in ocean science – has begun. However, there is a way to go before they are dismantled. The introduction to this chapter set out a number of qualities associated with feminist leadership: to be transformative, reflective, caring, responsible, transparent, non-violent, inclusive, courageous and with zero tolerance for discrimination and the abuse of power (Action Aid, n.d.; Fair Share of Women Leaders, n.d.; Centre for Feminist Foreign Policy, 2023). While these qualities have mostly not been obvious in the three research projects, some evidence of reflective, caring, responsible, inclusive, and transparent practice has been noted in connection with the instances of feminist leadership described. It is too soon to determine how transformative this will be, and how much more widely these isolated examples will penetrate the ocean sector as a whole.

In terms of practical recommendations, systematic and participatory strategies need to be developed and implemented with monitoring and accountability actions. Institutions need long-term strong and committed leadership, which recognises that gender data is ocean data, and therefore promotes gender equality as an indivisible aspect of ocean sustainability.

Note

1 The project coordinated with focal points from the Intergovernmental Oceanographic Commission (IOC – UNESCO), the Food and Agriculture Organisation (FAO), the Division of Ocean Affairs and Law of the Sea (DOALOS), the International Seabed Authority (ISA), the Joint Group of Experts on the Scientific Aspects of Marine Environmental Protection (GESAMP),* and with non-governmental organisations, including the International Union for Conservation of Nature (IUCN) and the High Seas Alliance (HSA).

References

Action Aid (n.d.) *Ten Principles of Feminist Leadership* https://actionaid.org/feminist-leadership (Accessed 24 Jan 2024)

Boyle, Alan and Redgwell, Catherine (2021) *Birnie, Boyle and Redgwell's International Law and the Environment* (4th edn., Oxford University Press), 504–505.

Buckingham, S. and Kulcur, R. (2017) 'It's not just the numbers: Challenging masculinist working practices in climate-change decision-making in UK government and environmental non-governmental organizations' in Griffin-Cohen, M. ed *Climate Change and Gender in Rich Countries*. London and ~New York: Routledge, pp. 35–51.

Burns, William C.G., (2007) 'Potential Causes of Action for Climate Change Impacts under the United Nations Fish Stocks Agreement', *Sustainable Development Law and Policy* 7(2), 34–38, 36.

Caron, D. (1990) 'When Law Makes Climate Change Worse: Rethinking the Law of Baselines in Light of a Rising Sea Level' *Ecological Law Quarterly* 17(4), 621–653, 629 and 635–636.

Centre for Feminist Foreign Policy (2023) Feminist Foreign Policy Summit (Accessed 26 January 2024)

Charlesworth, H., Chinkin, C. & Wright, S. (1991) 'Feminist Approaches to International Law', *The American Journal of International Law*. 85 613–645.

Doelle, Meinhard, (2006) 'Climate Change and the Use of the Dispute Settlement Regime of the Law of the Sea Convention' *Ocean Development & International Law* 37, 319–337, 322.

Fair Share of Women Leaders (n.d.) *Feminist Leadership*. Heinrich Boll Stiftung. The Green Political Foundation https://www.boell.de/en/feminist-leadership (Accessed 24 Jan 2024)

Goettsche-Wanli, G.(2019) 'Gender and the Law of the Sea: a Global Perspective', in Irini Papanicolopulu (ed), *Gender and the Law of the Sea* Brill, 39.

Harrison, James, 'Litigation under the United Nations Convention on the Law of the Sea: Opportunities to Support and Supplement the Climate Change Regime', in Ivano Alogna, Christine Bakker and Jean-Pierre Gauci (eds), *Climate Change Litigation: Global Perspectives* (Brill, 2021), 415–432, 421.

Intergovernmental Oceanographic Commission (IOC) of UNESCO, United Nations Decade of Ocean Science for Sustainable Development 2021–2030, Implementation Plan, Version 2.0, July 2020.

International Maritime Organisation (IMO), Resolution MEPC. 377(80), adopted 7 July 2023, '2023 IMO Strategy on Reduction of GHG Emissions from Ships'.

International Maritime Organisation (IMO) LEG/MISC.8, 30 January 2014, Implications of the United Nations Convention on the Law of the Sea for the International Maritime Organization.

International Seabed Authority (ISA), (2023) Advancing Women's Empowerment in Marine Scientific Research, https://www.isa.org.jm/isa-voluntary-commitments/enhancing-the-role-of-women-in-msr/

IPCC (2019) Summary for Policymakers. In: H.-O. Pörtner, D.C. Roberts, V. Masson-Delmotte, P. Zhai, M. Tignor, E. Poloczanska, K. Mintenbeck, A. Alegría, M. Nicolai, A. Okem, J. Petzold, B. Rama, N.M. Weyer (eds.) *IPCC Special Report on the Ocean and Cryosphere in a Changing Climate*. IPCC.

IPCC, (2021) Summary for Policymakers. In: *Climate Change 2021: The Physical Science Basis. Contribution of Working Group I to the Sixth Assessment Report of the Intergovernmental Panel on Climate Change* [Masson-Delmotte, V., et al. (eds)]. Cambridge University Press, pp. 3–32, doi:10.1017/9781009157896.001 (IPCC SPM, 2021).

ITLOS (2024) Request for An Advisory Opinion submitted by the Commission of Small Island States on Climate Change and International Law, Advisory Opinion of 21 May 2024, ITLOS List of cases: No. 31.

ITLOS/Press 314, 25 March 2021, Tribunal Amends Rules with a View to rendering them Gender Inclusive. The following Rules were amended in the English text: articles 4, 5, 6 and 7 (The Members); articles 11, 12 and 13 (President and Vice-President); articles 16 and 19 (The Composition of the Tribunal for Particular Cases); article 31 (Special Chambers); articles 36 and 39 (The Registry); article 42 (Internal Functioning of the Tribunal); article 45 (General Provisions); article 76 (Oral Proceedings); and article 136 (Advisory Proceedings). https://www.itlos.org/en/main/resources/media-room/calendar-of-events/#ar492

Johannesen, E. (2024, forthcoming). Participatory approaches in support of institutional actions to advance gender equality in ocean science in the IGO context. Manuscript submitted for publication.

Johannesen, E., Barz, F., Dankel, D. J., & Kraak, S. B. M. (2023). Gender and early career status: Variables of participation at an international marine science conference. *ICES Journal of Marine Science*, 80(4), 1016–1027. https://doi.org/10.1093/icesjms/fsad028

Johannesen, E., Ojwala, R. A., Rodriguez, M. C., Neat, F., Kitada, M., Buckingham, S., Schofield, C., Long, R., Jarnsäter, J., & Sun, Z. (2022). The Sea Change Needed for Gender Equality in Ocean-Going Research. *Marine Technology Society Journal*, 56(3), 18–24. https://doi.org/10.4031/MTSJ.56.3.6

Kenya Human Rights Commission (2018). Ethnicity and politicization in Kenya. Available at: https://www.khrc.or.ke/index.php/publications/183-ethnicity-and-politicization-in-kenya/file (Accessed on 24 January 2024).

Kitada, M., Johannesen, Ellen, Ojwala, Renis Auma, Buckingham, Susan, Sun, Zhen, Rodríguez-Chaves, Mariamalia, Neat, Francis, Long, Ronán, and Schofield, Clive, (2023). 'Barriers to the Collection of Gender-disaggregated Data in Ocean Science' *Human Factors in Management and Leadership* 92, 71–80, https://doi.org/10.54941/ahfe1003736

Koh, Tommy T.B. (1982). 'A Constitution for the Oceans', Remarks by Tommy T.B. Koh, of Singapore, President of the Third United Nations Conference on the Law of the Sea, United Nations Conference on the Law of the Sea https://www.un.org/depts/los/convention_agreements/texts/koh_english.pdf

Kulp, S.A. and Strauss, B. H. (2019). 'New elevation data triples estimates of global vulnerability to sea-level rise and coastal flooding', *Nature Communications*, 10, 4844. https://doi.org/10.1038/s41467-019-12808-z

MARPOL (1997). Protocol of 1997 to amend the International Convention for the Prevention of Pollution from Ships, 1973, as modified by the Protocol of 1978 relating thereto (MARPOL PROT 1997), Annex VI: The Prevention of Air Pollution from Ships, adopted 26 September 1997, entered into force 19 May 2005.

Ojwala, R. A. (2023). Status of gender equality in ocean research, conservation and management institutions and organisations in Kenya, *African Journal of Marine Science*, 45(2), 105–115, https://doi.org/10.2989/1814232X.2023.2213724

Ojwala, R. A. (2024). Unravelling gender and ethnic bias in higher education: Students' experiences in access to ocean science education and career opportunities in Kenya. *Higher Education*, 87(3), pp. 1–22. https://doi.org/10.1007/s10734-024-01198-x

Ojwala, R. A., Buckingham, S., Neat, F., & Kitada, M. (2024). Understanding women's roles, experiences and barriers to participation in ocean science education in Kenya: recommendations for better gender equality policy. *Marine Policy*, 161 (106000), 1–12. https://www.sciencedirect.com/science/article/pii/S0308597X2300533X?via%3Dihub

Ojwala, R. A., Kitada, M., Neat F., & Buckingham, S. (2022). Effectiveness of gender policies in achieving gender equality in ocean science programmes in public universities in Kenya. *Marine Policy*, 144, 105237, 1–12. https://doi.org/10.1016/j.marpol.2022.105237

Papanicolopulu, I. (2019). 'Introduction – Gender and the Law of the Sea: Oceans Apart?', in Papanicolopulu, I. (ed), *Gender and the Law of the Sea* Brill, 19–20.

Redgwell, C. (2019). 'Treaty Evolution, Adaptation and Change: Is the LOSC 'Enough' to Address Climate Change Impacts on the Marine Environment?' *International Journal of Marine and Coastal Law*, 34, 440–457, 444–445.

Sun, Z., Auma Ojwala, R., Johannesen, E., Rodriguez-Chaves, M., Long, R., Kitada, M., Buckingham, S., Schofield, C. and Francis Neat (Eds.), (2023). *Scaling up Actions to Empower Women for the Ocean Decade-Joint 2022 United Nations Ocean Conference Side Events Report*, Malmö: World Maritime University.

Sun, Z., Long, R., Rodriguez-Chaves, M., Kitada, M., Neat, F., Schofield, C., Buckingham, S., Jarnsäter, J., Barjandi, E., Auma Ojwala, R., and Johannesen E. (Eds.), (2021). *Empowering Women for the United Nations Decade of Ocean Science for Sustainable Development*, Malmö: World Maritime University.

Taaliu, S. T. (2017). Ethnicity in Kenyan universities. *Open Journal of Leadership*, 6(2), 21–33.

United Nations (2023). Global Sustainable Development Report: Times of Crisis, Times of Change. Science for Accelerating Transformations to Sustainable Development. https://sdgs.un.org/gsdr/gsdr2023 (Accessed 24 Jan 2024).

United Nations DOALOS (2023). Regular Process for Global Reporting and Assessment of the States of the Marine Environment, including Socioeconomic Aspects. Third World Ocean Assessment Annotated Outline 2023. Available at: https://www.un.org.regularprocess/sites/www.un.org.regularprocess/files/2_clean_ver_edited_annotated_outline_of_third_assessment_final_clean.pdf (Accessed 19 April 2024).

United Nations Framework Convention on Climate Change (UNFCCC), (1992). adopted 9 May 1992, entered into force 21 March 1994, 1771 UNTS 107.

United Nations Framework Convention on Climate Change (UNFCCC), (2015) Paris Agreement adopted 12 December 2015, entered into force 4 November 2016, Article 4(2), UNTS I-54113.

United Nations Framework Convention on Climate Change (UNFCCC), (2023). Oceans. Available at: https://unfccc.int/topics/ocean#COP-28-2023 (Accessed 23 June 2025).

Weinreb, A. A. (2001). First politics, then culture: Accounting for ethnic differences in demographic behaviour in Kenya. *Population and Development Review*, 27(3), 437–467.

Wellbeing Economy Alliance (n.d.) Wellbeing Economy Governments https://weall.org/wego (Accessed 24 Jan 2024).

5

ENSURING JUSTICE THROUGH GOOD PRACTICE

Establishing the context for change across organisational scales

Seema Arora-Jonsson

Introduction

The European Green Deal as well as the mechanism for Just Transitions both speak of the need for justice in climate transitions. Feminist scholarship that has been at the forefront of thinking and the politics of justice, is sorely needed in times of change. Feminists have insisted on addressing the organisation of power relations in relation to environmental and climate questions, often through gender mainstreaming projects in organisations, with grassroots groups working and living in the environments and more recently in relation to the climate emergency. Echoing the overarching theme of the book, this chapter asks, is it possible for policymakers – even assuming that they are sympathetic – to overcome institutional barriers and conventions in pursuit of the radical changes necessary for a gender-just climate emergency response? What lessons can be drawn from "good practice cases"? I go on to problematise what this might mean in practice in relation to gender mainstreaming cases in Sweden that I identify as potential good practices.

Gender mainstreaming entails paying attention to how gender intersects with other dimensions of power and how attention to these dimensions of power would help to challenge discrimination. Dealing with this complexity in policies and projects has been far from easy. In recent years, in light of the Sustainable Development Goals (SDGs) and other international conventions, there has been a reaffirmation of the importance of gender and climate – both in policy and in academic literature. In 2014, in an overview of how research and policymaking had addressed the gender/power and environment nexus internationally,[1] I analysed the disjuncture between what feminist scholarship advocated and what was translated into policies. Gender scholars spoke

DOI: 10.4324/9781003461005-8

of the need for structural change, for example, through establishing clear ownership in favour of women while policy did not always address this. Gender research demonstrated how change happened through collectives and in particular through collective organising while policy, hoping to address marginalisation, turned largely to individuals. Importantly, gender research brought attention to how women are not a homogenous group, to how intersecting dimensions of power order people's lives and that disadvantage is produced as a result of these differently in different contexts. This too was rarely addressed in policy or projects (Arora-Jonsson, 2014).

While Sweden has had an important ongoing discussion on gender-equality nationally, attention to gender and power in relation to the environment has been minimal. The mainstream view in policy continues to be that while questions of gender and power are relevant in environmental governance in countries in the Global South, Sweden, with its good governance and overarching goal of gender-equality in society, does not need this. Moreover, environmental policymaking is regarded largely as a natural science or technical issue, or at most an economic one, separated from societal questions (Arora-Jonsson, 2017a, 2018; Arora-Jonsson & Wahlström, 2023). The attention given to the climate crisis has, however, galvanised a great deal of research on gender/power and environment/climate (much of it by the authors in this book) and some attention to this nexus in policymaking. Given the highly conflictual relations and contestations over land in relation to the green transition in Northern Sweden (Cambou, 2020; Össbo, 2023) feminist perspectives on gender and power are highly pertinent in this context.

Thus, ten years after the 2014 review that I mentioned above, I examine how Swedish policymaking on the environment and in relation to the issues of the Green Deal have addressed the nexus of gender/power and environment/climate – in relation to structural inequalities, collective organising as well as intersecting dimensions of power. To do that I focus on organisations in Sweden that have sought to work towards gender-equality in relation to the environment and climate in various ways and reflect on the potential of these 'good practices' for gender justice.

I reason that projects on gender mainstreaming (albeit depoliticised) can contribute to making a space for, although not necessarily bring about, change. By making dissonances apparent as well as providing a platform, they are a vital piece of the puzzle for establishing a context for change.

Identifying good practices: The cases and the Swedish context

The 'good practices' were identified as a result of having worked for 26 years in the field of environment and gender and more specifically from four research projects on gender and environment. This includes an analysis of regional climate adaptation strategies in Sweden (Arora-Jonsson & Wahlström, 2023), of Swedish bureaucracies' work with questions of gender and power in environmental governance, both within the country and in their development work in the Global South (Arora-Jonsson, 2017a, 2018), research on community and forest relations and the forest sector in Sweden (Arora-Jonsson, 2013; Holmgren & Arora-Jonsson, 2015) and work on the Farmers' Federation Union and its gender-equality academy (Arora-Jonsson & Leder, 2020; Pettersson & Arora-Jonsson, 2009).

Based on this work, I identified three areas where government agencies and private organisations within what is called the green sector are making efforts to mainstream gender and in some cases, intersecting dimensions of power in farming, forestry and related environmental fields and industries. Three cases from these fields were studied as part of a European Institute for Gender-Equality (EIGE) project on *Collection of Good practices on Gender Mainstreaming in the European Green Deal* that was carried out in 2023.[2] The project was intended as support for the advancement of gender mainstreaming strategies and activities in the European Green Deal policy areas. It was part of ElGE's efforts to support EU policymakers and practitioners to increase practical know-how on gender mainstreaming in the context of the European Green Deal, ensuring a gender-just and inclusive green transition. The material analysed is based on a reading of policy documents and interviews with ten officials and others (consultants, members of associations/federations) from different organisations involved in working with the three cases: gender-equality in the forest sector in Northern Sweden, gender-equality and the green transition in Norrbotten and the Gender Academy of the Farmers' Federation in Sweden.

Gender and power in the green sector in Sweden

Historically, the green sector in Sweden has been dominated (in relation to ownership, decision-making and in forest education), by white males. For example, 83 percent of agricultural entrepreneurs (owners of agricultural companies) are men (Jämställdhetsakademi, 2019). The forest sector is also highly gender segregated. Approximately 400 women and 2,000 men worked within large-scale forestry in 2016 in Sweden. For entrepreneurs, the corresponding figures were 400 and 14,000, respectively. The percentage of employed women has varied between 2–6 percent in entrepreneurial companies since 1993. In large-scale forestry, the percentage has increased somewhat over time. In 1993, it was 12 percent and in 2016 it reached 16 percent (Länsstyrelsen Västernorrland, 2020). There are no statistics on ethnicity but insights from 25 years of working in the field as well and that a large part of forest ownership is inherited suggest that there are almost no non-white or non-ethnic Swedish owners, although a lot of forest work is done by immigrant populations, many of whom are from Eastern Europe. Indigenous populations dependent on the forests have relied on customary access to forest land. They are now in the midst of protracted conflicts in relation to large-scale governmental projects such as mining, wind power, national parks and/ or military installations today (Klocker Larsen et al.).

In 2021 the Swedish labour force consisted of 5,547,000 persons, of which 53 percent were men and 47 percent women. In contrast, the environmental sector or what is called the green industry/sector (including agriculture, forestry and fishing) in Sweden, employed 87,000 persons, of which 75 percent were men and 25 percent women (SCB, 2021 cited in Hansson et al., 2023). This does not include work outside formal labour statistics such as collective work for the upkeep of common spaces (Arora-Jonsson, 2013, 2017b) or indigenous care of reindeer herding areas and so on which is normally unpaid and not documented.

Policies seeking to ensure gender-equality have often come up against strong patriarchal norms. Within agriculture, public policies that advocate equal inheritance, for example, have not had the desired effect as can be seen by the skewed ownership of land in favour of ethnic Swedish men. Although many women work within the agricultural sector, they are less represented in leading positions and normally earn less than men in the sector. These inequalities are also related to difficulties in measuring unpaid care work, more often carried out by women than men (Åström, 2010).

In forest policies, a discussion on gender had been strikingly absent for a long time despite forestry being an extremely gender segregated sector. In a 2011 policy overhaul, a gender strategy for forestry was published by the government. Similarly, over the years there have also been several attempts within the Swedish University of Agricultural Sciences to raise awareness, establish new practices and recruit women to forestry education (Powell & Arora-Jonsson, 2021). An analysis of the documents revealed that both climate change and gender-equality were regarded as business opportunities and a means to secure growth and employment in the green sector. Women were represented as potential employees and active forest owners who would fulfil the needs of the industry, rather than as active citizens with a voice in forest decision-making. Climate change and gender-inequality were displaced from the political to an economic sphere, linked to industrial needs, private forest ownership and profit rather than to public and collective decision-making (Holmgren & Arora-Jonsson, 2015).

In national climate policymaking, there are minimal references to gender and power. Studies indicate that where government agencies do discuss power, they largely paid lip service to the issues in broad and vague ways (Singleton et al., 2022). The Regions, on the other hand, have been more active.[3] In the course of our work on climate adaptation, we found two Regions, Västra

Götaland and Norrbotten, that had actively taken up questions of gender and/or indigeneity in their climate adaptation plans (Arora-Jonsson and Wahlström, 2023). However, during our research and interviews for the EIGE project, the new conservative Swedish government cut all funding for climate adaptation making it difficult for the officials who had pushed for gender and power in climate adaptation to even address climate adaptation. Overall, in the strategies, climate like the environment is regarded as a technical and natural science issue where ignorance on issues of gender and power was consciously constructed in order to keep professional identities intact (ibid.).

In parallel, a feminist discourse driven by feminists and others scattered across the sector has permeated organisations working in the sector. The #MeToo movement that started in 2017 galvanised the green sector (or industries) in Sweden to act. Stories of gendered inequalities and sexual harassment in the forest sector exploded into the limelight during the #MeToo years (see Johansson et al. 2023; Grubbström and Powell 2022). This led to a number of initiatives from the farming and forest sector in the form of awareness raising campaigns and seminars and gender-equality projects in the industry as well as in forestry education at the Swedish University of Agricultural Sciences. They also prompted funding for gender research financed by the forest industries. The Minister of Rural Affairs brought together the organisers of the #MeToo movement in the sector, representatives of the forestry industry and the Swedish Forestry Agency, to discuss how the industry would address issues of harassment and gendered inequalities.[4] Research since then has indicated that although explicit sexism is no longer tolerated in the sector, masculine structures and cultures continue to prevail (Johansson et al., 2023).

One of the most important consequences of #MeToo was that it made visible a widespread affective dissonance (Hemmings, 2012 in Hansson et al., 2023). This dissonance was made more apparent by a 'feminist infrastructure' (ibid.), that is, previous work against sexual harassment, feminist networks, and digital and feminist literacy in the country. The dissonance also had to do with the existence of various formal incentives in society, such as educating and encouraging women and immigrants to work in environments dominated by white men. The #MeToo movement made it clear how such formal incentives collided with, and were counteracted by, informal power structures where strong norms of masculinity are maintained through sexism and sexual harassment.

My previous research on international environmental organisations showed me that even a depoliticised and disciplined narrative on gender in organisations, by providing a platform from which to destabilise dominant debates on sustainability and the environment, can open up a space for counter discourses and is vital for bringing about change (Arora-Jonsson & Basnett, 2018). This has a correlation to what happens on the ground. Broader discourse on democratic participation and local management in forest relations have helped to open the space for women's agency and resistance in forest governance and democratic rural development on the ground (Arora-Jonsson, 2013). In cases of sexual harassment in organisations, researchers have pointed to the importance of activating or training bystanders, that is people not directly involved in cases of discrimination or sexual harassment, to provide the support needed for change (Fenton and Mott, 2017 cited in Grubbström and Powell, 2022). I now turn to the three projects to examine how these developments fed into the questions of structural inequalities, collective organising and intersecting dimensions of power in the projects below.

Gender-equality in the forest sector

The forests, their ownership, use, businesses and the manufacture of forest products make up the basis of Sweden's single largest industry. As part of the implementation of the Regional Forest Programme, some of the biggest private actors in the forest sector (the forest companies, SCA, Holmen och Norra Skog) and the County Board in Västernorrland, a county in northwestern Sweden highly dependent on forestry, put together a gender-equality project between 2020 and 2023. The project was led by the County Administrative Board in Västernorrland (the administrative authority in the region) and financed equally by the region and private actors.[5] The project comprised actors from private businesses, government agencies, a university, an agricultural high school and a trade union. The aim was to create conditions for a long-term competitive, sustainable and profitable forest sector. The project participants met once a year and were divided into six working groups depending on their interest. These were: 1) communication, 2) individual forest ownership, 3) recruitment and education, 4) work-life, 5) contracts and procurement, and 6) wildlife management. Following Sweden's National Forest Programme's vision of the forests that they dubbed 'the green gold', they hoped "to contribute with employment and sustainable growth throughout the country as well as to the development of a growing bioeconomy". The National Forest Programme encourages actors in the sector to take measures to increase gender-equality and diversity (page 21) among a range of other measures (such as innovations, creative product development and collaboration across sectors) that it regards as indispensable for developing the forest sector (23).[6]

A total of 16,800 people, 7 percent of Västernorrland county's inhabitants are forest owners, of which 39 percent are women. Forestry and the forest industry accounts for approximately 4.5 percent of the county's total employment and for approximately 8 percent of the county's gross regional product. For the County Administrative Board, 'gender-equality work was key to increasing the attractiveness of the forest industry and that this meant that women and men needed to have the same opportunities to own and use forests and to work in or run companies within the forestry industry' (Länsstyrelse Västernorrland, 2020). This was to be achieved by 1) designing strategies that would lead to work with gender-equality in the forestry sector to progress to the next stage, meaning that a real and lasting change occurred and where the change-oriented work had an impact; 2) being brave in the design of goals and actions; 3) identifying and managing resistance; and 4) focusing on the norms that associate forestry knowledge and competence with men and masculinity. Based on this, it was decided that gender mainstreaming would be the strategic foundation of the project.

The goals were intended to ensure that project participants: 1) were able to identify masculinity norms within their own organisation and know what the organisation was doing for gender-equality work; 2) had developed measurable goals for implementing the action plan (entitled *This is how we strengthen the attractiveness of the forestry industry based on an equality perspective*) (Länsstyrelse Västernorrland, 2020), within each organisation and within the sector as a whole, and had decided on how these goals were to be implemented, monitored and reached; 3) integrated equality goals into the steering documents. Further, 4) the Swedish Forest Agency's proposal for measures for an equal forest sector (in relation to equal work-life, recruitment and certification) were set as industry standards for Västernorrland and intended to be used as good examples/practices for other regions; and 5) criteria for equal opportunity procurement and contracts for services in the forest sector were designed in a joint working group in which both clients and contractors participated.

The project invited a wide array of stakeholders in the forestry sector in Västernorrland, including workers, students, executives, forest owners and companies, to join the project. Through trainings and webinars on topics such as 'the current state of gender-equality in the sector', 'when can we be equal', 'myths and resistance', 620 people (49 percent men and 51 percent women) from 128 different organisations participated in the project.

Gender-equality training was also conducted in autumn 2021 for leaders and employees of forest companies. These focused on gender-sensitive recruitment, on the importance of activities and meeting places for forest owners being consciously equal, on integrating common measurable goals into the operations, developing business relationships and advisory services to reach wider groups of forest owners, on developing strategic communication in each organisation and jointly to strengthen the forest sector from an equality perspective. Management teams were offered a workshop to make visible how masculinity norms were expressed within their own organisation, how they affected the organisations' ability to pursue gender-equality and what they could do to enhance the attractiveness of the sector. The participation rate was, unfortunately, low. The project organisers felt that it would have been desirable if the steering group, as well as management and working groups' participants, would have participated to a greater extent in the project's activities.

Nevertheless, the organisers felt that the project was transformative in that they were able to incorporate gender mainstreaming not only as a strategy within the project and for the participating actors, but to test how national gender-equality goals and the Swedish Forest Agency's goal of gender-equality could be put into practice within different organisations and how gender could be mainstreamed throughout the Swedish forest sector. In other words, it was a scaling up of gender-equality in the forest sector that had not been done before. They pointed to how the method promoted collaboration across company boundaries. This was appreciated by many, otherwise always in competition with each other. In the organisations, the work was promoted by a senior manager within each area. This raised the status of the gender mainstreaming and made it easier for the gender officers to work with the issue.

The gender mainstreaming broadened and created awareness in new areas in the companies about gender-equality. The project also enabled collaboration on gender-equality between different actors within the organisations and between actors of different sizes, such as companies and government agencies. This was regarded as positive by many. Despite the work that still needs to be done, the project organisers believe it to have succeeded in creating a common agenda on gender-equality and for having initiated a conversation throughout the sector, thus laying the ground for the newly established Council for Gender-equality at the Swedish Forest Agency.[7]

The decision to work with norms around men and masculinity shifted the focus from recruiting women (that is adding women to existing structures) to dealing with leadership and organisational culture. Working with gender-equality and gender mainstreaming meant that privilege was made visible. It also entailed that those who were privileged risked losing those privileges. With that in mind, it came as no surprise that there was resistance in all organisations represented in the steering group. The organisers felt that there should have been a plan for handling this. Similarly, there should have been a plan in place for how to manage situations where individuals in working groups acted in ways that made it impossible to achieve the tasks of the working group and the goals and purpose of the project.

But perhaps the most important achievement was that the project provided a context and platform to address gender inequalities within the organisations. It was a common arena in which to test tools and methods for gender mainstreaming, and to adapt the national and regional ambitions regardless of the size or type of organisation.

A just green transition for Northern Sweden

Previous research on climate adaptation strategies in Sweden had shown that the region of Norrbotten had been one of two exceptions when it came to including questions of gender and power in their regional climate adaptation strategy (Arora-Jonsson and Wahlström, 2023). As a result, we chose to interview the officer responsible for climate adaptation who had actively sought to include gender-equality in their work with climate change for the EIGE project. Gender mainstreaming had not received specific funding in the work for climate adaptation strategies. The officer at the County Administrative Board had funded this work from the climate adaptation budget, but that had now been stopped by the government in the budget proposition for 2024. In the course of our study, however, we came upon a major gender mainstreaming project, meant to lead to a just green transition being undertaken by the two county boards of Norrbotten and Västerbotten in Northern Sweden, that we studied instead.

In close collaboration with actors within the energy sector, the gender-equality officers in the County Administrative Boards from both Norttbotten and Västerbotten regions had put together a plan of action for gender-equality signed by several actors within the energy sector and by regional and local government agencies. The project, meant to promote gender-equality and loosely modelled on the forestry project in Västerbotten described above, is designed to run from 2023 to 2030.

Northern Sweden is the site of the much-vaunted green transition in the country where industries such as green steel and electric batteries as part of 'smart energy systems and circular economy' are intended to replace fossil fuel-based production. The two chairs of the regional authorities and the gender-equality officers in the county administration felt the need to launch a collaborative initiative to ensure that a gender-equality perspective would permeate the transition in the region and the expansion of green industry. Northern Sweden is characterised by low population density and the planned transition is expected to involve growth for businesses and to require new forms of labour. The project organisers believed that this made it important to change the current image of the male-oriented industry and to use the project to present a new narrative that would be gender-equal and create opportunities for both women and men. When talking about the ongoing societal (industrial) transition in the north of Sweden, one of the interviewees stated that a gender lens (which involves analyses of power) can help answer questions such as what a safe society is and for whom it is safe, thus putting questions of gender in a wider intersectional perspective.

The gender officers and the heads of the regions invited leading industrial business actors to roundtable discussions to talk about gender mainstreaming the transition. Facilitated by the gender experts, the various actors (industry as well as regional and local agencies) formulated goals and objectives of the project that were included in the letter of intent that aimed to achieve a gender-equal and just transition in Norrbotten and Västerbotten. The objectives in the letter of intent were to: 1) ensure management-driven gender-equality work; 2) nuance the image of the industry and the counties by challenging and changing structures and notions that limit women and men, respectively; and 3) be an attractive employer by enforcing a gender-aware work environment. The letter of intent did not mention power inequalities explicitly or refer to other dimensions other than men and women as homogenous entities.[8]

The project focuses specifically on just and equal payments for both women and men and the promotion of economic independence. The project also works to avoid reproducing negative masculinity norms and combats sexual harassment at the workplace. The various actors agreed to work together to change the male-dominated image of the industry and make the region lucrative for all those interested in working there. The target group for the project were companies involved in the green transition and public and private actors interested in the project who wanted to be part of the region's transition.

The overall strategy was to focus first on actors in management roles with leadership courses who could promote gender mainstreaming in their company/organisation. Roundtable conversations were meant to ensure participation and ongoing evaluations. At the time of the interviews (2023), the County Administrative Boards were still funding the project though they hoped that the companies would take this over. One officer remarked that it was difficult to run such a large project from two relatively small units and rather than the gender experts leading the projects, they believed that it would be good if they were led by the management of the County Administrative Board with the gender experts facilitating the process. They agreed and, as has been shown by gender experts in international environmental organisations elsewhere, internal work with gender-equality within the organisation is vital in order to be able to promote gender mainstreaming externally in the sector (see Arora-Jonsson and Basnett, 2018). Whether or not male heads of the organisations unaware of gender issues should be leading gender mainstreaming projects needs further discussion and research. It is an issue that recurs in the next example and this approach can also have its own drawbacks.

The project is still in its infancy. An interesting reflection in the interviews was about how resistance to gender mainstreaming seemed to have changed, from being openly expressed to being more hidden. Colleagues and managers associated with the project were generally positive and enthusiastic about gender mainstreaming, yet, resources and help to continue to develop the work was lacking. As one person mentioned, "people who work with gender questions need to be better at identifying that resistance can take on many different facets." The "kill them with kindness" attitude is often harder to counteract than open resistance.

The gender-equality academy

The third and last example is the Federation of Swedish Farmers' Gender-equality Academy (*LRF's Jämställldhetsakademin*). The Federation (LRF) is a politically independent interest and business organisation that operates on national, regional and local levels. The Federation (and the Academy) works within agribusinesses and green industries, that is, in agriculture, forestry, the horse industry and related fields. The organisation has 128,000 members[9] and, according to the interviewees at the Academy, the members of the Federation represent more than 90,000 businesses within the green sector. These include some of the largest food and agriculture businesses in Sweden as well as independent farmers. Apart from individual farmers, almost all cooperatives within Swedish agriculture and forestry are members. The Federation finances its activities through membership fees combined with asset investments and business operations.

The Academy came into being in 2009 as a response to arguments for greater gender-equality within the Federation and as a result of a history of somewhat contentious gender relations in the past (personal communication with previous female members). At the time, the Swedish government had also been urging private companies to implement a gender perspective. The Academy was seen as a way to reach women and to increase the profitability of the sector. It is meant for all members of the Federation, business owners and entrepreneurs within the green-agricultural sector and companies working within or related to the sector. The boards of the various cooperatives represented in the Academy are dominated by men. This is often put down to the persistence of tradition. As a former chairman of the Academy expressed it, a certain amount of tradition leads to the green sectors not being as gender-equal as they could be (Arora-Jonsson and Leder, 2020:21).

Key priorities and objectives of the Academy are to highlight and promote women's roles within the green-agricultural sector, women's economic rights, in particular in family businesses and in making women's work within the sector visible in statistics. The Academy is often described as a think-tank whose mission is to spread awareness and inform on matters of gender mainstreaming. The Academy commissions gender reports in various fields, raises awareness on gender through discussions and seminars and through the promotion of gender-sensitive data collection and analysis (gender statistics and sex-disaggregated data) across the national, regional and local levels.

The Academy has presented a series of reports on gender inequalities within the green-agricultural sector. After #MeToo, a report presented testimonies from three #MeToo campaigns from within the forestry, agriculture and the horse industry. A report also demonstrates how the media fails to represent the reality of Swedish agriculture while reproducing gendered and other stereotypes. The question of parental leave was the subject of one report where men taking parental leave were presented as good examples that led to overall better contexts for the family and the farm.

The Academy's most important contribution has been to illustrate the structural inequalities in agriculture. It helped to point out that women in agriculture and forestry often ended up outside official statistics, because men more often were registered as company and/or landowners. This resulted in women in agriculture/forestry/farming becoming invisible in official statistics. In the wake of the #MeToo movement, the Academy commissioned several reports that highlighted women's hidden contributions to the running of farms as part of their everyday work.

Thus, the Academy's most concrete achievement has been in influencing the official methodology for agricultural statistics in Sweden. It has changed so that women and women's work is now visible through sex-disaggregated data and the focus is not merely on the landowners or the owners of the business. In the statistics that the Board of Agriculture gathers on farms, the Board began to count not only those who owned the farm but also began to ask who else was involved in its running. Through these new surveys, they have shown that although only 17 percent of women are listed as owners in the green sector, 44 percent actually manage the farms (Swedish Board of Agriculture, 2010 cited in Jämställdhetsakademi, 2019). This has been extremely important in recognising that formal ownership does not have to correspond with the work put into farming (Arora-Jonsson and Leder, 2020). By changing the way that statistical data is measured, agricultural knowledge builds on more accurate data that better reflects the actual work being carried out in farms. Whether this is likely to affect the structure of ownership is as yet difficult to say.

Since the Academy was started in 2009, its members have varied. In 2023, 11 people were engaged in its work. One person coordinated the work of the Academy at about 10 per cent of a full-time position. The chairman of the Farmers' Federation heads the Academy, something called for by the gender officers in the case above. The interviewees emphasised the importance of having the Chair of the Federation as the Chair of the Academy as they believed that it sent a clear signal to everyone in the Federation that gender-equality is a priority area. It is also intended to ensure support to the Academy from the Federation. It was stated during the interviews that having the Chair of the Federation at the forefront of the Academy has made it difficult not to participate for those less interested or positive towards gender mainstreaming.

However, the lack of paid staff and a communications person or coordinator at only 10 percent appears paltry given the size of the Federation. In a previous study, it also appeared that the leadership position, while giving legitimacy to the Academy, was more symbolic than actual (Arora-Jonsson and Leder, 2020). This is in contrast to the cases above where gender experts heading the projects were the driving force behind the projects.

Although some of the members of the Academy showed an interest in addressing dimensions such as age, urban/rural residence or disability in relation to gender, such attention to intersecting dimensions of power is a somewhat unprioritised area of gender mainstreaming. Much like the policy focus in general (for example, in forestry and agriculture), the Academy has a strong focus on business owners who are a fairly homogenous group – predominantly white and male. The diversity is in the labour force in forestry and agriculture where migrant labour is vital. This is, however, not yet a subject of discussion for the members of the Federation or the Academy. One of the persons interviewed for the EIGE study remarked that the work of the Academy has become stuck on equality between women and men and while sexism in the field was urgent to address, as brought out by the #MeToo movement, other dimensions such as racism were not discussed.

One of the main challenges in the work of the Academy has been to make sure that everyone is on board and working towards gender mainstreaming and that there is a general support for the Academy's work. Creating lasting engagement across an entire organisation has been difficult. One representative mentioned that the Academy could be perceived as a top-down institution that lacked grounding among the members of the Federation, underscoring the importance of inviting members to participate in the reports, share their experiences and evaluate the reports. An important lesson has been that it can be difficult to ensure that the knowledge trickles throughout the organisation. During an internal meeting, it was concluded that the Academy sometimes "feels like its own satellite". The chairman is clear that he wants the Academy's work higher up on the agenda within the Federation.

As a result of the Academy, the Farmers' Federation also decided to add a gender perspective to their overall vision, exhorting the entire organisation to work towards gender-equality and to include a gender perspective in all its work. The Federation now has its own gender-equality strategy that was developed in 2016 and which was being revised at the time of the interview. The strategy explains that the organisation is yet to be gender-equal (including representation of different sexes, ages and occupations) and that the use of sex-disaggregated data shows the difference in representation between men and women among the elected officials in the organisation, highlighting the need for an equal distribution of power.

Conclusion: A potential platform for change

The projects studied here take an overarching gender mainstreaming perspective that involves organisations from regional to local levels. As I stated in the introduction, feminist research has emphasised the importance of policies that seek to address structural inequalities such as the skewed ownership of land between men and women, the importance of supporting women's collectives and, importantly, for the need to address intersecting dimensions of power that divide women and men as groups. In this last section, I reflect on these issues and on how gender mainstreaming attempts might provide the space for bringing about change in organisations and across different scales at the nexus of gender, environment and climate.

In relation to structural change, all three cases demonstrate that they are very much in the mould of what has been called a 'business case for gender-equality' where gender is used to promote other objectives, in this case, the profitability and sustainability of the sector as such. For example, in the forest as well as the transition projects, it felt important to justify gender mainstreaming as a means to better management as well as better collaboration across different organisations that were otherwise unnecessarily competitive – although it was also important for the organisers to challenge the male-dominated norms in the sector. For the Academy, gender mainstreaming is justified by the potential of gender mainstreaming to promote the Farmers' Federation's vision of growth and profitability within the green sector. And for all, the hope is that gender mainstreaming would lead to a better and new labour force.

The driving force of the projects acceptable to all project participants is that gender mainstreaming would pave the road to better and more sustainable growth and not necessarily on the focus of much of feminist scholarship – justice or democracy. As the Gender-equality Academy demonstrates, policies that have promoted structural change have not been enough to bring about widespread change. For example, despite policies and laws promoting equitable inheritance and ownership, ownership of land and businesses within the green sector continues to be skewed in favour of white men.

Thus, justice is considered served because gender-equality is on the agenda. The use of gendered terminology can render various activities legitimate without fundamentally changing the systems and practices in an organisation. This can backfire on gender-equality and also on women. As the cases demonstrated, a lack of discussion on power and privilege beforehand means that resistance to gender mainstreaming can come as a surprise and be difficult to deal with. At the same time, these conversations beforehand are not easy. There is need for more attention to this facet of gender mainstreaming that has not been highlighted much before.

Yet, the work of the projects in highlighting gender inequalities and providing a potential platform for organising for change helps to chip away at the structure. The Academy's work in highlighting how statistics are compiled is vital in bringing to light structural inequalities in the first place and making them more difficult to ignore. An earlier analysis of climate adaptation policies in Sweden showed a construction of ignorance about questions of gender and intersecting dimensions of power. Questions of gender and power were ignored and made difficult to work with (Arora-Jonsson and Wahlström, 2023). In contrast, all the mainstreaming projects bring gender (and power?) into the public domain. The Academy's work with raising awareness (albeit its challenges in reaching out to all members) and importantly the push for discussing masculine norms within forestry and transitions are important in this regard.

The structure of leadership has been felt to be important among those we interviewed. The gender experts felt that the transition project should ideally be anchored with the management. At the same time LRF is anchored in the management but has few gender experts driving the Academy in the same way. Clearly both are needed or perhaps, ideally, gender experts within the management driving these issues so that it goes beyond the symbolic.

If there is one thing that feminist research has taught us, it is that justice is truncated without a proper gender perspective and that entails studying how different dimensions of power intersect. This makes the question of intersectionality important. However, it appears that bringing in questions of multiple dimensions of disadvantage is clearly unsettling. While the organisers for the transition project were clear on the importance of working for human rights and with inequalities in power, the focus in the letter of intent has been mainly on gender and in particular on women. For example, those most likely to be affected by the green transition are the Sami indigenous groups and they are not mentioned. At LRF, the officer felt that they needed to deal with women's disadvantage first and then come to the other dimensions. This sidesteps the entangled nature of disadvantage as well as the many different people beyond landowners who work in the sector. As class and ethnic advantage are ignored, gender-equality projects risk becoming focused on a narrow group of women and not having the base they need to ensure change.

Feminist research has brought attention to how change has been possible through collectives and coalitions. The #MeToo movement as well as organising in relation to environmental and development questions on the ground but also internationally is a testimony to this (Hansson et al. 2023; Arora-Jonsson, 2013, 2017b; Arora-Jonsson and Basnett, 2018). In the projects at the moment, this has been absent. The Farmers' Federation itself is a collective bound together by ownership and assets. However, it needs to adapt to meet changing conditions as is made clear by the Federation's attempts to reach out to women who are excluded. But this outreach has been to individual women entrepreneurs and the Academy is yet to become a site for collective organising for different women that would be needed for change. As research has shown, it has often been in the interests of marginalised groups to organise themselves separately – in relation to mainstream organisations – in order to bring about change for equality (c.f. Arora-Jonsson, 2013; see overview in 2014). The Academy could potentially provide that platform though the little time devoted to its infrastructure is not conducive for that. It would need more time and gender experts such as those present in the other two projects.

Feminist research has shown how inequalities permeate across scales and that these connections need to be addressed. The sheer scale of the other two gender mainstreaming projects open up potential space for change. The projects potentially provide the infrastructure for change by making evident gendered connections across different scales and also make visible gendered outcomes that are otherwise hidden. They provide the space by which to bring issues of power into the limelight as well as an infrastructure that extends beyond the immediate local context. Perhaps an added focus also on 'bystanders' (Grubbström & Powell, 2022), not necessarily directly involved in the companies' work could be useful to strengthen that infrastructure.

As I discuss earlier, the established feminist infrastructure meant that #MeToo activism, by articulating a widespread affective dissonance, pushed open doors that enabled change. The projects provide such a potential platform and infrastructure for solidarity across different levels. As an organiser pointed out in the forest project, the project provided a context and a platform for those who do want to transform the current structures. Importantly, it also provides legitimacy to bring up questions of gender and power, especially in face of resistance in their own contexts. This wider context and feminist infrastructure is vital for change. While the projects are perhaps far from changing structures they are certainly establishing the context for it.

Notes

1 At the time, the literature that did exist was overwhelmingly on the Global South and although there has been more attention to this nexus in the Global North, most is still from the Global South.
2 I was assisted in the EIGE project by the invaluable work of two research assistants, Anna Berg Grimstad and Emil Planting who interviewed officials in the organisations. Anna also compiled the templates for the project.
3 Sweden is divided into 21 regions and each region is administered by the County Administrative Board.
4 This was also the case in other sectors such as in the male-dominated engineering universities where # call for change was heard and acted upon.
5 https://www.lansstyrelsen.se/vasternorrland/om-oss/vara-tjanster/publikationer/2023/slutrapport-for-jamstalldhet-i-skogsbranschen.html.
6 https://www.regeringen.se/informationsmaterial/2018/05/strategidokument-sveriges-nationella-skogsprogram/.
7 https://www.skogsstyrelsen.se/om-oss/organisation/radgivande-grupper/skogssektorns-jamstalldhetsrad/.
8 https://www.lansstyrelsen.se/norrbotten/om-oss/vara-tjanster/publikationer/2023/avsiktsforklaring-jamstalld-industri.html.
9 https://www.lrf.se/om-lrf/.

References

Arora-Jonsson, S. (2013). *Gender, Development and Environmental Governance: Theorizing Connections*. Routledge.

Arora-Jonsson, S. (2017a). Blind spots in environmental policy-making: How beliefs about science and development may jeopardize environmental solutions. *World Development Perspectives*, 5, 27–29.

Arora-Jonsson, S. (2017b). The realm of freedom in new rural governance: Micropolitics of democracy in Sweden. *Geoforum, 79*, 58–69.

Arora-Jonsson, S. (2018). Across the Development Divide: A North-South Perspective on Environmental Democracy. In T. Marsden (Ed.), *Sage Handbook of Nature* (Vol. 2, pp. 737–760). Sage Publications.

Arora-Jonsson, S., & Basnett, B. S. (2018). Disciplining Gender in Environmental Organizations: The Texts and Practices of Gender Mainstreaming. *Gender, Work & Organization, 25*(3), 309–325. https://doi.org/10.1111/gwao.12195

Arora-Jonsson, S., & Leder, S. (2020). Gender Mainstreaming in Agricultural and Forestry Institutions. In C. Sachs, L. Jensen, P. Castellanos, & K. Sexsmith (Eds.), *Routledge Handbook of Gender and Agriculture* (pp. 15–31). Routledge.

Arora-Jonsson, S., & Wahlström, N. (2023). Unraveling the production of ignorance in climate policymaking: The imperative of a decolonial feminist intervention for transformation. *Environmental Science and Policy, 149*. https://doi.org/10.1016/j.envsci.2023.103564

Åström, G. (2010). *Trygghet ger balans: Företag, familj och fritid i de gröna näringarna*. LRFs Jämställdhetsakademi.

Cambou, D. (2020). Uncovering Injustices in the Green Transition: Sámi Rights in the Development of Wind Energy in Sweden. *Arctic Review on Law and Politics, 11*, 310–333. https://doi.org/10.23865/arctic.v11.2293

Grubbström, A., & Powell, S. (2022). Från hashtag till handling Skogsutbildningens #metoo- rörelse i spåren av #slutavverkat. In H. Ganetz, K. Hansson, & M. Svenningsson (Eds.), *Maktordningar och motstånd. Forskarperspektiv på #metoo i Sverige* (pp. 279–302). Nordic Academic Press.

Hansson, K., Ganetz, H., & Sveningsson, M. (2023). The significance of feminist infrastructure: #MeToo in the construction industry and the green industry in Sweden. *Gender Work and Organization*. doi:10.1111/gwao.12994

Holmgren, S., & Arora-Jonsson, S. (2015). The Forest Kingdom – with what values for the world? Climate change and gender-equality in a contested forest policy context. *Scandinavian Journal of Dental Research, 30*(3), 235–245.

Jämställdhetsakademi, L. (2019). *Jämställdhet i det gröna näringslivet: En guide från LRFs Jämställdhetsakademi*.

Johansson, K., Johansson, M., & Andersson, E. (2023). All talk and no action? Making change and negotiating gender-equality in Swedish forestry. *Foresty Policy and Economics, 154*. https://doi.org/10.1016/j.forpol.2023.103013

Länsstyrelse Västernorrland (2020). *Så stärker vi attraktionskraften i skogsbranschen utifrån ett jämställdhetsperspektiv Handlingsplan*.

Össbo, Å. (2023). Back to Square One. Green Sacrifice Zones in Sápmi and Swedish Policy Responses to Energy Emergencies. *Arctic Review on Law and Politics, 14*, 112–134. https://www.jstor.org/stable/48722458

Pettersson, K., & Arora-Jonsson, S. (2009). *Den osynliga entreprenören: Genus och företagande i de gröna näringarna*. LRF.

Powell, S., & Arora-Jonsson, S. (2021). The conundrums of formal and informal meritocracy: dealing with gender segregation in the academy. *Higher Education*. https://doi.org/10.1007/s10734-021-00719-2

Singleton, B. E., Rask, N., & Kronsell, A. (2022). Intersectionality and climate policymaking: The inclusion of social difference by three Swedish government agencies. *EPC: Politics and Space, 40*(1), 180–200. doi:10.1177/23996544211005778

Interview 2

THE HON MARAMA DAVIDSON, CO-LEADER OF THE GREEN PARTY OF AOTEAROA NEW ZEALAND: THE IMPORTANCE OF GRASSROOTS AND COMMUNITY ACTION

Interviewed by Dory Reeves and Julie MacArthur, 31 August 2023, New Zealand

Note: Māori terms are translated in brackets and there is a glossary at the end so that readers can follow up on the more detailed meaning of the words.

Introduction: The Aotearoa New Zealand context

Aotearoa New Zealand is a developed country in the Pacific, to the south and east of Australia. It has a population of 5.25 million, 16% of whom identify as Māori. Although a relatively small country in terms of landmass, it is a world leader in many respects. It is home to the world's first national Green party – the New Zealand Values party (est. 1972), the predecessor of the NZ Greens. New Zealand was also the first country in the world in which women got the vote (in 1873) and is one of only six countries in 2023 where women make up more than 50% of elected parliaments (UN Women 2023). Due to vast hydroelectric and geothermal resources, New Zealand electricity generation is more than 80% renewable, one of the highest shares in the world, and ranks as one of the leading countries in the world in the World Energy Council Trilemma Report (WEC report 2024). Former Prime Minister Jacinda Ardern said climate change is her generation's 'nuclear free moment' and on December 2, 2020 declared a climate emergency in New Zealand, committing NZ to a carbon-neutral government by 2025.

However, the country's 'clean, green' image hides much, particularly in terms of the fraught colonial relationships between the colonial settler state and Indigenous Māori peoples. New Zealand contributes 0.09% of the world's total CO_2 equivalent emissions. As of 2019 though, New Zealand's CO_2 emissions were 6.8 metric tons per capita, considerably higher than the world average of 4.6 metric tons per capita and higher than many countries

DOI: 10.4324/9781003461005-9

with larger populations. Despite the high percentage of renewables in the energy mix, fossil fuels continue to fuel the majority of the country's primary energy supply (MacArthur and Bargh, 2022, pp. 259–261), and road transport emissions have increased more than 100% from 1990 levels. Agriculture in New Zealand is responsible for 49% of emissions in the country (MfE, 2024).

The New Zealand government has taken a range of policy steps to address the climate challenge. Its key tool since 2008 has been the Emissions Trading Scheme (ETS) (MFE, 2023). New Zealand has a target of net zero emissions by 2050 enshrined in law, under the Zero Carbon Act. In May 2022, New Zealand released its first emissions reduction plan, outlining a series of policies and incentives to decarbonise all sectors. Aotearoa New Zealand is using a system of three multi-year emissions budgets to meet the 2050 target. Progress is monitored by the Climate Change Commission.

The Honourable Marama Davidson was interviewed for this chapter. Between 2020 and 2023, she was Green Party co-leader with James Shaw who was Minister for Climate Change. Marama Davidson was the Minister outside Cabinet for the Prevention of Family and Sexual Violence, as well as holding the Associate Housing portfolio. After the 2023 New Zealand Election, Davidson was returned to parliament but the previous governing coalition of Labour and Greens was replaced by a right-wing coalition of National, ACT, and New Zealand First. Marama describes herself as follows: 'I am from Aotearoa, the country called New Zealand. My people are *tangata whenua*, (the Indigenous people of this land). We have come to be known as the Māori people. I descend from two different kinship groups in the North called Te Rarawa and Ngāpuhi, and also from the peoples of the East Coast region who are Ngāti Porou. 'One of her marae is Matai Ara Nui Marae in Whiriniaki. Marama describes herself as an environmentalist and human rights advocate. She brings a particular political lens as a parliamentarian into the discussion.

What motivates you in your climate policy leadership work with the Green Party of Aotearoa New Zealand?

Mokopuna, [grandchildren] and Indigenous leadership sum it up, and are the main drivers. The Indigenous leadership around the world, not just here in Aotearoa, has been a voice I am particularly motivated to amplify in the climate space and environmental spaces since the dominant narrative is still quite white. From a political standpoint, a driver is also seeing what happens when activists for climate aren't listened to. By that I mean there are other parties and political leaders who are wanting to undo all of the painstaking work and all of the momentum and so keeping the momentum up is important.

As to specific movements, when Occupy Wall Street went around the world there were Indigenous peoples everywhere wanting to make sure that we included a colonial analysis in to that. The whole genesis of the oppressions that Occupy was raising stemmed from the mates of colonisation and imperialism, and capitalism. Not to have that analysis rooted in that movement was a massive flaw; especially since the very land that Wall Street is on is occupied territory for the First Nations peoples, and so here in Aotearoa we worked really hard to try and make sure that there was a *tangata whenua* [people of the land] analysis underneath. This was a good opportunity to raise analyses of power, privilege, and oppression. Then when Idle No More happened in Canada rooted in the pipeline protests, we saw this as another opportunity for this joined-up thinking. But that's not the only movement that motivates me. The Greens have always been strong supporters of West Papua independence. We have long joined up with movements including Friends of Palestine, Australian Aboriginal, and Standing Rock. I actually organised an Aboriginal land *hikoi* [march] in Auckland.

Could you tell us the importance of your feminism and intersectionality to your work on climate and other related policy

For me it's *Mana wāhine* [Māori feminism] and the starting point is about understanding that colonisation violently impacted on the status of Māori women and that has led to ongoing and inter-generational impacts in the stronghold of *whānau* [family]. The *whānau* stronghold is particularly a *wāhine* leadership zone, which again filtrated out to all levels of governance and leadership for Māori women. The Western patriarchal lens completely eroded that and continues to be really harmful and this drives the reclamation work we are doing. Also, one of the key differences between a white feminism space and a *Mana wāhine* space is that ours is an understanding of whole of *whānau* well-being, which includes men, and all genders and children; it's a lot more holistic than anti-men approaches as well. I am being very simplistic here but those are very simplistic understandings. I work in a prevention of violence space. I have had to pull the roots of violence back, and we must continue to highlight the gendered nature of violence while at the same time broadening out our understanding that violence is basically 'power over' and that is why disabled people, rainbow, and trans people also have a unique experience of violence that deserves to be told; it's not a gendered lens alone that is going to help us to resolve violence. So, in the climate space, I'll give a story to capture it all.

Probably 10–12 years ago, I was on a political climate for women panel, before I became an MP and I talked about the link between the violence to *Papatūānuku* [Earth mother] the violence to our earth directly linking to the violence against women, and the patriarchal drivers of exploiting women, exploiting the planet, exploiting the earth for quick satisfactory gain, having power over the planet, women, and *Papatūānuku*, to benefit the few. I talked about why we need to understand the link between the role of women and the role of climate destruction. And a young white male in the audience [verbally] tore me down. He considered himself a lead climate activist and he just rubbished everything I said. How dare you, [he said] put something up so kooky in this discussion. How dare you feel you have any right to make those outlandish links and claims in this climate discussion space, and that just freaked me right out and I have been determined ever since to keep making those links.

Which of the policies that you have been instrumental in developing regarding climate would you consider the most successful? Why?

The reason why I am with the Green Party is because we continue to make the intersectional links and that is very much from an Indigenous lens of everything being connected, a lens that Kate Raworth (2017,) who wrote *Doughnut Economics* has herself acknowledged. An economic lens that is holistic and connected has been part of an Indigenous economic theory for thousands of years. In 2020, Māori created our own doughnut economic framework, putting the environment in the middle and sent it to Kate, (Shareef and Boasa-Dean, 2020). She loved it and said she has always acknowledged the circular and regenerative nature of the Indigenous economic lens, an inter-generational lens and acknowledges Indigenous peoples around the world who have held long-standing values and understandings of how the world connects. This was reflected in the second iteration (Raworth, 2022).

The question asked undermines the connected nature of and the interconnected nature of policies. In the election in 2023 and the last election, prioritising minimum incomes and wealth tax, is a climate policy. People have got no time to think about being involved in collective action and climate movements when they are literally struggling for food, shelter, and healthcare and freedom from violence, so we are all climate MPs. We are all climate ministers, all of our policies are climate justice policies and I also want to add that where I am really proud and it's not just me (I can claim nothing individually and that's another holistic way of looking at things), it's our collective Māori voice that has been able to support James [Shaw] as climate minister, to make sure he got his engagement with *tangata whenua* early and correct and I'll give this example. There are, as you know, nasty actors; there are mis/disinformation nasty actors and when we put up a nature proposal, a lot of nasty manipulation was happening among Māori communities to get Māori up and angry about it, and we knew what was happening so we decided to take the bull by the horns and support James to sit down with people and explain and walk through and by the end of those meetings, all up in the North [of Aotearoa New Zealand] I can remember we had the Māori community saying: "you need to go stronger". So, I want to point out that all our policies are interconnected and the process for good policy requires engagement and tapping into the leadership and the expertise in the sectors relating to those particular policies. And so I am really, really proud of our climate policies as a package that we have been working on and we have only one more to go and that is the oceans policy, keep an eye out, because oceans are sucking up carbon as well, our policies are all connected and inter-related under the banner of climate justice, environment, inequality, housing, income and underneath a bigger banner of Te Tiriti justice (Green Party, 2023).

What personal qualities do you think are needed for developing and implementing successful policies to address the climate emergency?

Sincerely, it's understanding you are never alone. And we have seen the downfall when that pressure, when you feel you need to hold it all up. It doesn't work. It's not human. It's actually dehumanising. And we are personally experiencing and coming from the trauma of when our leaders are isolated, which is a strategy to undermine leaders and voices and that's why you can never stand as an individual leader; you have to have a network of support, particularly as a brown woman in politics. I mean I'll just give you an example. When I talked about white cis male violence publicly, holy shit that mobilised external white supremacist movements around the world: the UK, the US, Australia, Russia, and possibly more, to actually mobilise troll farming from around the world that actually had an impact here in New Zealand and now that has been terrifying and I would not have survived any of that and would not still be here if it weren't for the collective support and my closest trusted circles; so leadership is actually about knowing that you are part of a collective. My instinct was to go into a corner and bubble into a foetal position and that is what that is designed to do. And so, because I was aware of that happening to others I had to go against my instinct and show vulnerability to my circles and reach out for help and only then were they able to push me back into place and out of the corner in the dark and that stuff is not meant to be survivable. Humans aren't supposed to withstand that amount of pressure, sustained pressure. That's not the only example; I've got tons.

But I guess what I'm saying is I wouldn't still be here, I wouldn't be sane, I wouldn't be well if it wasn't for understanding my role as part of a collective and I also want to say that my only other point is the extra pressure and strength and resilience that is required for brown women in any leadership role, in any public-facing role and I'll give you the final example of that. Recently [2023] David Seymour (leader of the ACT Party) gave some absolutely violent rhetoric towards the Minister of Pacific Peoples. It got massive media coverage (Newshub, 2023). We were appalled and had to figure out how we could make sure we had a voice and put our hands up for the community to see us not standing for that rubbish. We worked with the disinformation project (TDP) down here [in Wellington] who had been monitoring the attacks and targets at me (TDP, 2023). They are the ones who released a report that said that the targeting I got after the white cis male comment was made [and that] the only other time they had seen that much was with Jacinda [Ardern, then PM of New Zealand] through the [COVID-19] lockdown stuff; but it all came in a very tight time frame. So, it was the most the country had ever seen basically, and so when David [Seymour] did his thing, we worked with the disinformation project to say, 'what's the best and safest way for us to have a voice'. The first thing they said was, 'Make James [Shaw], the white guy in a suit do it, make him be the lead not Marama, they will just target her and it will mobilise white supremacy around the world. David [Seymour] is linked into that. He is doing this to do that. It is an intentional strategy. It can't be Marama, it has to be James. Marama can amplify James but it has to be James leading. He will not get the target anywhere near'. What the fuck. Like what the fuck. So I'll just give you that. A woman in power is triggering but being a brown woman in power, that is even more triggering.

What networks and relationships have you found most useful in developing and implementing policies to address the climate emergency?

It goes to an enduring role in politics. I have been here [in the NZ parliament] nearly eight years now but I have the privilege of being connected to the activist community and because of that and because of where I come from, I have been able to maintain those links and those relationships and they are who I am accountable to. So, our policies and our policy analysis and our political positioning, what we vote for and against in the House, that's who I am directly accountable to, and that's why our policies have been bold and strong because of who we remain accountable to.

There are so many activist groups. It has been really important for us to support directly the 'School Strike for Climate' and in turn the 'Make it 16' and lowering the voting age, and we also visited Ihumātao and all of the Māori land occupations (BWB, Forthcoming), Shelly Bay in Wellington and Māori-led occupations all around the country and that's been a really cool part. As you know, even among Māori there is no all-encompassing political thought; Māori vote for every single party, including ACT, including the most conservative right-wing parties. Our accountability in our political work, as the Green Party, is with Māori who are carrying the progressive values of the Māori activist movements; the genesis of *Mana Motuhake* [self-determination] and *Tino Rangatiratanga* [leadership], and those have been my particular priorities, but also I basically lean on the mandate of the broad, broad, broad prevention of violence response sector and those groups and advocacy groups – and basically, they are my particular political cover because everything I do is from their leadership and no one can dare, dare question them.

What do you find to be the main obstacles to developing a socially and economically just climate policy in your jurisdiction?

Disinformation and political will. Labour could have pulled so many levers. So, it's what your theory of change is. We are very clear that our theory of change relies and leans on the progressive grassroots movements. That's where we see our basic power. Exciting and inspiring progressive grassroots, community-driven movements and, in a nutshell, our role is to amplify their leadership. Even Labour's theory of change and everyone else's theory of change the political will only come with what they sense to be the broader public mandate, but they're even reading that wrong. Hello, a wealth tax? A tax of some sort to redistribute wealth is growing in popularity and we are very proud in having a key role in that, but the other political parties have still not taken up that mantle or are trying to look like they are taking it up but not really. They still funnel wealth to their rich mates. So political will, built by populous public mandate instead of 'progressive moving public mandate' has been one of the biggest barriers; it means that Chris Hipkins [PM New Zealand after Jacinda Ardern 2022–2023] ruling out a wealth tax, so we just continue to resist that and say, "no politician can rule anything in or out; that power is for the voters".

The barriers have been political will; the barriers have been a theory of change that relies on populous conservatism and the lobbying power of people in powerful corporations, farmers – but not all farmers. I want to be really clear; it's the few tiny vocal but angry farming voices in particular, the Federation of Farmers and those sorts of corporations who aren't really good representatives of the really good work that many farmers are doing.

So it's powerful lobbies, political will, and flawed theory of change as well as a lack of bravery. In the political system, big money is influence and when the political system is set up to favour big power and big money influence, the system is flawed. When you look at the National [Party], etc. and the money they are raking in. We have always targeted the $3 donations, because that aligns with the spread of power theory and so when the democratic system is rigged to favour powerful political wealthy influence it's going to be real tough to get change.

What climate related projects/policies are you considering/ would like to introduce in the future, and how might these contribute to gender equality in your country?

A more important follow-on from policy is community story building. So, what I am trying to get to here is we have a big anti co-governance tour happening and it's toxic, and racist, and yucky. And many, many *tauiwi* [non-Māori] are standing with Māori to decry and it is beautiful to see that. What I also know is that co-governance and the benefits of it have been happening for a long time already; people just don't realise and cannot see the benefits and it's really simple, like the *Ngāti Whatua marae*, [in Tāmaki Makaurau Auckland], inviting people to plant native trees and reclaim Indigenous forests and *iwi* and *hapū* (sub-tribes) building the community to restore their rivers and return the tuna (eel) to their rivers by river-cleanups and also working together to resist big corporate water take; all of that *mahi tahi* [collaboration]. This is all that co-governance is and yet it's been made into this very scary thing, and so in terms of policy, we need to ignore the noise and just make it happen and in policy, in housing policy and in environmental policy, absolutely in climate policy, we just need to ensure that Māori have a governance role in all of it and just do it. Our ability to cut through to a better understanding is on the ground. Media won't, we can't rely on media to do it for us. The conservative white will continue to whip up fear and mis/disinformation so we have to carry on through the noise and be in touch at the ground level so that we are building immunity, we are building immunity to the rubbish and the bullshit. And so, yes, that is policy but policy alone is not going to fucking change anything.

Who do you draw inspiration from in your role as co-leader of the Green Party of Aotearoa New Zealand, and why?

Honestly, it really is the grassroots leaders: their messaging, their clarity, their focus, their approach, their strategy. And I am talking about teenagers as well, like honest to God, they have just got it and they inspire me, I mean I learn from them and grassroots organisations around the world. For example, when you talk about Black Lives Matter, when you look at the origins and the black women particularly and their clarity of purpose, focus, and messaging, it's there, they've got it. So that is who I take my inspiration from. In my political circle there are obviously the Green MPs, staff, and my senior staff; they have really kept me alive in this job and it has been their support... and I'll give you an example: When I got a major, major shit ton of bricks thrown at me for how dare I meet with gang women and that really ignited the white supremacist movement again; that was really harsh, that was pretty brutal and I thought I wasn't going to survive that; I thought I was going to go home and not get out of bed for three months. It was actually Eugenie Sage, one of our most senior *pākehā* MPs [who retired in 2023] who came up to me on one of those mornings when I was struggling and said 'don't you dare apologise, don't you dare stop, don't you dare apologise to us'. Because I was at that point when I was feeling apologetic for what I had done and apologetic to my team, she was like we won't have a bar of that. You will not apologise to us. You'll keep going. We are here. We have got you. And that was just, oh my God, that was so inspiring and they continue to inspire me.

Māori Glossary

Aotearoa   the land of the long white cloud, New Zealand
Kaitiaki   guardian, stewardship
Mahi tahi   to work together as one, partnership. In the context of Te Tiriti it means Māori and the Crown working together for mutually beneficial outcomes.
Mana   authority, power, status
Māori   the Indigenous Polynesian peoples of New Zealand
Mana Motuhake   self-determination, independence and sovereignty *mana* through self-determination and control over one's own destiny
Mana Wāhine   Māori feminist
Mokopuna   grandchildren
Pākehā   New Zealander of European descent
Papatūānuku   the Earth mother
Tangata whenua   the people of the land
Tauiwi   supporter
Te Tiriti   the Māori version of the Treaty between the Crown and Māori
Tino Rangatiratanga   chiefly leadership, the right to exercise leadership
Wāhine   woman
Wāhine   women
Whānau   family

References

Bridget William Books (Forthcoming) Ihumātao – BWB talk filmed onsite. [Online] Will be available on https://www.youtube.com/channel/UCTUZ1pwxgEqLyAffc3LuHAw

Green Party New Zealand (2023) *Healthy Oceans Act*. [Online] Available: https://www.greens.org.nz/healthy_ocean_act (accessed 9 May 2024).

MacArthur, J., and Bargh, M. (2022) *Environmental Politics and Policy in Aotearoa New Zealand*. Auckland: New Zealand.

Ministry for the Environment (2023) *About New Zealand's Emission Trading Scheme*. [Online] Available: https://environment.govt.nz/what-government-is-doing/areas-of-work/climate-change/ets/about-nz-ets/ (accessed 9 May 2024).

Ministry for the Environment (2024) *New Zealand's Greenhouse Gas Inventory 1990–2022*. Wellington: Ministry for the Environment. [Online] Available: https://environment.govt.nz/publications/new-zealands-greenhouse-gas-inventory-19902022/ (accessed 9 May 2024)

NewsHub (2023) '*It's a joke*': *Seymour doubles down on 'unacceptable' Ministry of Pacific Peoples*. August 18, [Online] Available: https://www.youtube.com/watch?v=qvX6PuLZt88 (accessed 9 May 2024).

Raworth, K. (2017) *Doughnut Economics: Seven Ways to Think Like a 21st-Century Economist*. Vermont, USA: Chelsea Green Publishing. 320 p. ISBN-13: 978-1603586740.

Raworth, K. (2022) *Doughnut Economics, 7 ways to think like a 21st century economist*, Milton Keynes: Penguin Books.

Shareef, J. and Teina Boasa-Dean (2020) *An Indigenous Māori View of Doughnut Economics.* [Online] Available: https://doughnuteconomics.org/stories/24 (accessed 9 May 2024).

The Disinformation Centre (2023) *Race and Rage, Examining Rising Anti-Māori Racism and White Supremacist Ideologies in Aotearoa New Zealand.* [Online] Available: https://static1.squarespace.com/static/65c9ceb1a6a5b72d6f280d67/t/6 5cc1beeb543cd42bdfa3792/1707875314859/Race-and-rage-Examining-rising-anti-Maori-racism-and-white-supremacist-ideologies-in-Aotearoa-New-Zealand. pdf (accessed 9 May 2024).

UN Women (2023) *Women in Politics.* [Online] Available: https://www.unwomen. org/en/digital-library/publications/2023/03/women-in-politics-map-2023 (accessed 9 May 2024).

World Energy Council (2024) *World Energy Council Trilemma Report.* [Online] Available: https://www.worldenergy.org/publications/entry/world-energy-trilemma-report-2024 (accessed 9 May 2024).

PART II
Initiatives

6

GENDER SMART MOBILITY FOR ALL

Lessons Learned from Encounters with Danish
Municipalities

Hilda Rømer Christensen and Michala Hvidt Breengaard

Introduction

In this chapter we present the concept of Gender Smart Mobility, and the
lessons learned while we introduced this concept to Danish municipalities
and related communities of experts and citizens. We demonstrate how Gender
Smart Mobility can enable stakeholders, notably technical and planning staff,
to address green and Smart Mobility in new and just ways that go beyond
dominant notions of economy, growth, and motorised mobility.

Gender Smart Mobility is a notion aligned with the current UN Sustainable
Development Goals linking green and climate friendly city developments
with gender and equality. The notion of Gender Smart Mobility outlines a set
of principles for gender-just and diversity-inclusive mobility through reflections on five specific and aligned dimensions: inclusivity, affordability, effectiveness, attractiveness, and sustainability. While Gender Smart Mobility is a
visionary approach, it is also a complex concept to translate into practice. In
this chapter we present the potentials as well as the challenges of implementing the vision of Gender Smart Mobility.

Gender and diversity in transport includes a focus on how different social
categories affect people's transport needs and how they shape people's transport patterns. The notion of Gender Smart Mobility places particular emphasis on the conceptual approach of intersectionality which provides a
methodology for working with social differences in mobility. The focus on
diversity within and across gender is captured in the term 'intersectionality'
(Crenshaw, 1989; Phoenix, 2006; Angeles, 2017). Working with an intersectional approach in transport includes recognising not only the differences in
people's mobility needs but also their opportunities to access different modes

DOI: 10.4324/9781003461005-11

of transport. We argue that if planners and politicians ignore these differences, we get a transport system that simply marginalises some parts of the population's mobility needs and enhances social inequality.

During the one-year project *Green Transport, Technology and Diversity*, financed by the Danish Ministry of Research, we connected the application of Gender Smart Mobility in a range of Danish municipalities and activities. The motivation for the project was that Danish municipalities play an important role in terms of providing the framework for daily practices and changes in transport customs and traditions. This also assumes that municipalities are vital actors in the revision and introduction of green transportation as stated in the current climate plans in most municipalities.[1] However, in daily policy-making, transport and mobility have a low priority in municipal policy interventions and often compete with the provision and finances of social services and local schools, and especially in the present time of growing assignments and budget constraints.

Danish municipalities have a weak track record in terms of addressing gender and diversity in local affairs. While gender mainstreaming forms part of the Danish Equality law, it has only been subject to a bleak and uneven implementation in municipal governance and administration. Municipalities are required to submit reports on gender equality every third year for analysis at the central Ministry of Equality. Yet, the lack of local commitment as well as efficient and specific goals and sanctions in the field means that these reports have little or no effect. In particular, the field of mobility and transport is often omitted or given a low priority in these annual reports. This echoes the general debate on gender equality and diversity, where transport and mobility are hardly recognised as an issue of relevance.[2]

In this chapter, we present how the notion of Gender Smart Mobility came about and was developed as part of a recent EU Horizon project. We show how it can assist in creating new horizons of mobility for all with a particular focus on local municipality practices. This includes the presentation of five guiding principles for a gender and diversity indicator as well as the data it requires. Also, we will elucidate the potential in current municipal practices and how they can be aligned and strengthened to include gender and diversity. We conclude with critical reflections on the use and meanings of central concepts and terminologies in Gender Smart Mobility.

What Is Gender Smart Mobility?

The concept of Gender Smart Mobility was developed in the EU Horizon 2020 project TInnGO, which aimed at creating frameworks and mechanisms for developing a sustainable European transport system for all. Gender Smart Mobility was set up as a transformative concept and strategy to address (in) equality of gender and diversity in the field of transport and mobility, with particular emphasis on development and planning of Smart Mobility. Gender Smart Mobility aims to include and widen the range of modalities and social groups in processes of transport decision-making and planning. Accordingly, Gender Smart Mobility has been specified as an indicator with five vital dimensions, each of which make up the practical implications of the strategy (Christensen, Breengaard and Levin, 2023; Christensen and Breengaard, 2023). In technical terms, this type of indicator is referred to as a composite indicator, which means that it is composed of several dimensions, which can be illustrated in the matrix and list below.

The five dimensions of *inclusivity, affordability, effectiveness, attractiveness, and sustainability* are meant as steppingstones for creating a paradigm shift in transport policy, planning, and employment towards an all-inclusive Smart and green transport system. Each of the dimensions works as guidelines for the integration of gender and diversity perspectives. In practice, it includes assessments of a particular modality or device in relation to the indicators and whether they can be seen as inclusive, affordable, attractive, effective, and sustainable.[3] Besides, the Gender Smart indicator includes multiple intersecting agents across gender, age, ability, locality, and sexuality where relevant. It is an exploratory indicator – which differs from so-called tick box indicators. In the current projects, this explorative indicator spurred both findings and new questions, rather than definite answers to inadequate survey questions which are often used in transport behaviour studies.

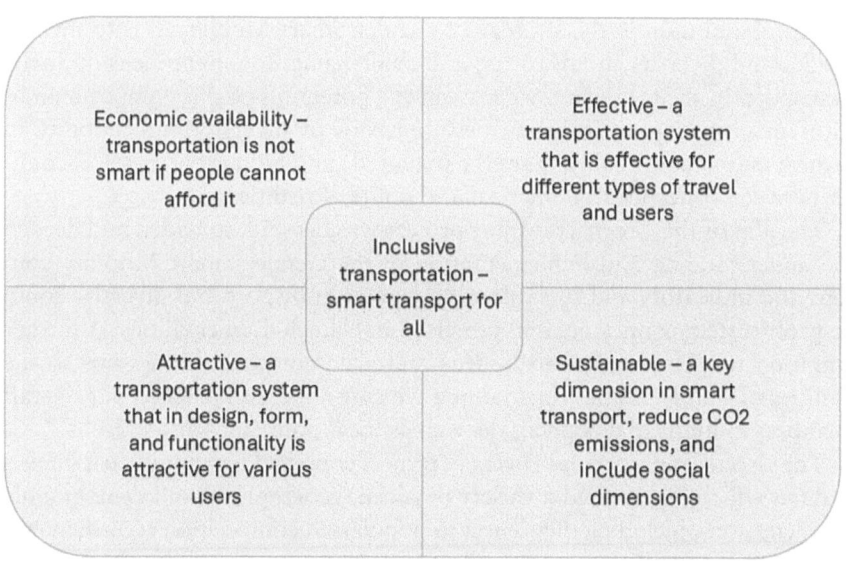

FIGURE 6.1   Gender Smart Mobility dimensions.

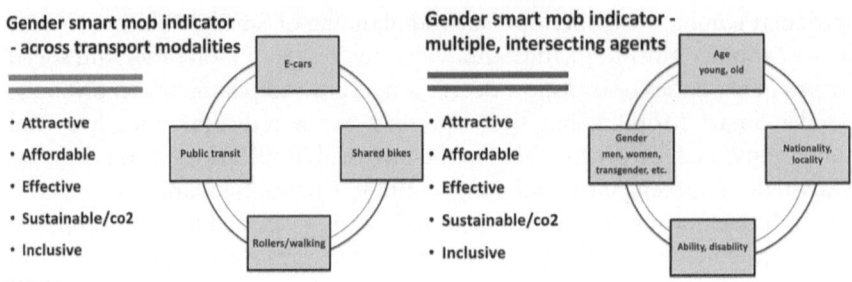

FIGURE 6.2 Gender Smart Mobility indicator.

Source: TInnGo 2021 www.tinngo.eu

To consolidate the Gender Smart Mobility indicator to include both multiple transport modalities and multiple groups, we emphasise the intersectional perspective within the understanding of the notion. As seen in Figure 6.2, we point to how intersections are present between and among both various agents (gender, age, disability, etc.) and various means of transportation (public transport, bikes, etc.). Paying attention to these intersections highlights how applying the Gender Smart Mobility indicator might lay out a variety of definitions in what Smart Mobility is for different social groups.

The Green Transport Project

A central ambition in the strategy of Gender Smart Mobility was to involve gender and diversity in advancing and challenging dominant ideas of Smart Transport. In the following, we present the potentials of this composite indicator in inspiring and creating new horizons of diversity and mobility in Danish municipalities. What are the potentials and limitations in the encounter between concepts and practical and political realities?

The aim of the Green Transport project was, as said, to widen and deepen the understanding and implementation of the Gender Smart Mobility concept and indicators and to explore how green transport and diversity could be proliferated, connected, and practiced in Danish municipalities. A further ambition was to generate new ideas and accommodate the notions to the realities of municipalities. In so doing we consulted and reached out to staff members in municipal councils, as well as local politicians and citizens.

The Green Transport and Diversity project consisted of mapping ten municipalities which represented a variety of social, geographical and demographic characteristics, including differences in population composition regarding ethnicity and levels of income, as well as urban and rural settings. To cater for geographical diversity a selection of diverse municipalities was included, including characteristics such as high and low income, city and suburban, as well as three towns in Jutland, with specific urban and rural challenges.[4]

The ambition was to align green transport and new technologies with perspectives on diversity and to transpose the notion of Gender Smart Mobility into hands-on approaches and tools to be used by stakeholders and practitioners in municipal practices.[5] More specifically the purpose was to:

- Develop education tailored to employees within the transport, environment, planning, and engineering fields in the Danish municipalities.
- Increase practice-oriented knowledge about how green measures in transport can meet and change the transport habits and mobility needs of the municipality and various groups of citizens.
- Contribute to a more sustainable, inclusive, and equal transport system and society by involving different groups in strategies and initiatives.

The Green Transport project connected strategically to the Danish DK2020 climate action plans. According to the DK2020 plans, municipalities must actively engage their citizens to achieve their climate targets. The inclusion of citizens is found to be particularly important in creating change through encouraging people to practice new sustainable lifestyles. DK2020 also means to ensure that all citizens have a platform to raise their concerns and wishes. Municipalities are increasingly aware that citizen involvement is key in climate strategies and action plans. Politicians and planners – both in Denmark and globally – often lack the necessary knowledge to involve gender and diversity perspectives in transport. While the DK2020 plans seemed promising, they were, however, mainly inspirational and aspirational, and the reality we encountered in the municipalities appeared rather different to the images and outreach of the plans.

Through consultations with politicians and planners in the various municipalities, we learned that reflections of diversity in the mobilisation and needs of the citizens were either absent or random. The municipal strategies seemed to be guided by more general and overall aims and priorities. The various levels and ambitions of addressing green transport were also influenced by the wider political culture and traditions of addressing sustainable solutions. While the limited urban space and a more just provision of space for non-motorised mobility – cycling and walking – was a pressing issue in Copenhagen, the alignment of urban and rural mobility needs was a more pressing issue in the Jutland municipalities, which included provincial towns and rural areas. Moreover, suburban and provincial towns with similar challenges and conditions seemed to handle the issues of green transport and inclusion in rather different ways. While some of them had a long tradition of prioritising car drivers, others were more focused on making green transport a real possibility in the provision of accessibility and alternative modes of transport.

Methods

The methods of the Green Transport project consisted both of data collections and consultations with municipal stakeholders and other relevant stakeholders. We aimed to create a new and more dynamic type of project of co-creation with communication and interactions with the staff and municipality council members in consultations and at webinars. A final conference in Copenhagen in spring 2023 included the municipalities involved in the project and high-level politicians, along with Danish and Nordic researchers and experts. A seminar at the University of Copenhagen in October 2023 was set up as a follow-up event, with an invited panel of municipal and transport experts to assess the concept and strategy of Gender Smart Mobility and its potential impact.

It was assumed that the usefulness of the project's end product – the educational material – would be strengthened if the knowledge, interests, and practices of the recipients were included early in the process. To involve stakeholders as much as possible, as well as to ensure the participation of these crucial actors in the project, an initial Living Lab (see Figure 6.3) was designed as a one-to-one consultation. Concretely, data was collected among the ten Danish municipalities by means of qualitative interviews with the municipalities' employees, predominantly at municipal town halls as well as telephone and web interviews. The consultative interviews addressed knowledge and experiences, availability of data and data collection, relations between politicians and municipal staff members, as well as good practices, plans, and proposals. To qualify the consultations with the municipalities, we mapped available data beforehand in relation to general conditions such as demography and availability of climate and transport action plans. Also,

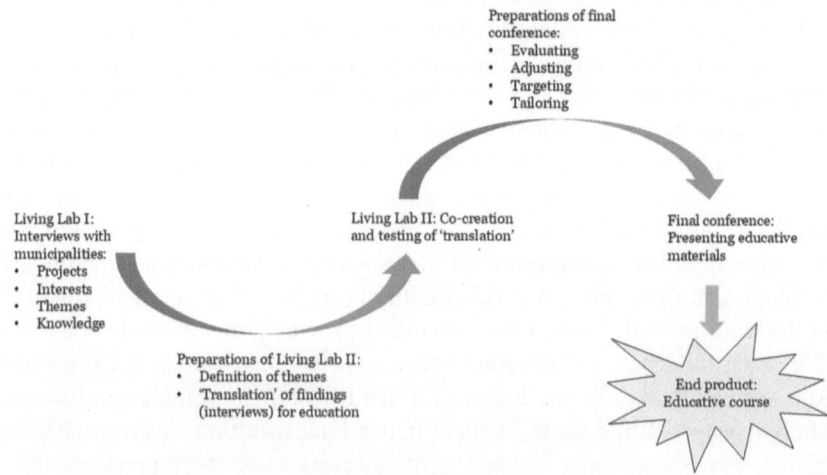

FIGURE 6.3   VEU project development model.

data on gender, ethnicity and age composition in city councils and in technical and environmental committees were collected, analysed, and visualised.

The interviews and consultations were guided by a semi-structured list of questions. They included queries about personal and professional backgrounds and position in the municipality – politician or/staff member. The municipalities were selected according to geographical diversity and people for interviews and consultations were located through various channels. For instance, we identified staff members related to problems and initiatives addressed in local newspapers, through direct contact with the planning offices in some municipalities, and through professional contacts from earlier projects. Some consultations were conducted only with politicians, some were a mix of a politician and a staff member, and some with one or several staff members in a group. All in all, five politicians and 13 municipal staff members were consulted. The interviews took between one and two hours. Three of the municipalities were subsequently selected to present their experiences in short case videos, which included a second round of short and focused interviews. In addition, we conducted interviews with relevant researchers and staff members in related institutions, for instance, the Nordic Hub for Sustainable Urbanisation – BLOXHUB, Copenhagen City Museum, The Danish Town Planning Institute, as well as the climate director in the Association of Municipalities in Denmark (KL). The interviews were made anonymous in the dissemination of findings and insights, except for the staff and politicians who were involved in the videos and therefore visible. We generalised and summarised, where feasible, the findings from interviews and consultations in a list of needs and experiences. Besides, we anonymised quotations from various municipalities and used them as illustrations and elucidating examples in the textbook which accompanied the videos.

Local Knowledge and Green Transport Challenges

Interviews with municipal staff and council members in the Green Transport project drew our attention to the lack of systematic indicators for working with diversity in the current decision-making process related to transportation and mobility at local levels. Several also underlined the need to include diversity in the plans and strategies that form the basis of political decisions about transportation and mobility as well as the need for more knowledge-based and considerate decision-making in the field.

Range of Needs Expressed by Different Municipal Politicians and Staff Members

- We would like to be updated on the latest research and methods.
- We would like, to an even greater extent, to consider and incorporate specific groups into the municipalities' transport strategy.

- We need data related to mobility which includes gender, age, ethnicity, sexuality, and disability.
- We are already spearheading strong initiatives but still lack long-term strategies for sustainable transportation – e.g. cycling and walking strategies.
- We would like to get inspiration and ideas for solutions that can be put into practice without any substantial extra expenditure.
- We would like to communicate our plans in new ways.
- Our work would benefit from concrete instructions and guidelines on how to integrate diversity.

We approached the municipal staff and politicians as 'experts' in their field with knowledge about the local opportunities and challenges of transport policy and planning. Their knowledge was presented to us and in the next step connected to the concept of Gender Smart Mobility. The interpretations focused on localising good practices and clarifying challenges which could enrich the Gender Smart indicator. How could the municipality solve challenges of rural and urban mobility in their setting? Which strategies were feasible in suburban communities to get more people to use public transport and bikes? And how did successful green transport such as cycling create new challenges for planning and exclusion of non-cycling groups? All along, the notion of Gender Smart Mobility was seen through the lens of context, structures and political culture which all fed into the interpretations and were reflected in the dissemination. The aim of the interviews and consultations was first and foremost to add local specificities to the general stage of gender, diversity and green transport. The aim was not to dive deeper into individual interviews or localities, but rather to provide a general overview of the stage and potentials of linking of gender, diversity and green mobility.

Challenges in Local Transport Policymaking

This section presents some of the findings of the challenges in including gender and diversity in transport policymaking, based on interviews with municipality staff and council members.

As for the procedure of transport policymaking, several council members pointed to the random ways in which priorities and policies are made. Several current and former council members openly admitted that transport had a low priority and that decisions were made opportunistic, comparing those to "sticking your finger in the air". Some mentioned climate and CO_2 emissions, welfare and employment, wellbeing and health as points of attention. When asked about gender and diversity in transport policymaking, the council members and staff voiced that these topics were handled as fragmented precautions primarily related to disability and age.

The loose priorities at the local level echo the limitations in the Danish Ministry of Transport's manual of socio-economic analysis of the transportation sector (2015). This manual at the national level prioritises dimensions of travel time, traffic safety, and economic growth with a focus on motorised transport. The manual does not explicitly include other modalities beyond motorised transportation. Also, it does not present reflections on who benefits from the actions, such as who will get to their destination at a faster pace or for whom the safety and economic advantages are intended. The lack of social perspectives leaves the municipalities with challenges and troubles of dealing with problems of mobility, safety, and accessibility for all. As laid out in the DK 2020 plan, municipalities are called to work closely with citizens and their needs in everyday life. Proximity to the citizens is central to the work of the municipalities, from the safety of schoolchildren, the needs for seamless public transit, to the safety and security of cyclists and the mobility of elderly and disabled, to mention a few (see short videos from Herning and Sønderborg in particular).[6] Several of the council and staff members in the municipalities responded to the multiple challenges of meeting the variety of transport needs among their population by asking for a breakdown of the current municipal silo system and urged for coordination across sectors. Particularly, they called for a closer proximity of health and social services and (active) transport and mobility initiatives. Some municipal staff members also asked for more influence in the municipal governance system to empower their position as project leaders and bridge-builders between citizens and the middle and high management of the municipalities. As said by one of the staff members:

> We know that decisions are made at a high level in the organisation, by the ones I call the blue shirts [managers]. If we want to change that, we need to focus on the operative levels, at my level, where we work directly with public transit in daily life. It is me who citizens turn to, it is me that the drivers turn to, and it is me who the management in the transport companies turn to if they have problems. There is a need for creating a national network, preferably at this operational level, and with participation of politicians, because they have the money, or they anyway decide how to spend the money.
>
> *(Municipal staff member in an old industrial and*
> *entrepreneurial area in mid Jutland)*

The interviews and conversations with municipality employees and municipal legislatures throughout the Green Transport project illustrate that problems and potential solutions are interlinked. Despite differences, they addressed similar themes and challenges across the municipalities. Congestion and prioritisation of non-motorised transport, lack of data and the lack of holistic and comprehensive approaches to transport planning and policy were noted as challenges. Regarding matters of diversity and reflections on different groups of citizens, the municipal consultations showed that these are more implicit than explicit. Also, it was stated that the lack of knowledge and strategies on diversity seemed to contribute to misunderstandings and non-successful initiatives.

> At one point we introduced small buses instead of the big diesel buses for reasons of climate and environment. But the small buses were very unpopular because they had poor accessibility. The municipality thought they were doing something good, but many citizens were frustrated by the lack of accessibility and space in the small buses, e.g. for walkers. It was not thought about well enough. We thought we were ordering a smart product – but that was not the case.
>
> *(Politician, North Zealand municipality)*

The relations between democracy, transportation, climate politics, and diversity are relevant when we look at municipal strategies and political priorities. In general, women in Danish municipal councils consider public transportation to be more important than men, irrespective of party affiliation (Hermansen, 2019). Moreover, research shows that municipalities with a greater diversity in gender and age tend to have greater engagement and focus on the broader context in technical committees (Kronsell et al., 2020).

Data from a recent Danish transportation survey and the municipalities' DK 2020 plans reveal, not surprisingly, that there are differences between municipalities in relation to current transportation practices. These differences can, in part, be explained by municipalities' size, geographical location, and the social composition of residents in terms of income, education, ethnicity, age, etc. In addition, there are varying political and cultural traditions, which means that municipalities that seem to be strikingly similar may have different compositions of employees as well as very different strategies and priorities to the subject of green transport.

Disseminating the Green Transport Project

Findings and deliverables of the Green Transport project were set out as educational knowledge and tools, which were disseminated in creative forms. To optimise and ensure a wide outreach we used visuals and short texts in a booklet and video, assuming that visual forms were more enticing than a report text. These forms of dissemination presented immediate and accessible understanding of the concept and strategy of Gender Smart Mobility and good practices and challenges in specific contexts. We presented insights in short videos which included introductions of why and how to include perspectives of gender and diversity in transport, short interviews with municipal staff and a politician as well as a selection of good practices. Furthermore, we published *Green Transport, Technology and Diversity. An inspirational guide for Danish Municipalities* (Christensen & Breengaard, 2023). This publication contained introductions to the field of gender and diversity in transport as well as statistics, guidelines, and examples of challenges and good practices from Norway and Sweden.

To achieve a broad outreach, the publication was sent to the technical and environmental committees of all 98 municipalities in Denmark. We also distributed the publication at various major conferences, such as a national conference on public transit, which was a hot topic in Denmark in the autumn of 2023 due to budget restrictions both at regional and municipal levels.[7] Some of the municipalities, such as Copenhagen and Aarhus, along with projects and green foundations, reacted almost immediately and invited us to attend seminars and other local staff events to introduce findings and discuss methods and challenges. Here we met a broad community of municipal staff, consultants, and other experts, who found the green transport and diversity approach relevant and with refreshing perspectives. All along, a range of critical and explorative questions were raised by the participants. They form the structure and content of the following elaborations and reflections which include and problematise the concept of Gender Smart Mobility.[8]

Critical Reflections and Lessons Learned

'Why do you focus so much on gender in your approach to Gender Smart Mobility?' was an issue that was raised time and again during our presentations, mainly in discussions with expert audiences. While municipal staff and counsellors found the issue of gender new and relevant, several experts and researchers saw it as a problem. The problem was often formulated around the potential essentialising approach in the focus on gender, explicit in the term Gender Smart Mobility. This problematisation addresses a range of issues and reflections which are briefly summarised below.

Gender Neutral

Even though there are ongoing changes in gender roles and divisions of work, there are still differences that have an impact on women's and men's travel patterns as seen below in Figure 6.4. These differences are due to continuing structural imbalances in society, but also in the transport sector itself. Historically, employees' travel needs have been privileged in relation to the domestic sphere, and this hierarchy continues in the planning of transport, although often in an invisible form. By 'invisible' we mean that transport research, policy, and planning assume that transport modes and needs are gender neutral, despite the diverse use of these modes.

Women are more likely to...	Men are more likely to...
Take journeys with more trips on the way, such as home – childcare institution – work – childcare institution – supermarket – home	Make direct trips between A and B (home – work)
Make more non-work-related trips	Travel more kilometres
Travel in suburban areas	Travel longer hours
Travel outside peak hours	Travel more often
Travel in more sustainable ways, e.g. by walking and using public transport	Stop driving due to health reasons
Stop driving car due to social-economic conditions, such as lack of finances or little driving experience	Stop driving later in life than women

FIGURE 6.4   Gender differences in transport.

Gender neutrality is reflected in terms such as 'users' or 'passengers', which are often used to describe actions in the field of transport. The question is what constitutes the understanding of a 'passenger' or 'user' and their transport needs and travel patterns? And how can we move from gender neutrality to gender-inclusive modes?

Intersectionality

Throughout the project, we emphasised that we applied an intersectional approach to the notion of Gender Smart Mobility which means that we see gender in the context of other social categories. Intersectionality expands the binary understanding of men and women as two homogeneous groups and works with a more complex understanding of gender as a multifaceted social category. Incorporating an intersectional perspective into transport policy, planning and design means paying attention to how different social categories interact with different transport patterns and needs.

Applying an intersectional approach to gender implies identifying the many factors that lead to differences in various social groups' transport behaviour, their choice of transport mode, and barriers of accessibility. For instance, we have been met with the argument that cycling is problematic to promote as a green solution, because cycling excludes many people, including ethnic minority women who do not know how to cycle. A sidetrack in the Green Transport project was to raise the need of promoting cycling courses for women with ethnic minority backgrounds. This focus on a specific group spurred discussions among various audiences. While students tended to problematise and see the focus as stigmatising, politicians and practitioners such as social workers, saw this as a strength, and acknowledged a challenge which has been ignored up till now.

Gender before Sustainability

One of the experts, an experienced researcher in the field of sustainable transport and indicators, problematised the unclear, and what he considered weaker, position of sustainability compared to gender. His suggestion that sustainability should take centre stage urges us to clarify what we mean by sustainability. Sustainability is the backbone in the strategy of Gender Smart Mobility. While the term Smart Mobility most often describes technological innovations of transport to solve problems of, not least, pollution and environmental challenges, we critically use the term by addressing the uneven resource allocation and attention given to public transit and non-motorised mobility compared to individual car mobility. We see the relevance in addressing and clarifying critical issues related to cars and gender, in particular e-cars which are often promoted as the dominant sustainable and climate friendly solution. Recent analysis of the e-car market in Denmark has demonstrated that the roll out of electric and hybrid cars has first and foremost benefited the upper middle class which presumably consists of a majority of male drivers (Statistics Denmark, 2023).[9] Implying a perspective on gender and diversity in Smart Mobility – Gender Smart Mobility – is a way to raise the question of the quality of sustainability as 'smart for whom?' which includes problematising the car-centric society (Christensen et al., 2022).

Why Gender Smart – and Not Gender Slow – Mobility?

One of the issues that this chapter critically assesses is how gender and diversity are entangled in the concepts and practices of Smart Mobility. During the dissemination process of the Green Transport project the critical question of the use of Smart was raised. Why do you use the 'Smart Mobility' terminology which connects to instrumentalisation and technology? Why not use a counter concept, such as 'Slow Mobility' or the like?

The reason why we chose to use and connect gender with the concept of Smart Mobility is our ambition to connect gender to the mainstream climate and transport agendas; currently 'Smart' is a buzzword with many meanings. With the introduction of the concept Gender Smart Mobility, we intended to address the hype and attention of the *Smart* agenda with a new dimension – and to relocate and suggest *Smart* as a more social-oriented and inclusive term. Our choice was based on the following considerations:

- *Smart as a travelling concept*: For a brief historical outlook, the close alignment of technology with the Smart City and Smart Mobility, accentuated by the critical question above, dates to the 1990s where the focus was on the significance of new technology, ICT and its application in urban infrastructure (Albino et al., 2015: 3).
- *Smart as problem solving*: The idea of the Smart City was invented as a solution to multiple urban problems. Initially, Smart Cities addressed a variety of escalating problems, including traffic congestion, inefficient services, and economic stagnation in the 1990s and proliferated as a solution to the financial crisis in 2008 (Hollands, 2014). Today in the 2020s the Smart City, including Smart Transportation, is identified as both hegemonic and seductive, yet also with aspirations of prosperity and growth on the one hand, and the promotion of healthy lifestyles for all on the other.
- *Smart as a means*: In general terms, the Smart City and Smart Mobility discourses have become part of ongoing conceptual and practical dynamics, and this also creates a thriving field of contrasting views.[10] Various agents, among them the consultants that are engaged in carving out future directions, have pointed to the need to include more human agency in Smart City strategies. One example is a McKinsey report from 2018, saying that Smart cities 'need to focus on improving outcomes for residents and listing their active participation in shaping the places they call home' (McKinsey 2018 Executive Summary).
- *Smart in sum*: By applying the term Gender Smart Mobility, we intended to join forces with other social scientists to offer a new take on how to wrestle the definition of a Smart City from the fixation on technology, the pure profit interests of big business, and the associated weakness of public services and disempowerment of citizens. The urge to continuously discuss and improve the understanding and strategies of what Smart is, also the reason and motivation behind the development of the composite indicator of Gender Smart Mobility presented above.

Conclusions

The Gender Smart Mobility indicator is vital to include in mobility and transportation, not only locally but in Europe and in global perspectives, to enhance and extend strategies of welfare and wellbeing as well as the green transition. We have demonstrated that Danish municipalities play an important role in terms of daily practices and changes in customs and traditions. This makes municipalities vital actors in the revision and introduction of green transportation. A focus on the connection between diversity, transportation, and mobility in the Green Transport project was intended to enhance the Danish climate politics agenda and help Denmark achieve its green vision. However, as demonstrated, there is still a long way to go.

The Green Transport project was meant as a wake-up call aimed at advancing the relevance of including diversity as part of the green transition and transportation. Even though we do not have a follow-up survey to quantify the effects of the project, we have seen a substantial interest in the video films and publications which were the outcome of the project.[11] Moreover, the Green Transport project took place at the same time as perspectives of gender and diversity in transport and mobility; emerged after years of ignorance; it was made a priority of The European Institute of Gender Equality (EIGE) as well as The European Green Deal. This includes the campaign: Towards a green and gender equal Europe with slogans such as Unite for a Green and Fair Tomorrow. Along with the campaign, Carlien Scheele, Director of EIGE, unveiled a powerful vision to accelerate the green transition while advancing gender equality. "Now you too can be part of this transformative journey. Together we can shape a sustainable, inclusive future for everyone in Europe."[12]

Notes

1 Through the DK2020 project, led by the green think-tank Concito, 95 Danish municipalities have committed to developing climate action plans that are in line with the objectives of the 2015 Paris Agreement. As a major CO_2 emitter, transportation is included as one of the most significant parts of these action plans.

2 A recent report on gender mainstreaming in Denmark provides evidence of the low level and omission of gender mainstreaming in transport. The Ministry of Transport out of a total of 93 new laws ended up with zero equality assessments. See https://kvinfo.dk/wp-content/uploads/2023/02/Ligestillingsvurdering-af-lovforslag_KVINFO_Februar-2023.pdf: 29.

3 See examples of the implementation in the TInnGO Hubs and GaDAPs, https://transportgenderobservatory.eu/resources/gender-diversity-action-plans/.

4 Outline of the Danish municipalities included in the project: Copenhagen was chosen as the capital well known as a cycling friendly city, while at the same time confronted with an influx of new residents and a neglect of walking and accessibility in public transit. We also included several suburban municipalities north of Copenhagen. One dominated by high income and a high level of car ownership (also electric cars which require both public and private charging stations) and neglect of policies to improve conditions for cyclists, pedestrians and public transit. This compares with a couple of mixed middle- and low-income municipalities which were eager to address cyclists and non-motorized mobility with various experiments at the stations and in terms of good biking facilities and rental bikes. We initially included several low-income municipalities south and west of Copenhagen; unfortunately, a couple of them left the project due to new political priorities after the local elections in 2021. We were left with one municipality with low-income earners (the lowest average in Denmark) where around 40 % of the residents have another ethnic background than Danish. The most pressing problem here as reported by the municipal staff was how to provide decent parking lots for the residents in rental housing where parking had not been included from the outset. The shortage meant that cars were parked all over, since tenants owned several cars including company cars; an indication that low-income areas are catching up in terms of car ownership and that cycling is not seen as a tempting alternative. Green transport in this low-income area was mainly seen as public transit and not

as cycling or walking. To achieve geographical diversity three municipalities in Jutland were included. They presented provincial towns with between 73.000 and 97.000 residents. A major reform in 2009 reduced the number of Danish municipalities from around 300 to 98, meaning that many rural communities have been enrolled in such urban structures. Some of the former rural communities had even been split, so that they now belong to separate urban municipalities. Conditions as these clearly influenced the stage and ambitions of green mobility in the various locations, along with specific cultures and traditions and often leaves the rural population with bad or absent alternatives to the private car.

5 Green Transport, Technology and Diversity was a one-year project financed by the Danish Ministry of Education and Research in 2022–23 and conducted by a team of researchers and students at the Coordination for Gender Studies, University of Copenhagen.

6 The videos can be found at https://koensforskning.soc.ku.dk/projekter/groen-transport/.

7 Kollektiv Trafik Forum og Transportøkonomisk Forening Kollektiv Trafik Konference mandag d. 2. oktober, 2023.

8 We were invited to give talks in various think tanks and projects, e.g. Concito – the Green Think Tank in Denmark, The Nordic Council of Ministers and the Nordic green transition project, the Danish Road Directorate, and a couple of municipalities. In addition, the TInnGO and Green transport project(s) have been presented at major academic and outreach conferences, the Regional Studies conference in London October 2023, The Nordic Green Transport Forum June 2024 and the annual RSA conference in Florence in June 2024. Finally, presentations have been made at various University of Copenhagen seminars and at the summer school: Gender and Diversity in Public and Private Organizations in August 2023 and at project meetings in the GILL project/Horizon Europe and I project in dec. 2023.

9 According to Danish Statistics there was a surplus of high-income earners (at the highest level) and top managers among the electric car owners in 2021. Top leaders consisted of 3% of the population but owned 10% of the electric cars. The highest wage group makes up 10% of the population and 28% of car owners. DKs statistik. 25.10. 2023. https://www.dst.dk/da/Statistik/nyheder-analyser-publ/bagtal/2023/2023-25-11-portraet-af-elbilkoeberen.

10 No less than 43 smart city definitions were found in recent literature (Hollands 2014; Vanolo 2013; McFarlane and Söderström, 2017). The multiple definitions also form dynamics aimed at widening, displacing, and twisting the meaning of 'smart city' and associated practices.

11 Our website: https://koensforskning.soc.ku.dk/english/projects/green-transport/ had over 3.000 visits from October to December 2023.

12 Green Deal and Gender Equality | European Institute for Gender Equality (europa.eu) https://eige.europa.eu/newsroom/green-deal-and-gender-equality?language_content_entity=en.

References

Albino, Vito, Umberto Barardi and Rosa Maria Dangelico. 2015. Smart Cities. Definition, Dimensions, Performance, and Initiatives. *Journal of Urban Technology*, 22 (1): 3–21. https://doi.org/10.1080/10630732.2014.942092

Angeles, Leonora 2017. Transporting difference at work. Taking gendered intersectionality seriously in climate change agendas. In: Griffin Cohen, M. (Ed.) *Climate Change and Gender in Rich Countries: Work, public policy and action*, 103–118. London, Routledge.

Christensen, Hida Rømer and Michala Hvidt Breengaard. 2023. *Green Transport, Technology and Diversity. A Guide for Danish Municipalities.* University of Copenhagen. https://koensforskning.soc.ku.dk/english/projects/green-transport/green-transport.pdf

Christensen, Hilda Rømer, Michala Hvidt Breengaard and Lena Levin. 2023. *Gender Smart Mobility: Concepts, Methods, and Practices.* London, Routledge. https://doi.org/10.4324/9781003191025

Christensen, Hilda Rømer, Louise Anker Nexø, Stine Pedersen and Michala Hvidt Breengaard. 2022. The lure and limits of smart cars: visual analysis of gender and diversity in car branding. *Sustainability* 14 (11), 6906. https://doi.org/10.3390/su14116906 - 06 Jun 2022

Crenshaw, Kimberle. 1989. Demarginalizing the Intersection of Race and Sex: A Black Feminist Critique of Antidiscrimination Doctrine, Feminist Theory and Antiracist Politics in University of Chicago Legal *Forum*: Vol. 1989: Iss. 1, Article 8. http://chicagounbound.uchicago.edu/uclf/vol1989/iss1/8

Hermansen, Amalie Dahlerup. 2019. *Køns betydning for politisk repræsentation i danske kommuner.* MA Dissertation. Department of Political Science, Aarhus University. https://surveyselskab.dk/wp-content/uploads/2020/03/Speciale_Amalie-Dahlerup_2019.pdf

Hollands, Robert 2014. Critical interventions into the corporate smart city. *Cambridge Journal of Regions, Economy and Society.* 8. 61–77. https://doi.org/10.1080/13604810802479126

Kronsell, Annica, Christian Dymén, Lena Smidfelt Rosqvist and Lena Winslott-Hiselius. 2020. Masculinities and femininities in sustainable transport policy: a focus on Swedish municipalities. *NORMA* 15 (2), 128–144. https://doi.org/10.1080/18902138.2020.1714315

McFarlane, Colin, Ola Söderström. 2017. On alternative smart cities. *City* 21 (3–4), 312–328. https://doi.org/10.1080/13604813.2017.1327166

McKinsey 2018. Global Institute: Smart cities: Digital Solutions for a more liveable Future. June 2018. Accessed October 25, 2024. https://www.mckinsey.com/business-functions/operations/our-insights/smart-cities-digital-solutions-for-a-more-livable-future

Phoenix, Ann. 2006. Interrogating intersectionality: Productive ways of theorising multiple positioning. *Kvinder, Køn & Forskning* (2–3). 21–30. https://doi.org/10.7146/kkf.v0i2-3.28082

Statistics Denmark 2023. Portræt af elbilkøberen. Accessed October 25, 2024. https://www.dst.dk/da/Statistik/nyheder-analyser-publ/bagtal/2023/2023-25-11-portraet-af-elbilkoeberen

Interview 3

ADA COLAU, MAYOR OF BARCELONA 2015–2023: ADDRESSING THE CLIMATE EMERGENCY IN COLLABORATIVE WAYS AT THE CITY LEVEL

By Inés Novella Abril, 21 February 2024 in Ada Colau's office at Barcelona City Hall.

Could you tell us about your political career and how you got to this important position?

I never thought that I would be the mayor of Barcelona, because I am a girl from the suburbs who studied philosophy and who has been a social activist for most of her life, especially for the right to housing, but not only. I've been a social activist for decades, I've done many different jobs, I've known precariousness – because I come from a humble family – and I didn't know anyone from the Barcelona elite. But despite this background and origin, something extraordinary happened. A context of great social mobilisation, such as the movement for the right to housing in which I was involved and which later joined 15M, coincided with a social upsurge and a political and institutional crisis provoked by very serious cases of corruption. It was then suggested collectively that it might make sense for those of us who had never felt represented to take a step towards the institutions. So a different political proposal emerged, which was not a classic party. I took the step into institutional politics not because I wanted to individually, but because there was an exceptional situation where many people who did not feel represented decided to undertake this political innovation. We built a new political platform which was not a classic party, but which brought together many different people; people who had never been in a political party and people who came from political parties but who understood that the way of doing politics had to be updated. And all these people who came together made a programme for the city of Barcelona that ended up being called *Barcelona En Comú*.[1]

DOI: 10.4324/9781003461005-12

In your first electoral programme, as a candidate for mayor of Barcelona, how important were climate issues and feminism in your proposal for local politics?

They were very important, which was strange at the time. When we came to the elections with Barcelona En Comú, we showed that institutional politics had been stagnant for too long and far removed from social debates. We were the first to talk openly about the climate emergency as a central issue. We were the first to talk about feminist politics, not just equality, but feminism. We were the first to talk about confronting speculation from within the institution, something that no one else in the institutional sphere was saying.

And, to tell the truth, I am happy because later, when we came to the city council, we were able to do part of that programme and, beyond what we did, we managed to change the agenda. In other words, now in the institutions, not only in the Barcelona City Council, they are talking about feminist policies and climate emergencies, issues that were not talked about before. For example, we created the Regidoria de Feminismes and a few years later the Generalitat de Catalunya created a Conselleria de Feminismes.[2] Beyond our policies, the political agenda has really changed, with the climate emergency, feminism, and the fight against speculation at the centre. I think all this is here to stay.

What do you see as the role of local government in the fight against climate change? Could you describe the main measures you have taken during your eight years in office?

In these eight years, despite all the constraints we have faced, we have shown that much more can be done on climate change at a local level than has been done before. We have placed environmental issues at the heart of the city's policies and implemented measures that had not been done before. Apart from traffic calming and the *superilles* model,[3] which is perhaps the best known, we have increased the amount of green areas, doubled the number of cycle paths and made a much greater commitment to public transport. These are the most classic initiatives, but we have also built climate shelters, protected school settings, and renaturalised school playgrounds. We were told at the beginning that we would slow down the economy, but we have left it better than we found it, with fewer people unemployed and good economic indicators. We have shown that climate policies do not work against the economy.

Municipalities can do a lot about climate change by raising public awareness. It is important for citizens to be aware and more demanding of all administrations, because many things do not depend on local authorities. In Barcelona there has been a change in awareness in recent years. It used to be something "for environmentalists" and now it is a majority opinion of the citizens, who want to reduce pollution, noise and emissions, and want more recycling and more public transport. It is a priority, they are conscious and they are demanding. And this is essential, because climate change policies face resistance from big economic interests.

A mobilised and aware citizenship is essential so that demands can then be made to other levels of government that have more competences. The question of competence is one of the great limits we have. We often hear that metropolitan areas produce most of the world's emissions. But a lot of those emissions are the result of an economic model that is governed by rules that do not depend on cities, but on national, European, or even global rules. That is always a difficulty. Cities need to do more, but we are not given the budgets or the powers to act. And yet, I believe that in these years as Barcelona's government, we have shown that it is possible to innovate, to do many things beyond the classical ones.

For example, we created the public energy company "Barcelona Energía", which allows all the lighting in Barcelona to be powered by 100% green energy. We started with public lighting because it was directly dependent on us, then as a trader we reached other councils in the metropolitan area, public markets, and more and more small and medium-sized companies. In addition, as a public supplier with no profit motive, it is possible to adjust prices and advise users on which tariff is best for them. At the same time, we help reduce emissions by buying 100% green energy.

Would you say that the difficulties in developing climate policy stem from different aspects? Which ones would you highlight?

Yes, there are general issues, such as administrative times and bureaucracy, which affect all administrations. The question of competences is also an additional difficulty, as you often have to reconcile your local policy with regional, national, or European policies. We are now in a period of permanent instability with constant elections. Before, with absolute majorities, there was more stability, but that has come to an end. In addition to these structural administrative problems, there are also the problems of conflicting interests.

There are business lobbies that refuse to lose a single euro of profit, so they wage legal and media wars. We have seen real subsidised smear campaigns and "fake news". There are non-transparent digital media that we know are funded by large corporations and that are dedicated to saying outrageous things about me. These are proven lies, but they are out there and they are circulating. We denounced it publicly, they corrected it in their digital media, but fake news is much more attractive and it has been proven that it circulates 10,000 times more than a correction. The same thing happens with the lawsuits, they all ended up being filed; but when they are filed, you have to declare and the damage is already done, because the filing of the lawsuit that follows reaches very few people.

Based on your experience as a mayor, what would you recommend to people who are now in charge of local government if they want to develop transformative and socially sensitive climate policies?

One lesson learned is the need for planning. Reviewing traffic, emissions and noise, investing more in public transport or making the city calmer and healthier are measures that need to be taken in every city. It is not an ideological political programme, it is a question of survival. But it is true that it is important to plan these actions so that they are not seen as a paternalistic or moralistic imposition by the institutions, but as a collective advance by the citizens. That is a different approach.

There is sometimes a temptation on the left to impose a set of policies that are necessary only in terms of rational argument. This is one way, but it is important to think about how these policies are communicated, prioritised, and elaborated. For example, when working with schools, children are very clear about the need to fight climate change. Giving children a leading role, which then influences their families, is an indirect way of creating a civic culture on climate. It is a way for people to demand that governments develop policies. It is more effective than having the administration act in a paternalistic way.

We started with the low emission zone, an unavoidable measure that involved restricting traffic and most polluting cars. This is necessary and is being done all over Europe. But when you arrive and this is the first measure you take, when there is a large part of the public that sees it as an imposition by the administration, it ends up generating a lot of resistance. We were pioneers in Spain and now it is being done all over the country. So, it is necessary to find a balance between this pioneering character and the pedagogy that must accompany these measures. If the low emission zone had been introduced at the end of the mandate, it would certainly have met with less resistance because much more public debate has been generated.

Participatory budgeting is another measure that indirectly helps these measures to be seen as collective progress and not as an imposition. We allocated 70 million euros to investments decided by citizens through participatory budgeting. Many schools, families, and neighbourhood associations were mobilised. In the end, most of the measures were aimed at improving school settings, cycle paths, and high-quality green areas. But it is different when it is a citizens' initiative and when it is imposed by the administration.

We learned this over the years. Maybe now I would do some things differently at the beginning, in a different order or with a different pedagogy. The process may be a little slower, but in the end it is worth it because it is seen as a civil process and not an imposition.

Where did you find support? Who would you say were your allies in developing environmental and feminist policies?

The allies were citizens in general, not just mobilised and aware organisations. The key was to make the city debates public and to set up different mechanisms for citizen participation, such as working with schools or participatory budgeting.

We proposed a new participation regulation to be able to carry out citizen consultations. But AGBAR[4] managed to stop this by putting a lot of resources into lawsuits against us, thus preventing a public, non-binding debate on public water management. But the debates that are generated around something like this end up activating citizens. This is important because we have been able to make more progress where citizens have been mobilised.

Confronting privileges that no one has confronted before touches interests and provokes reactions. This is what happened with the car industry, real estate speculation, and water management. The way to be strong as an institution in the face of these big lobbies, which reach the media and have a lot of power, is to mobilise citizens.

On housing, we achieved things that seemed impossible because there was a mobilised citizenship that put pressure not only on us but also on the other political parties. For example, we proved that any new housing development had to include 30% social housing. This was a way of making the private sector co-responsible for making housing more affordable. The same happened with measures on sustainable mobility or the regulation of tourism. Without an active citizenry demanding the connection of the tramway along Avinguda Diagonal, we could not have achieved it. Without social mobilisation in favour of controlling and limiting overcrowded tourism, we would not have been able to implement a series of measures in this area.

On the other hand, we also find an alliance in the strengthening of networks between cities. This is also feminist politics. Networking serves to defend cities together against common problems. For example, against multinationals like Airbnb or the car lobby, we networked with Paris, Milan, Amsterdam, Berlin, and London, among others. This gave us more capacity to lobby the European Commission or national governments. It gave us visibility, so that our proposals no longer looked like madness from Barcelona or just my own ideas. It was a network of European capitals saying the same thing, and that was also an important support.

What qualities do you think a politician needs to develop successful policies to tackle the climate emergency? Is there anything special about a woman doing this?

I believe that everyone has something to contribute to politics. But one thing we represented is the opportunity for people who have never been in positions of power before to exercise it for the first time. Many people on the street told me that it was the first time they had felt represented. It is clear that we were different from the people who usually occupy these positions: men from a certain social class, with a certain cultural capital, and political, and economic connections. We were an anomaly, not only because I was the first woman mayor, but because we all came from activism, or from the working and humble classes.

And I think that is what we need in general. It's not that people have to look like us, but that social diversity should come more into institutional politics for the common good and to make politics better for everyone. So there should be more women, but also more LGTBI people and especially more migrants. In Barcelona, between 25% and 30% of the population was born outside Spain, and they are not represented in the different spheres of power. We will have a serious problem of social fragmentation if this entire population is not represented in decision-making.

Diversity will make it easier to face the current major challenges to democracy, such as the rise of the extreme right or the questioning of politics. Diversity in politics also means having more and more diverse talents.

In your role as mayor of Barcelona, have you been inspired by anyone? Who were your role models?

There was not just one person, but if I had to choose, I would say my mother. She is one of the people I admire and love the most. She is a woman who has done things that I could not do, she has raised four daughters, she has worked all her life and fought against precariousness. She has faced very difficult situations without losing her good humour or her empathy. For me, she is a reference point in terms of always knowing what is most important and taking care of the people you love. She always encouraged me and my sisters to follow our dreams.

After that, I have many extraordinary women as role models. When I came to the city council, there was hardly any representation of women from the city. So, on the first 8 March that I was mayor, in 2016, I placed a photo of Federica Montseny[5] because she is a reference for me. Not just because she was the first woman minister in Spain, but because of the way she faced her contradictions. She was an anarchist, she didn't want to join any institution or hold a government position, but in the end she took office and faced her contradiction because she understood that there was a more important cause than her own coherence. Fascism was coming and a popular front had to be organised. Confronting her own comrades, she took over the Ministry of Health and was a good minister, introducing for the first time in Spain the issues of abortion, single mother families, and caring for sex workers. These were revolutionary things at the time. For me, Montseny represents the noblest kind of politician, one who overcomes her own contradictions and does what she believes is most useful for the citizens as a whole.

When this book began to take shape, there was a group of women in different parts of the world who were leading transformative political projects on climate change: Jacinta Arden, Sanna Marin, Nicola Sturgeon, Marianne Borgen, and you. Within a few months, most of you resigned or were not re-elected, and almost all of you were replaced by men. What do you think about this?

As long as we are exceptions, it is difficult to be sustainable. We are interviewed as exceptions; even if we are not one or two, everyone identifies us as 'the first woman minister', 'the first woman and ecofeminist mayor'. For me, the lesson is that in order for this to be something more widespread and sustainable over time, there needs to be many more women. It's a question of numbers, that's why I defend quotas. For me, feminism is much more than parity quotas, but parity is essential for there to be many more women in decision-making. So, taking into account this question of numbers, there must be opportunities for many more women so that there continue to be women in these and other positions of power. But of course, as long as we continue to be the few exceptions, it is easy for this to stop happening.

Notes

1 "Barcelona en Comú" ("Barcelona in Common") is a left-wing political party active in the city of Barcelona. It was founded in 2015 and Ada Colau was one of its main promoters.
2 'Regidoria' and 'Conselleria' refer to the department responsible for sectoral policies within local and regional governments respectively.
3 *Superilles* (superblocks in English) is the main street transformation programme for the city of Barcelona. Through a redefinition of Cerdà's urban grid, this model reclaims for citizens part of the public space that has been occupied by private vehicles. Aiming at creating a healthier, greener, fairer and safer public space, the *superilles* model improves urban biodiversity, supports sustainable mobility, encourages social cohesion, and promotes local economy.
4 Formerly known as "Aigües de Barcelona" and with a history of 150 years, AGBAR is a Spanish water utility company operating in Spain and Latin America.
5 Federica Montseny Mañé (1905–1994) was the first woman minister in Spain. An anarchist and intellectual, she was Minister of Health and Social Welfare during the Second Spanish Republic. Despite her short time in office, she brought about far-reaching changes, including the abortion law and the creation of social facilities for women such as soup kitchens for pregnant women, children's homes, and centres for the liberation of prostitutes.

7

WHAT DOES DEGROWTH SAY ABOUT GENDER EQUALITY AND SOCIAL JUSTICE?

Bipasha Baruah and Andrea Burke

Introduction

Degrowth is an emerging field of research and a social movement in industrialised countries founded on the premise that infinite economic growth is incompatible with the biophysical limits of our finite planet (D'Alisa et al., 2015; Asara et al., 2015). Central to the degrowth philosophy is "imagining and enacting alternative visions to modern growth-based development" (Kallis, 2015: 1). Within the degrowth literature, new ways of working and new forms of employment (such as part-time work, work-time reduction, remote work, flexible work and job sharing), alongside interventions such as universal basic income, universal social protection, free post-secondary education, maximum salary caps, and reform of financial institutions, have been proposed as ways by which industrialised countries might balance economic security with environmental protection (D'Alisa et al., 2015; Kallis, 2011).

The fact that the current economic growth paradigm perpetuates existing gender and environmental injustices is well established in the scientific literature. A reliance on economic growth cannot be the solution to environmental injustices since it is, as Anguelovski (2015: 35) points out, "part of the process that creates environmental injustices". An evolving body of scientific evidence has emerged in the last decade that endorses the role degrowth in industrialised countries can play in balancing global economic needs with environmental concerns, thereby enabling humans to live within the Earth's carrying capacity. However, within this literature, the link between degrowth and environmental sustainability is much more clearly established than the link between degrowth and gender equality. The link between degrowth and environmental sustainability is more straightforward since decreasing production and consumption, as evidenced by industrial and air travel

DOI: 10.4324/9781003461005-13

shutdowns in the early months of the COVID-19 pandemic, directly allevi-ates environmental pressures. When it comes to issues of gender equality, the link to degrowth is not as clear-cut although feminist scholars have long asserted that "environmental and gender concerns taken together open up both the need for re-examining, and the possibility of throwing new light on, many long-standing issues relating to development, redistribution, and insti-tutional change" (Agarwal, 1992: 152). Other feminist scholars have long emphasised that economic growth, measured simply in terms of income and production of goods, fails to factor in environmental harms as well as the costs of social reproduction. For example, Marilyn Waring (1988) was one of the first authors to criticise gross domestic product (GDP) for being structur-ally blind to women's unpaid care labour as well as ecological processes.

Within the degrowth literature that has emerged in recent years, there does not appear to be consensus on the relationship between planned economic contraction and gender equality. Some theorists have emphasised that degrowth may emancipate women and other groups that are marginalised within the capitalist system from the limitations imposed by the current eco-nomic model and generate new forms of access, participation, and representa-tion (D'Alisa et al., 2015). Others have suggested that current forms of gender and other social inequalities within the capitalist system may be exacerbated under degrowth and lead to further gendered marginalisation unless appropri-ate social reforms and public policies are implemented simultaneously (Dengler and Strunk, 2018). When it comes to gender equality, the only point of agree-ment in the degrowth literature seems to be that "further elaboration is needed on the gender dimensions of degrowth" (Demaria et al., 2013: 206). Several scholars have asked concrete questions about the social equity and justice dimensions of degrowth. For example, as a follow-up to Paech's (2010) pro-posal for a 20-hour work week, scholars have asked that if middle-class pro-fessionals are able to reduce their childcare needs by working less, what would happen to the livelihoods of those whose economic mainstay is caring for the children of others (Eicker and Keil, 2017; Perkins, 2010)? Similarly, if hotels start (as many already have) reducing their ecological footprint (and cutting maintenance costs) by offering guests incentives to forego daily housekeeping and laundry services, would the jobs held by cleaning staff (among whom women, immigrants, and racialised minorities are overrepresented) not be ren-dered even more precarious and poorly paid than they already are (Lipsig-Mummé and McBride, 2015)? If our economies were to contract, what types of policies and programmes might we implement to reduce social inequality? These are the types of questions we hoped to find evidence for in this research.

In this chapter, we draw upon domestic, international, and cross-sectoral evidence on emergency response benefits, basic income schemes, childcare, eldercare, and new ways of working to understand how degrowth might be reconciled with gender equality and social justice to guide public policy and practice agendas in Canada. We reviewed the existing literature and evidence published within the past 14 years (2010–2024) on degrowth, gender equality, and social equity, including within the context of the COVID-19 pandemic (2020–2022), and sought to identify potential impacts that degrowth in industrialised countries may have on women, gender relations, gender equality, and social equity. Since degrowth as a concept was developed by and for the Global North, this research focused primarily on industrialised countries. We assembled and analysed scientific research and practitioner literature on these topics from countries that are members of the Organisation for Economic Cooperation and Development (OECD). Additionally, we selectively identified promising programmes, policies, and practices from other industrialised as well as emerging and developing economies for reconciling degrowth with gender equality and social justice. We did so because Canada may benefit from learning about relevant initiatives in international settings.

Methodology

The peer-reviewed scholarship on degrowth, gender equality, and social justice is currently quite limited but there is a significant amount of "working knowledge" available from practitioner sources. We drew upon and amalgamated both scientific and practitioner literature to better enable policymakers and other end users of this research to fully appreciate the nature, magnitude, nuance, and complexity of the issues involved. To carry out this project, we used three online databases and search engines, namely, EBSCOhost, JSTOR, and Google Scholar, to locate peer-reviewed scientific literature. We also collected and analysed relevant professional reports, policy reviews, position papers, and survey results from governmental and non-governmental organisations in Canada that work on environmental protection, economic security, and social justice (Smart Prosperity Institute, Ecotrust Canada, the Canadian Centre for Policy Alternatives, the David Suzuki Foundation, Climate Justice Saskatoon, and Solidarity Halifax, for example). Collected data were analysed using the *Codebook for Standards of Evidence for Empirical Research* (SoE) (Heck and Minner, 2009). The SoE and their process of application result in a careful review of the claims of individual studies and reports based on six categories: adequate documentation, internal validity, analytic precision, generalisability/external validity determination, overall fit, and warrants for claims.

Findings and Discussion

COVID-19 was declared a global pandemic just as we received news that the grant application that enabled us to carry out this research had been successful. The global economy shrank dramatically soon after, in effect offering opportunities to study degrowth and its implications for gender equality in real time. Of course, it is important to clarify that what we had planned to study was the relationship between planned deliberate degrowth, gender equality, and social justice, not unplanned economic contraction triggered by a pandemic. Proponents of degrowth envision the economies of industrialised countries shrinking by design, not disaster or disease. Nevertheless, it was impossible not to notice that interventions endorsed by degrowth proponents – basic income support, reduced working hours, flexible work, and remote work, for example – were rolled out in industrialised, emerging, and developing economies around the world within weeks of the coronavirus being declared a pandemic. We discuss these briefly in the next few pages.

Canada Emergency Response Benefit (CERB) and Universal Basic Income (UBI)

In Canada, the Canada Emergency Response Benefit (CERB), although introduced as a temporary income support programme, has reinvigorated an existing debate about a guaranteed basic level of income (see, for example, Hennessy, 2020; Ibbitson & Parkinson, 2020). In May 2020, fifty senators sent a letter to Justin Trudeau calling for the CERB to evolve into a minimum basic income programme. News stories and op-eds spanning the ideological spectrum have appeared in the media over the past three years. Although CERB has not been unanimously endorsed, even Conservative think-tanks like the Frasier Institute (Clemens et al., 2020) and newspapers like the *Financial Post* have emphasised the need for a genuine public discussion about replacing existing provincial and federal income support and related programmes with one programme that reduces administrative costs, better co-ordinates assistance and increases resources available to Canadians.

The deployment of CERB, albeit as a temporary income support programme, has also led to a broader conversation about the possibilities and benefits of providing social protection within a human-rights framework and delinking social security from employment status. There have been significant advancements globally in expanding and strengthening social protection policies in recent years, as more countries transition toward developing welfare systems. Some strategies that are being tried in European, African, Asian, and Latin American countries include basic income schemes (see, for example, Hiilamo, 2020, for findings from Finland's 2017–2018 basic income experiment) as well as conditional and unconditional cash transfer programmes that enable poor women to make priority decisions for themselves and their dependents. Programmes like Brazil's Bolsa Familia, Mexico's Prospera, Mali's Social Cash Transfer initiative, and India's basic income pilot are hopeful developments given that structural inequality constrains individual ability to exercise rights and demand entitlements (Campello and Neri, 2014; Davala et al., 2015; Mary Robinson Foundation, 2016).

Within the context of COVID-19, Togo's Novissi digital scheme for social protection has been reviewed very positively for reaching the poorest and most vulnerable members of society, among whom women and informal sector workers are the majority (World Bank, 2021). Togo has a population of 8.1 million (2019) and a gross domestic product (GDP) per capita of $679. The Novissi platform delivers contactless, emergency cash transfers based on machine-learning techniques and mobile money. In order to expand Novissi's platform coverage, the poorest villages and neighbourhoods are selected through high-resolution satellite imagery and nationally representative household consumption data. Within those villages and neighbourhoods the poorest individuals are then prioritised through machine-learning algorithms using mobile phone metadata and phone surveys. Machine-learning algorithms trained on mobile call detail records predicted consumption patterns for 5.7 million individuals (70% of the population). Between November 2020 and March 2021, 57,000 new beneficiaries were prioritised for contactless social protection payments using predictive algorithms (ibid.).

Globally, 92% of women and 87% of men work in the informal sector (ILO, 2020). Almost 1.6 billion informal workers have been affected by the lockdowns and other COVID-induced containment measures and/or are working in the hardest-hit sectors (ibid.). In the past year, only 30 out of a total of 190 countries (for which we were able to find information) introduced some form of income support aimed at informal sector workers. Algeria, Pakistan, and Togo are notable examples of developing countries that have attempted to extend social protection programmes to informal sector workers during the COVID-19 pandemic (ibid.). There is much evidence to suggest that large numbers of Canadians – gig and contract workers, migrant workers, informal sector workers, and sex workers, for example – did not benefit from income replacement measures such as CERB. Canadian policymakers could benefit from looking at and learning from programmes like Novissi that succeeded in reaching large numbers of informally and precariously employed workers.

Alongside coverage about the CERB and the possibilities of a universal basic income programme, several reports have been published in Canada about how unevenly the burden of essential work and caregiving has been shouldered during this pandemic, notably by women and communities of colour (see, for example, Public Health Ontario, 2020, Desjardins et al., 2020). Identifying the types of policy responses that might alleviate these inequities may also help us understand how degrowth (in its planned iteration) might be reconciled with gender equality and social justice to guide public policy and practice in Canada.

The following is a discussion of policy conversations and possibilities other than CERB and UBI that have emerged in Canada in the past four years due to COVID-19 that present the possibility of reconciling planned degrowth with gender equality and social justice.

What Is "Essential" Work and Who Carries It Out?

The coronavirus pandemic has created an unprecedented reckoning of what constitutes essential work and essential service in Canada. This has led to calls for comprehensive reforms of sectors such as long-term care, home care, and childcare as well as the need for stronger societal valorisation (via higher wages, more secure employment, paid sick leave, affordable childcare, housing and student debt assistance, and other stronger social protections) of those who work in such sectors. The lessons learned and evidence generated from the COVID-19 pandemic have the potential to inform responsive social and economic policies that could benefit large groups of essential workers who are presently underpaid and often precariously employed (such as personal support workers, health aids, sanitation workers, grocery clerks, meat-packers) and among whom women, immigrants, racialised minorities, and documented and undocumented migrants are overrepresented.

Gender has played an important role in determining how the costs and burdens of essential work have been borne. Globally, men are more likely to die of COVID while women are more likely to lose employment (Ravanera, 2020). Almost twice as many women as men have lost their jobs worldwide because of COVID. From March 2020 to February 2021, women accounted for 53.7% of year-over-year employment losses in Canada (Statistics Canada, 2021). Women are more likely to work in industries hardest hit by COVID-19 – hospitality, retail, tourism, health, and social services, for example. These industries are also less well-suited for physical distancing and remote work. This also explains why sectors like construction, manufacturing, and transportation, which tend to employ larger numbers of men and include more jobs that are easily adapted for physical distancing, recovered much earlier than sectors in which women are more likely to be employed. Women in Canada are also twice as likely as men to work part time (26% versus 13%) (Patterson, 2018). These part-time rates have been relatively stable over the last 20 years (ibid.). When employers must resort to cost-cutting measures, part-time workers are often the first to be let go. The high proportion of women working in small-sized firms in service industries combined with the greater employment losses among small firms in the service industries exacerbated the gender difference in loss of employment (Statistics Canada, 2021). The confluence of such factors has resulted in higher levels of employment loss for women, and to the pandemic being described as a "she-cession" (Trichur, 2020).

Another reason why women have fared worse than men during the pandemic is that men tend to be employed in much higher numbers than women in professional, scientific and technical fields (science, technology, engineering, mathematics, architecture, actuarial sciences, logistics, project management, for example) which tend to be easier to carry out remotely or with physical distancing. Ability to work remotely correlates with other entrenched social hierarchies based on race and ethnicity; for example, only 16% of Hispanic and 20% of Black Americans were found to be able to work from home, compared to 30% of Whites and 37% of Asian Americans (Ravanera, 2020). The "gender of COVID" intersects predictably with "the colour of COVID", to result in women and people of colour making up most of both essential workers and the unemployed. The emerging stay-at-home economy in Canada reveals a two-tiered society: "non-essential" workers who can work from home, and "essential" workers—not only health care workers and other first responders but also childcare workers, long-term care workers, gender-based violence service providers, cashiers, cleaners, grocery clerks, delivery workers, bus drivers, mail carriers, and warehouse workers, among whom women and racialised minorities are the majority.

Women are more likely than men to be healthcare workers, on the frontline of the coronavirus pandemic. According to the World Health Organization, across 104 countries, women comprise 70% of health and social care sector workers (Boniol et al., 2019).[1] At 81%, women are even more disproportionately represented among health care and social assistance workers in Canada (Statistics Canada, 2019). This makes women more vulnerable to the disease. Given the underrepresentation of women in leadership positions, women also do not often have their voices heard in decision and policymaking (Wenham et al., 2020). Women in general, and particularly women of colour, are also concentrated in jobs in the services and hospitality industries. These jobs involve high contact with people, and often do not offer paid sick leave (Scott, 2020). This increases not only women's risk of contracting COVID-19, but also puts their families and communities at higher risk. For trans and other gender diverse populations, existing experiences of employment, health, and social service discrimination and stigma have been compounded by the pandemic (ibid.). Even a simple activity such as hand-washing, endorsed widely as an important sanitation strategy during the pandemic, poses a bigger challenge for a transgender than a cisgender person since public restrooms and change rooms in Canada remain overwhelmingly marked as single gender spaces in which non-binary persons regularly experience harassment and intimidation.

New Ways of Working and New Forms of Employment

The unprecedented global expansion of remote and flexible work brought on by COVID-19 in 2020 may also lead to longer-term changes in ways of working in many sectors of employment, with important consequences for gender equality. Countries with more equitable gender norms tend to have a better-established tradition of flex-time policies. For example, Finland offers the most flexible working schedules on the planet, with 92% of companies allowing workers to adapt their hours, compared to 76% in the UK and the US, 50% in Russia and just 18% in Japan. Finland's Working Hours Act, which ironically came into effect in 2020, gives most full-time employees the right to decide when and where they work for at least half of their working hours (Savage, 2019). However, it is important to recognise that even if such flexible work policies were in place in more countries, we would be left with the more significant challenge of changing the perception of caregiving as a gendered responsibility carried out solely or mostly by women. Until more transformative social change takes place in gender relations, flexible working schedules may just reinforce existing gender imbalances in employment and care (Partridge, 2021). Early evidence generated during the COVID-19 pandemic suggests that the wider availability of flexible and remote work will not be adequate to disrupt the entrenched gender division of labour within the household (Lewis, 2020; Medina and Lerer, 2020). This was evidenced most poignantly by the fact that spring 2020 witnessed the rolling back of three decades of women's participation in the Canadian workforce despite the widespread transition to remote and flexible work. And women in the 20–24 and 35–39 age groups constituted the two largest cohorts exiting the workforce. Women with children below the age of six constituted the single largest group exiting the workforce (Desjardins et al., 2020). The uncertainty of the length of school closures and online learning in Canada meant that large numbers of women remained unable to return to their old jobs or to start new positions even when the economy started to recover in the later months of 2020. Even temporarily leaving the workforce or being demoted to a position with less responsibility, as many women have during the past 18 months (Partridge, 2021), may produce longer-term negative career outcomes for those who were establishing themselves professionally or in mid-career positions before the pandemic. Potential consequences include the negative impact on career progression, the impact of work interruptions on earning potential, negative implications for pensions and retirement, a remaining gender-unbalanced applicant pool for promotions, significant challenges for workforce re-entry, and the loss of key talent and experienced women who decide not to return. Due to this disconnection from the workforce and/or loss of professional momentum, there may also be fewer potential women professionals who would eventually qualify for executive and board positions, which already tend to be held disproportionately by men in Canada (Baruah and Biskupski-Mujanovic, 2021). In other words, even temporary withdrawal from the workforce or professional setbacks can produce lifelong implications for individuals and for society at large.

Single parents face unprecedented challenges during the pandemic. Balancing caregiving and work is the norm for single parents, but during a pandemic they bear an even heavier load. For example, if single parents become ill and must isolate, the consequences for their children may be dire.

Single parents often depend on extended networks – including grandparents –but are cut off from these circles of support during the pandemic (Powell, 2020). The pandemic brings to the fore the importance of public support for childcare (Ravanera, 2020). It has become increasingly evident during this pandemic that caregiving status and support availability is more important than just gender in determining who has lost most ground in employment in the past 15 months (ibid.). Single fathers and other male caregivers – especially those with lower levels of income, education, and precarious employment – have also borne significant social and economic burdens during COVID-19.

For flexible work and remote work to become a policy prescription that advances gender equality, some structural changes to employment norms and gender norms are necessary. Strengthening long-term investment in the care sector (including care for seniors, children, and adults with special needs) is essential to enable workers to benefit optimally from the availability of remote and flexible work options. It is also crucial to make flexible work hours available at all levels of employment, including in executive and management positions. In present practice, employers that offer workers the option of working remotely often end up creating a two-tier system in terms of pay and career advancement, with those choosing remote work (still predominantly, although not exclusively, women) experiencing both lower wages and slower career advancement rates. For remote work to be effective as a means for reducing career advancement gaps, a re-definition of who an "ideal employee" is, based on performance in measurable terms, over "presenteeism", i.e. number of hours spent in the workplace, is necessary. This will benefit all employees since balancing work and family responsibilities is emerging as a concern for all workers irrespective of gender identity (Gunderson, 2016; Baruah and Biskupski-Mujanovic, 2021).

The Care Sector and the Care Economy

In Canada, COVID-19 has provided the strongest impetus in decades for a national childcare strategy. In September 2020, Justin Trudeau's federal government made a big promise in the throne speech[2]: the creation of a national childcare strategy, a response to the fact that women have been harder-hit than men by job loss due to COVID-19 (Saba, 2020). Advocates have been calling for an accessible childcare system dating back to 1970, when the Royal Commission on the Status of Women recommended a national childcare policy. More than 20 years later, a federal task force recommended the same. In 2004, Paul Martin's Liberals promised $1 billion a year to build a national childcare system. However, Canada has yet to see the creation of a national childcare programme (ibid.). But this time, advocates believe it could happen, buoyed by the government promise of not just one-time funding but sustained funding for the future. In the throne speech, the government promised to "build on previous investments, learn from the model that already exists in Quebec, and work with all provinces and territories to ensure that high-quality care is accessible to all".

Importantly, it is not only childcare advocates who have been asking – the call for affordable childcare has been echoed by other groups such as the Chamber of Commerce and numerous banks, after the pandemic made it clear that not fixing this problem would have widespread economic implications. A pan-Canadian system that provides childcare for all would require creative and sustained collaboration with the provinces and territories to create services in different communities tailored to the needs of children in those communities. At the time of completing work on this chapter in April 2024, the Canada-wide affordable early learning and childcare programme had reached all ten provinces and three territories. Although access remains uneven due to long waitlists in some provinces and territories, there is cautious optimism that a childcare system that is affordable and accessible for all families across Canada is possible and imminent.

In addition to childcare, COVID has provided unprecedented impetus for strengthening long-term investment in other important areas of the care sector, including care for seniors and adults with special needs, groups whose care also tends to fall disproportionately on women. For example, calls for a national plan for safe seniors' care, with long-term, dedicated funding and effective enforcement mechanisms, have been made by several associations of healthcare professionals in Canada even before the pandemic (see, for example, Armstrong et al., 2015). Since March 2020, there have been more than 16,000 deaths in senior care homes across Canada, which relies heavily on long-term care to house its ageing citizens (Fieber, 2021). With the COVID pandemic highlighting systemic neglect toward Canada's elders, a national eldercare strategy also seems imminent. In Canada, there are 400,000 people living in institutional care, one of the highest rates in the world, but inadequate attention has been paid to home care and supportive housing as strategies that would enable elders to continue living and ageing at home (Picard, 2021). Within Canada, Quebec appears to provide the best domestic example to emulate for homecare, hiring 10,000 personal support workers since March 2020 and doubling their wages (ibid.). Canada can also look to Denmark for an example of a society that keeps people living at home, through a strong supportive housing system (ibid.).

Proponents of degrowth have emphasised that while energy and resource-intensive sectors of the global economy need to shrink in the future, some critical democratic infrastructures, such as infrastructures of care, must expand and flourish. The coronavirus pandemic has demonstrated not just the need for a structural reform of the care sector but also the fact that employment in the care economy (childhood education, primary education, social work, health and library services, for example) is already green and sustainable in ways that employment that relies on extractive energy-intensive sectors is not. The care industry, for example, is 30% less polluting (in terms of greenhouse gas emissions) than the construction industry and the education industry is 62% less polluting than the construction industry (Women's Budget Group, 2020). Investing in the care sector could create 2.7 times as many jobs as the same investment in construction (ibid.).

In the United Kingdom, the Feminist Green New Deal emphasises that caring work should be a cornerstone of a clean and green post-COVID economy:

> Most people think of green jobs as installing solar panels, working on wind farms, or planting trees. These jobs will help us take better care of the planet, but caring for people is also inherently low-carbon work. The average job in health and care produces 26 times less greenhouse gas than a manufacturing job, and nearly 1,500 times less than a job in oil and gas. We need to rebalance our economy away from energy-intensive sectors focused on producing more things for us to consume, and towards labour-intensive sectors that will help secure wellbeing for all. Care jobs could and should be made even greener, but investing in many more of them would both address the care crisis and offer meaningful, low-carbon work for people moving out of industries that need to shrink, such as aviation and retail. Of course, we also need more jobs in renewable energy, energy-efficiency technology, and low-carbon infrastructure, and these must be made accessible to all. But we also need a wider definition of green jobs to include low-carbon health, care, and education. Creating many more of these jobs and improving wages and working conditions would address gender inequality in these sectors and beyond and lead to better outcomes for everyone by making sure we are better cared for and educated. It would provide opportunities for workers leaving precarious work elsewhere and industries that need to shrink or wind down altogether because of their climate impact.
>
> *(Women's Budget Group, 2022)*

Policies aimed at expanding such sectors and improving wages and working conditions could emerge as a positive legacy of the coronavirus pandemic. In addition to demanding better working conditions and social protections for care economy workers, there are broader calls for a care income across the planet for all those, of every gender, who care for people, the urban and rural environment, and the natural world. The Green New Deal for Europe, for example, looks at what work is needed for social and environmental wellbeing, and what work is not, and proposes a care income as essential for climate justice (Global Women's Strike, 2020).

In Canada, the Trudeau government ratified the Paris Agreement in 2016 and launched a Feminist International Assistance Policy (FIAP) in 2017.

While Canada has made significant efforts to mainstream gender into its climate commitments, these commitments focus almost solely on women in the Global South. They do not consider the need for domestic action on gender and climate (Fawad et al., 2023). Although gender gaps are systematically greater in the Global South than in the North, gender inequalities are also still widespread in affluent countries (Cohen, 2017). But discussion of gender inequalities in the Global North is rarely examined in the context of climate change, or with reference to the connections between domestic climate and energy policies and their global effects. Canada would do well to learn from the types of feminist climate policy being advocated in countries like the UK and Sweden that grapple with the domestic and global challenges presented by gender inequality and the climate crisis.

Conclusion

Although COVID-19 has led to a downscaling of production, transport, and emissions, as recommended by degrowth proponents, it is important to emphasise that it does not represent degrowth. A degrowth transformation must be planned and democratic. Unlike the temporary and short-term changes triggered by an economic recession, degrowth also requires a permanent long-term commitment to the downscaling of production and consumption as well as the reorganisation of society in a more just way (Rilović et al. 2020). COVID-19 has so far disproportionately affected the poorest and the most vulnerable in society whereas a degrowth transformation would be intentional and proactively pursued and have redistributive justice and equality at its core (ibid.).

Before the pandemic, degrowth proponents had to work hard to convince people of the value of a planned contraction of industrialised economies alongside expansion of redistributive activities and policies. Amid tangible evidence that an economic system based on over-exploitation, over-production, and over-consumption is not sustainable, support for degrowth may continue to grow even as we put COVID-19 behind us. If one virus can cause such upheaval throughout our entire social and economic system, then it stands to reason that we should consider different ways of organising our societies. Our current political-economic system has shown itself to be incapable of responding to crises in a just and humane way. A democratically planned yet adaptive, sustainable, and equitable downscaling of the economy may lead to a future where we can live better with less. COVID-19 may have serendipitously provided the most convincing rationale for degrowth in industrialised societies. That said, it is important to acknowledge that at the time of writing this chapter, Canada appears to be mostly resuming its pre-pandemic growth trajectory. Air travel has returned to pre-COVID levels; the CERB has long been discontinued; and Bill C-273, aimed at establishing a national strategy for a guaranteed basic income was terminated when the 44th General Election was called in August 2021 (although the strengths and weaknesses of UBI as a social protection strategy continue to be discussed and debated in the public domain); and employers are increasingly insisting on a return to in-office work (although post-COVID many employers have also incorporated remote and flexible work possibilities into their longer-term recruitment and retention strategies). Whether the policy lessons that COVID-19 inadvertently provided, about how an intentional degrowth strategy might be implemented in Canada, are carried into the future remains to be seen.

As evidenced by its current climate plan, The Pan-Canadian Framework for Clean Growth and Climate Change, the federal government in Canada remains convinced that economic growth and technological fixes are adequate solutions for the climate crisis, although it is becoming increasingly clear that since over-production and over-consumption in industrialised countries are the biggest impediments to environmental sustainability, technological changes such as transitioning to clean energy sources will not be enough to prevent climate breakdown and address other environmental problems.

Degrowth is a particularly important strategy for Canada to consider since the average Canadian produces 22 tons of greenhouse gases (GHG) per year. This makes Canada the highest per capita GHG emitter among the G20 countries, which as a group is responsible for more than 80% of the world's annual GHG emissions. Policy prescriptions such as shorter work weeks, and remote and flexible work can play an important role in reducing the carbon emissions of Canadians. Recent research in industrialised countries has established that working less is good for the environment: that if we spent 10% less time working, our carbon footprint would be reduced by 14.6%, largely due to less commuting or grabbing high-carbon convenience foods on our breaks. A full day off a week would therefore reduce our carbon footprint by almost 30% (Smedley, 2019). And yet degrowth has had very little mainstream policy uptake in Canada. As a policy framework, degrowth has had very little purchase beyond activist and academic circles, largely because politicians do not win elections on platforms of scaling back consumption and shrinking the economy. Even the two most socially and environmentally progressive parties in Canada – the New Democratic Party and the Green Party – envision a future based on green growth and limited economic redistribution delivered via income support and expansion of health and social protection infrastructure, not degrowth.

The existing scientific evidence suggests that in order to remain within the 2°C threshold of warming agreed upon in the Paris Climate Accord, Canada must leave 75% of all coal reserves, 74% of oil reserves and 24% of gas reserves undeveloped and underground. This would require a massive transformation of the Canadian economy and society. Yet, Canada's current commitments put forward as part of the Paris Climate Accord puts the country on track to 5.1°C of warming by 2100. These alarming projections suggest that the concept of degrowth must mobilise politics, policies and alternative visions of reconciling economic security, environmental protection, and social justice that have broad-based public support in Canada, i.e. well beyond the academic, NGO, and activist circles in which the degrowth conversation has presently gained traction.

Notes

1 Globally, women working in the health and social care sectors also earn 11% less than their male peers.
2 The throne speech opens every new session of Parliament in Canada. The speech introduces the government's direction and goals and outlines how it will work to achieve them.

References

Agarwal, B. (1992). The gender and environment debate: Lessons from India. *Feminist Studies* 18(1): 119–158.

Anguelovski, I. (2015). Environmental Justice. In *Degrowth: Vocabulary for a new era*, edited by Giacomo D'Alisa, Federico Demaria, and Giorgos Kallis. New York: Routledge. pp. 33–36.

Armstrong, P. et al. (2015). *Before It's Too Late: A national plan for safe seniors' care*. Ottawa: The Canadian Federation of Nurses Unions.

Asara, V. et al. (2015). Socially sustainable degrowth as a social-ecological transformation: Repoliticizing sustainability. *Sustainability Science* 10(3): 375–384.

Baruah, B. & S. Biskupski-Mujanovic. (2021). Navigating sticky floors and glass ceilings: Barriers and opportunities for women's employment in natural resources industries in Canada. *Natural Resources Forum*. doi:10.1111/1477-8947.12216

Boniol, M., et al. (2019). *Gender equity in the health workforce: Analysis of 104 countries*. World Health Organization. Retrieved from https://apps.who.int/iris/bitstream/handle/10665/311314/WHO-HIS-HWF-Gender-WP1-2019.1-eng.pdf?sequence=1&isAllowed=y

Campello, T. & M. Neri (2014). *Bolsa Família Program: A decade of social inclusion in Brazil*. Brasilia: Institute for Applied Economic Research.

Clemens, J. et al. (2020). Poor CERB targeting wastes billions. *The Frasier Institute*, August 27. Retrieved from: https://www.fraserinstitute.org/article/poor-cerb-targeting-wastes-billions

Cohen, M.G. (Ed.). (2017). *Gender and climate change in rich countries: Work, public policy and action*. London, UK: Routledge.

D'Alisa, G.et al. (Eds.). (2015). *Degrowth: A vocabulary for a new era*. New York: Routledge.

Davala, S.et al. (2015). *Basic Income: A transformative policy for India*. New Delhi: Bloomsbury.

Demaria, F.et al. (2013). What is degrowth? From an activist slogan to a social movement. *Environmental Values* 22(2): 191–215.

Dengler, C. & B. Strunk. (2018). The monetized economy versus care and the environment: Degrowth perspectives on reconciling an antagonism. *Feminist Economics* 24(3): 160–183.

Desjardins, D. et al. (2020). Pandemic threatens decades of women's labour force gains. *RBC Economics*, July 16. Retrieved from https://thoughtleadership.rbc.com/pandemic-threatens-decades-of-womens-labour-force-gains/

Eicker, J. & K. Keil. (2017). Who cares? A convergence of feminist economics and degrowth. Gender and the Economy - Perspectives of Feminist Economics. Retrieved from https://www.exploring-economics.org/de/entdecken/who-cares/

Fawad, A. et al. (2023). A feminist climate policy? Examining Canada's climate commitments. *Environmental Politics* 32:5, 815–837, DOI: 10.1080/09644016.2022.2144011

Fieber, P. (2021). 'Neglected No More' exposes deplorable state of senior care in Canada. *CBC News*, March 11. Retrieved from https://www.cbc.ca/news/canada/calgary/wordfest-book-neglected-no-more-exposes-deplorable-state-of-senior-care-in-canada-1.5946420

Global Women's Strike. (2020). Open Letter to Governments – A Care Income Now! Retrieved from https://globalwomenstrike.net/careincomenow/

Gunderson, M., (2016). *A Baseline on Youth Employment in the NR Sector: A data analysis*. Ottawa: Natural Resources Canada.

Heck, D. & D. Minner. (2009). *Codebook for Standards of Evidence for Empirical Research*. Chapel Hill, NC: Horizon Research.

Hennessy, T. (2020). A just recovery: The pandemic is a call for personal and collective change. *Canadian Centre for Policy Alternatives*, September 1. Retrieved from: https://www.policyalternatives.ca/publications/monitor/just-recovery

Hiilamo, H. (2020). The basic income experiment in Finland yields surprising results. *Nordic Welfare News*, May 7. Retrieved from https://www2.helsinki.fi/en/news/nordic-welfare-news/the-basic-income-experiment-in-finland-yields-surprising-results

Ibbitson, J. & D. Parkinson. (2020). CERB and other coronavirus benefits won't last forever. Or will they? What a universal basic income could look like. *The Globe and Mail*, May 15. Retrieved from: https://www.theglobeandmail.com/opinion/article-cerb-and-other-coronavirus-benefits-wont-last-forever-or-will-they

ILO (2020). Extending social protection to informal workers in the COVID-19 crisis: Country responses and policy considerations. Retrieved from https://www.ilo.org/secsoc/information-resources/publications-and-tools/Brochures/WCMS_754731/lang--en/index.htm

Kallis, G. (2011). In defence of degrowth. *Ecological Economics* 70(5): 873–880.

Kallis, G. (2015). The degrowth alternative. *Great Transition Initiative* 1–6. Retrieved from https://greattransition.org/images/Kallis-Degrowth-Alternative.pdf

Lewis, H. (2020). The Coronavirus Is a Disaster for Feminism. *The Atlantic*, March 19. Retrieved from https://www.theatlantic.com/international/archive/2020/03/feminism-womens-rights-coronavirus-covid19/608302/

Lipsig-Mummé, C. & S. McBride (Eds.). (2015). *Work in a Warming World*. Montreal: McGill-Queen's University Press.

Mary Robinson Foundation (2016). *The Role of Social Protection in Ending Energy Poverty Making Zero Carbon, Zero Poverty the Climate Justice Way a Reality*. Dublin: MRF.

Medina, J. & L. Lerer. (2020). When mom's Zoom meeting is the one that has to wait. *The New York Times*, April 23. Retrieved from: https://www.nytimes.com/2020/04/22/us/politics/women-coronavirus-2020.html

Paech, N. (2010). "Vom grünen Wachstumsmythos zur Postwachstumsökonomie." Retrieved from https://www.entrepreneurship.de/summit/files/Paech-2010-Sammelband-Wiegandt-Welzer.pdf

Partridge, J. (2021). Switch to more home working after Covid 'will make gender inequality worse'. *The Guardian*, June 19. Retrieved from: https://www.theguardian.com/business/2021/jun/19/switch-to-more-home-working-after-covid-will-make-gender-inequality-worse

Patterson, M. (2018). Labour Statistics at a Glance: Who works part time and why? Statistics Canada. Retrieved from https://www150.statcan.gc.ca/n1/pub/71-222-x/71-222-x2018002-eng.htm

Perkins, P.E. (2010). Equitable, ecological degrowth: Feminist contributions. In *Second International Conference on Economic Degrowth for Ecological Sustainability and Social Equity*, Barcelona, Spain, March (pp. 26–29).

Picard, A. (2021). *Neglected No More: The urgent need to improve the lives of Canada's elders in the wake of a pandemic.* Toronto: Penguin Random House Canada.

Powell, C. (2020). The Color and Gender of COVID: Essential workers, Not Disposable People. Think Global Health, Council on Foreign Relations. Retrieved from https://www.thinkglobalhealth.org/article/color-and-gender-covid-essential-workers-not-disposable-people

Public Health Ontario. (2020). COVID-19 in Ontario – A Focus on Diversity. Retrieved from https://www.publichealthontario.ca/-/media/documents/ncov/epi/2020/06/covid-19-epi-diversity.pdf?la=en

Ravanera, C. (2020). *Primer on the gendered impacts of COVID-19.* University of Toronto, Rotman School of Management. Retrieved from https://www.gendereconomy.org/primer-on-the-gendered-impacts-of-covid-19/

Rilović, A. et al. (2020). A degrowth perspective on the coronavirus crisis. Retrieved from https://www.degrowth.info/en/2020/03/a-degrowth-perspective-on-the-coronavirus-crisis/

Saba, R. (2020). On the cusp of a national childcare strategy. *Toronto Star*, November 14. Retrieved from https://www.thestar.com/business/2020/11/14/on-the-cusp-of-a-national-daycare-strategy.html?rf

Savage, M. (2019). Why Finland leads the world in flexible work. *BBC Worklife*, August 8. Retrieved from https://www.bbc.com/worklife/article/20190807-why-finland-leads-the-world-in-flexible-work

Scott, K. (2020). COVID-19 crisis response must address gender fault lines. Canadian Centre for Policy Alternatives. Retrieved from http://behindthenumbers.ca/2020/03/20/covid-19-crisis-response-must-address-gender-faultlines/

Smedley, T. (2019). How shorter workweeks could save Earth. *BBC Worklife*. August 7. Retrieved from https://www.bbc.com/worklife/article/20190802-how-shorter-workweeks-could-save-earth

Statistics Canada (2019). Employment by class of worker, annual (x 1,000). Retrieved from https://www150.statcan.gc.ca/t1/tbl1/en/tv.action?pid=1410002701&pickMembers%5B0%5D=1.1&pickMembers%5B1%5D=3.15&pickMembers%5B2%5D=4.3

Statistics Canada (2021). COVID-19 impacts on productivity growth and gender differences in employment. Retrieved from https://www150.statcan.gc.ca/n1/daily-quotidien/210526/dq210526a-eng.htm

Trichur, R. (2020). 'It's a 'she-cession'. Governments must put women first during the recovery', *Globe & Mail*, May 1. Retrieved from https://www.theglobeandmail.com/business/commentary/article-legislators-must-prioritize-women-combat-workplace-gender/

Waring, M. (1988). *If women counted: A new feminist economics*. San Francisco: Harper

Wenham, C. et al. (2020). COVID-19: the gendered impacts of the outbreak. *The Lancet* 395(10227).

Women's Budget Group. (2020). What would a Feminist Green New Deal look like? Retrieved from https://wbg.org.uk/wp-content/uploads/2020/05/A-Feminist-Green-New-Deal.pdf

Women's Budget Group. (2022). A Green and Caring Economy: Final Report. Retrieved from https://wbg.org.uk/wp-content/uploads/2022/11/A-Green-and-Caring-Economy-Report-FINAL.pdf

World Bank. (2021). Prioritizing the poorest and most vulnerable in West Africa: Togo's Novissi platform for social protection uses machine learning, geospatial analytics, and mobile phone metadata for the pandemic response. Retrieved from https://www.worldbank.org/en/results/2021/04/13/prioritizing-the-poorest-and-most-vulnerable-in-west-africa-togo-s-novissi-platform-for-social-protection-uses-machine-l

Interview 4

GUÐMUNDUR INGI GUÐBRANDSSON, EX-LEADER OF THE LEFT-GREEN MOVEMENT, AND MINISTER OF SOCIAL AND LABOUR AFFAIRS IN ICELAND

Interviewed by Gunnhildur Lily Magnusdottir, September 2024

Context

Guðmundur Ingi Guðbrandsson was, when interviewed in autumn 2024, the party leader of the political party, the Left-Green movement, and Minister of Social and Labour affairs in Iceland. He served as minister of environment and natural resources from 2017–2021 when Iceland, led by Prime Minister Katrín Jakobsdóttir, launched the wellbeing economy governments partnership (WEGo) with First Minister of Scotland, Nicola Sturgeon and the Prime Minister of New Zealand, Jacinda Ardern. WEGo emphasised care for humans and nature and that wellbeing is based on several SDGs where the GDP of a country is only one of many indicators. Guðmundur Ingi Guðbrandsson is an environmental scientist and has previously worked as the CEO for the Icelandic Environment Association. In 2024, the Icelandic government issued gendered state bonds, and became the first country in the world to do so.

Could you tell us the importance of gender to your work on climate change and other related policies such as social and labour policies?

Gender mainstreaming has been important in all policy-making in Iceland including climate change, for example, in our foreign policy; and gender equality is an important part of Iceland´s international image. This also means that when new legislative proposals are developed, policy-makers are required to make a gender equality evaluation, thus to assess the possible gendered effects of the proposal.

DOI: 10.4324/9781003461005-14

Another example is the so-called gendered state bonds, which the Icelandic government issued in 2024. The gendered bonds are a part of an existing legal framework about sustainable, green state investments. The government had earlier issued green bonds within this framework, requiring investments to take green transition and biodiversity into consideration. The gendered bonds follow a similar logic where the condition for investments is that they should promote gender equality.

When it comes to social and labour politics, I would say that parental leave is the best gender policy we have. Parental leave is 12 months in Iceland, and in 2020 an important legislative change was proposed to parental leave, making an equal division between both parents mandatory. This change, which has now been implemented, is an extremely important change since women tend to take a bigger part of the parental leave, which affects their participation in the labour market, career development, personal economy, and their retirement.

What qualities do you think are needed for developing successful policies to address the climate emergency?

A broad participation of stakeholders is very important, meaning politicians, policy-makers, and the public. Related to the broad participation I would also say that a cross-sectoral approach and coordination between different policies is necessary for successful climate policy-making, where all sectors are involved in emission reductions and other climate work. This cross-sectoral approach was, e.g. central in the governmental climate action plan in 2021. However, broad participation and involvement of stakeholders still require leadership and courage from the government as it is the government that sets the goals and then different stakeholders can be approached. It is also important to keep in mind that different sectors can contribute differently to reach common international reduction goals, e.g. in the Paris agreement. In Iceland the fishery industry can, for example, reduce more and quicker with the use of biofuels than, for example, agriculture.

The care discourse is also very important for successful climate policy-making. The care discourse needs to guide us, thus focusing on diminishing human suffering and increasing quality of life, respect for nature and intergenerational justice and what has been called the wellbeing economy with the focus on future generations will be responsible for all the burdens.

What networks and relationships have you found most useful in developing policies to address the climate emergency?

Primarily NGOs and social movements since they are often the first ones to frame new mindsets and have bold ideas and proposals, although they are not responsible for the implementation. My background as a CEO in the biggest environmental organisation in Iceland has been very important and it has been valuable to be able to tap into that knowledge and experience found within other organisations in Iceland. Also being a part of (and party leader) of a green political party is obviously also a social movement and connected to environmental grassroot movements. Putting carbon neutrality on the party agenda of the Left-Green movement, for example, came from the environmental grassroots, especially the Icelandic Environment Organisation. So environmental grassroot movements are very important for policy-making but it is also important to have a dialogue with different companies.

I also want to mention the important effects of Greta Thunberg and the Fridays for Future movement, which we so clearly saw in the climate discourse and increase in news on climate change. In 2018–2020 my ministry mapped the coverage on climate change in Icelandic media and we saw a huge increase in news on climate change and that affected many companies in Iceland, which saw the relevance of climate change and changed their strategies.

It is also important to recognise that climate change needs to be addressed by different academic disciplines, e.g. ethics, economy and philosophy, just to name some examples, which can all demonstrate the importance of care in climate change and how to diminish unnecessary suffering of people and nature. Different academic disciplines can help us understand that decisions about climate change need not to be solely based on economic prioritisations but what we perceive as ethically right, which in the end will benefit future generations.

Sources of inspiration/who do you draw inspiration from, and why?

I have first and foremost been inspired by several scholars from different disciplines, including philosophy, ethics, and biology. One of my main inspirations is the environmental ethics scholar Aldo Leopold. I have a favourite famous quote from his work *Land Ethic* (1949); "A thing is right when it tends to preserve the integrity, stability and beauty of the biotic community. It is wrong when it tends otherwise". I try to remind myself of the principle in this quote and use it as a guideline in my work. As a party leader I have emphasised that in work on nature preservation we need to perceive nature as sacred with its own merits, not merely as useable for people. This means that we should, just like we always have a gender lens in our policy-making, also have a nature lens in all policy-making.

Another important inspiration is the marine biologist and conservationist Rachel Carson, who wrote the book *Silent Spring* (1962). I am also very inspired by nature, especially the highlands of Iceland where I sometimes say that; "the calmness reigns and the peace of mind resides" [*Þar sem kyrrðin ríkir og hugarróin á lögheimili*].

Which of the policies/strategies that you have been instrumental in developing regarding climate would you recommend to a policy-maker new to the area? And why?

What is complex regarding climate policy is how cross-sectoral it is and is not limited to one ministry, thus the environmental ministry. Traditionally the finance and prime ministries are more powerful and have more political weight than the environmental ministry, although that has gradually changed and I focused on increasing media coverage about environmental matters in national media, which increased awareness and the weight of environmental issues, including climate change.

It is also important to have a party leader leading the environmental ministry, to signal that it is an important policy and requires coordination between different ministries. This is sometimes challenging, thus the coordination and collaboration between different ministries and actors; but a good example of a successful coordination was a legislative proposal by the environmental ministry about so-called F-gases in industries, which required collaboration with the finance ministry (ministered by a different governmental party). This proposal resulted both in taxation and in a regulation further limiting the import of F-gases and is now an important cross-sectoral policy. Another example of an important strategy was the sale tax reductions and aids for buying electric cars which was very important as Iceland became at that point in time second only to Norway in ownership of electric cars (per capita) in the world.

What do you find to be the main obstacles for developing a socially and economically just climate policy in Iceland?

I perceive it as a challenge and have criticised the current discourse that climate change is primarily about producing more green energy. This is a problematic simplification, although the green energy transition is very important, but we need to have a cross-sectoral approach and include sectors that might not be directly affected by the green energy transition. Also, the risk with this discourse is that it does not fully recognise the diverse effects on different groups in society and it can even free certain groups or stakeholders from taking responsibility or showing interest in fighting climate change, if governing authorities have a narrow focus on green transition. A broad focus is also important to enable low-income earners and disadvantaged groups to participate in a just transition and really in the environmental revolution which is needed. An example of this from my ministry, thus the social and labour ministry, is that recently financial support for purchasing cars for people with disabilities was increased with the intention to cover the additional cost that comes with buying a clean energy car and thereby enabling people with disabilities to further participate in the green and just transition.

What projects/policies are you considering/would like to introduce in the future? recommendations?

"The polluter pays principle" needs to be taken more seriously and we need to use emission taxation more widely in other sectors than today – not only in, e.g. the energy sector with traditional taxation on fossil fuels. We therefore need to use environmental taxes as a main policy-making tool to a greater extent than today.

I would also like to see climate mainstreaming as a primary tool in decision-making at different levels, similar to gender mainstreaming.

Further intertwining of different environmental policies is also needed, e.g. between biodiversity and climate change, different types of pollution, land deforestation to diminish parallel policy-making.

8

CLIMATE CHANGE POLICIES AND GENDER EQUITY

What are the views of women who work in construction?

Coralie Guedes, Vivian Price and Linda Clarke

Introduction

In June 2022, a ten-day international delegation of over forty construction women from North America and Europe, all trade union members, took place in London. They advocated for women to be represented in constructing the housing replacing the former Holloway Prison, the largest women's prison in Europe, shared best practices for recruiting and retaining women in construction occupations, toured building sites and vocational education and training (VET) facilities, and participated in co-facilitated workshops and community events.

Women are severely under-represented in the European and North American construction industries (Clarke et al., 2015; ESCO, 2020; Schiano-Phan et al., 2023; Price, 2002). The barriers to their entry are attributable to the nature of VET provision, the sector's structure and labour market, employment and working conditions, and social and cultural aspects (Clarke and Sahin-Dikmen, 2021). The industry's fragmented nature, including self-employment and layers of sub-contracting, does not provide a work-based training infrastructure, impedes change, and discourages women's participation (e.g. Wright and Conley, 2018; Clarke et al., 2022). Socially and culturally, traditional stereotypes and racist and sexist attitudes in a male-dominated environment, combined with lack of knowledge about the sector, lack of formal recruitment practices, appropriate selection criteria, and mentoring and role models (Clancy and Feenstra, 2019) contribute to their exclusion (Fielden et al., 2000). Working conditions remain a central challenge, given the lack of work-life balance possibilities (Worrall et al., 2010), long working hours (Watts, 2009; Styhre, 2011), tolerance of harassment and discrimination,

DOI: 10.4324/9781003461005-15

health and safety concerns (Curtis et al., 2018; Cruz Rios et al., 2017), unequal treatment, and denial of training (Bilginsoy et al., 2022). Whilst colleges may welcome women and ethnic minority applicants onto construction courses, trainees need an employer to gain work-based experience and employment, and white, male employers may be reluctant to recruit those 'unlike' themselves (Byrne and der Meer, 2005). As a result, higher proportions of women are found in colleges training for construction than in workplaces in UK, and USA apprenticeship programmes face retention challenges.

Unions act as both gatekeepers and gateways for those under-represented in the industry (Clarke et al., 2005; Goldberg and Griffey, 2019). In the US there are notable attempts to diversify the construction union workforce, including building community partnerships and organising pre-apprenticeship/VET programmes directed at increasing candidates from under-represented groups (Isingizwe et al., 2023). Mentoring, parental leave for both partners, and inclusive language also improve retention. Construction programmes have piloted an approach to empowering Indigenous groups by reinforcing the community values of appreciating the land and working collectively (Cameron and Rexe, 2022).

Lack of equity exists in the context of severe challenges facing the industry, including labour shortages (McGrath, 2021; Suryadi, 2018) and climate change. In Europe and the US, the built environment accounts for 40% of energy consumption and 36% and 29%, respectively, of greenhouse gas emissions (EC, 2020; Shoemaker, 2023; Leung, 2018), whilst construction employment is set to increase significantly through zero carbon building, renewable energy measures, and retrofit programmes.

Approaches to environmental policy and action

Hampton (2015) identifies three dominant union approaches to combat climate change:

1 *The neoliberal model*, dependent on market forces and characterised by technological solutions, with minimal acknowledgement of climate change, the role of labour and the exploitation of nature and an emphasis rather on 'green growth'.
2 *Ecological modernisation*, likewise assuming technology is the answer to reducing emissions and tied to 'green growth', whilst conceding the need for government intervention, regarding equity as good for a diverse workforce, and calling for a gradual phaseout of fossil fuels.
3 *Social ecological transition*, requiring fundamental social change, altering the nature of production and employment relations and shifting away from the growth economy and reliance only on technological solutions to reduce emissions.

From this third perspective, climate change offers an opportunity to transform society away from a profit-driven economy and maintenance of patriarchal white capitalist supremacy. State intervention is critical, unions and communities have an active role, and nature is no longer a passive commodity, separate from labour, but interconnected with labour, understood as an active force (Räthzel, 2021).

Construction unions generally follow the ecological modernisation paradigm, embracing green growth and sometimes opposing fossil fuel phaseout that jeopardises current employment. As governments invest in sustainability measures, unions seek to prepare members with relevant VET, primarily focusing on technical skills. In North America, for instance, joint union and contractor training programmes incorporate modules for solar, geothermal, or wind technology, with some also building model climate and sustainable energy training centres, though US government policy and investments in renewables are recent and unstable. Leading the way, the Canadian Building Trades Union introduced the government funded *Building It Green* programme, seeking to embed climate literacy into the building trades, albeit with a technological focus (CIRT, 2024). The European Union (EU) too has a longer history of investing in sustainable development and embracing climate goals with some effect on union building practices (Clarke and Sahin-Dikmen, 2020). VET systems in countries such as Belgium, Denmark and Germany also foster labour agency, ensuring workers can act autonomously rather than relying on supervision (Clarke et al., 2024). However, with VET programmes, though energy literacy is increasingly incorporated into curricula and workshop facilities as demonstrated by, for instance, insulation and heat pump installation, rarely is climate literacy included to give trainees an understanding of their role and their occupation's impact in addressing climate change.

The significance of networks of women construction workers

Our research focuses on a group of women construction workers, linked in an international network designed to overcome the exclusion women often face from this majority white, male-dominated industry. Such a network reflects not only survival mechanisms, but also critiques of the industry, providing information, support, inspiration, and collaboration to its members. In presenting their perspectives, we illustrate a collective openness among women employed in construction occupations to understand how climate change impacts their industry and their lives.

Much research has been published on social exclusion (e.g. Richardson and Le Grand, 2002), including the importance of agency in overcoming this (e.g. Burchardt et al., 2002). Recent research has, however, argued that dichotomies posed, such as between insider and outsider approaches, fail to take account of connections between stakeholders. For instance, Wagner et al. (2023) emphasise the importance of networks of collaboration in advocating for a joint climate strategy, exerting what might be termed 'collective agency'. In this respect, women construction workers can be seen to transcend outsider status by overcoming their local isolation and connecting with and supporting each other on many levels. Developing an international network of women working in construction has also enhanced their agency, raising their profile in addressing climate-related issues in construction.

Our approach raises the question of the relation between individual and collective agency, between the individual women interviewed and their involvement in a collective, a network of collaboration seeking to improve the position of women in construction. Insight into the tensions between the individual and collective levels is given by Pelenc et al.'s (2015) search to define and link concepts of collective agency, capability, and action in the context of sustainability implementation at local level. They found cooperation between individual actors enabled them to develop collective capabilities, which resulted in creating a new social organisation bridging different levels.

The process of interviewing and recording the experiences of construction women can enhance their awareness and considerations of what can be done to address climate change in their working environment. As Wamsler et al. (2022) show, individual and collective values, beliefs, and paradigms are expressed through narratives and everyday social practices that shape perceived possibilities and alternatives and inform individual and collective agency and action. Thus, our research, in facilitating these narratives and developing awareness of how everyday social practices have implications for climate change, can have material consequences. Given such a possibility, Wamsler et al. (2022) recommend giving more explicit consideration to and investment in the human dimensions of climate change, mostly absent in dominant external, technological, and information-based approaches. This involves better communication to nurture human potential and the creation of communities of practice and platforms, strengthening links between individual, collective, and public climate efforts. Additionally, measures are needed to support, first, learning environments nourishing people's capacities and increasing their sense of interconnection and, second, nature-based solutions and planning addressing both climate change mitigation and adaptation.

Our research can, therefore, play an important role in strengthening the collective agency and capacity of women construction workers interviewed by offering an opportunity to discuss our interpretation of the experiences and views they shared (MacLean, 2006; LaTour, 2008). Tradeswomen Building Bridges organised the London delegation with unions supporting women's efforts to fight isolation and exclusion. Such networks exist in many countries, including Canada, US, various European countries, India,

Philippines, Australia, Nigeria, South Africa, Nicaragua. Analysis of the interviews on women's experiences in green construction can further considerations of future alternatives that women's networks may follow and the possibility of adapting and transforming their collective actions and collaborations. In this way, women construction workers' networks of collaboration can be a force within unions and in other circles to nurture more consistent approaches to sustainable climate technology and encourage social ecological expression and action.

Methodology

The interviews conducted were designed to stimulate tradeswomen to reflect on climate change and their experience with green construction. They are situated in a semi- participatory framework as they took place within an event jointly designed by tradeswomen and researchers, and tradeswomen considered and approved the interview component. Life-history interviews are a window into the material conditions and intellectual and political evolution of people's actions and consciousness. Leading scholars in labour environmentalism have employed this approach to study union leaders' perspectives on environmental policy (Räthzel and Uzzell, 2011; Räthzel, 2021; Lundström, 2018). Conducting life-history interviews offers an opportunity to apprehend intersectional relationships (Crenshaw, 1991) between, for example, national origin, ancestry, racialised and gendered identities, occupation, education, and involvement with social movements, contributing to the way people think, feel, and act. An intersectional approach in turn facilitates understanding how tradeswomen's outlooks manifest common concern about climate change yet differ in their critique of the way the industry integrates green construction practices.

Biographical research methods are shaped by an exchange between interviewer/s and interviewees (Portelli, 1998; Räthzel, 2021). The authors are all social science researchers invested in scholarship and activism around equity in construction and green transitions, one having been a union electrician in US and involved with organisations led by women construction workers. Interviews with tradeswomen contain what Bornat (2008) refers to as interactivity, subjectivity, and structuring. The interactivity of face-to-face interviews facilitates subjective storytelling, expression of self, and sharing feelings that provide insight into how people make sense of their experiences. Structuring the interview format through prior theorising lends the interview a cohesive flow and interview questions were designed to overcome any tendency simply to echo the dominant position of employers and unions.

A convenience sample based on a one-time event enabled contact with a small and difficult to reach group (Baxter et al., 2015), with a collective identity based on shared purpose. The study was explained, and volunteers requested, securing positive responses from most delegates, representing a range of demographic and other characteristics such as trade, region, race, nationality, sexual orientation, language, migration status, class background, age, position in the industry and in their unions, and years in construction. To retain anonymity interviewees are identified in quotes mainly according to occupation, except when other characteristics, like location, are salient.

Data collection and analysis

Thirteen participants were interviewed in person during the delegation, with consent to be audio-recorded, and three were subsequently interviewed over Zoom. A set of open-ended questions was used to guide conversations, each lasting from one to one and a half hours with transcripts then loaded into an AI transcriber and subsequently polished. Of those prepared to be interviewed, these sixteen were selected based on their familiarity with green construction, the range of occupations covered (carpentry, pile driver welding, sheet metal working, iron working, plumbing, painting, bricklaying, machine operating, electrical work, pipe fitting), and geographical spread represented, including from Canada, Norway, Poland, Sweden, and the US. All North Americans had been through joint union-management apprenticeships, while the Europeans were qualified through recognised VET programmes. Several identified with a non-Western perspective through connections in the Global South or Indigenous roots, or were ideologically critical of market economies. The construction industry everywhere consists of many occupations and a variety of work activities, so we anticipated the diverse backgrounds of the women interviewed would present a rich range of individual experiences regarding green construction. Given the women's network, and hence collective agency, the overall result represents more than the sum of the individual experiences but a collective voice, expressing and reflecting on measures taken and what is needed in future.

The responses were examined according to four themes that emerged through analysis of the interviews: education and training; workplace; drivers of change, including unions, worker autonomy, and women's empowerment; beyond the workplace, including relations to nature, concerns about public regulation, and the importance of networks.

Findings

Education and training for green construction

The majority interviewed stressed both the lack and importance of education around climate change and the need for union members to feel comfortable having such conversations with employers. Most had experiences of training related to green construction in different settings, although this was not necessarily labelled as such and varied in terms of approach, practice, and content.

Practices and approaches to training

Experience of green construction training related to using tools and materials or processes. For instance, for painters, certified training can involve the use of special instruments with filters to clean paint brushes, so protecting the water, or low volatile paint to avoid emissions and damage to workers' health. For plumbers, training can involve heat pump installation, district heating, geothermal energy, or grey water systems. For sheet metal workers, the focus can be on heating, ventilation, air conditioning, and air circulation, whilst for carpenters, keeping the building envelope intact, window and door installation, insulated concrete forms, and insulation more generally were cited. Finally, for concreters, training can be about working with low emission concrete and, for electricians, solar panel installation and vehicle charging stations. Training can be of different length and provided in different settings, ranging from technical school to monitored work experience.

A critical area identified as empowering workers is climate literacy, acquiring the awareness and ability to understand how work undertaken impacts on climate change. Participants often point to lack of awareness and climate change not being taken seriously in their workplace because of the slow pace of change and people not suffering directly, though geographical location can have an impact. Climate change awareness is also seen by both North American and European interviewees as more common among younger workers, who discuss it in the union even though it is not part of workplace training.

Several women emphasised what is missing, above all a coordinated systems approach that includes cross-occupational training so that people understand one another's jobs and can be more mindful. It was, for instance, suggested that plumbers might undertake mechanical engineering training and become more involved in drawing up plans or supervising, sheet metal workers might obtain auditing certification, or sustainable building science could be introduced.

A carpenter from an Indigenous community stressed the need to change people's mindsets, learning from Indigenous traditional knowledge to understand, respect, and appreciate the land as part of VET, for instance about medicinal plants and not littering:

> At our toolbox meeting, my boss was talking about, 'you have to throw your cigarette butts in the proper receptacle, not on the ground'. And it sounds petty, but we do have these agreements with Indigenous communities. And if

you're doing something so simple and disrespectful like that, then that just means we're not standing by who we say we are.' And I was like, wow, he first of all acknowledged the littering. And then he acknowledged the Indigenous people. And he acknowledged our values.

(Carpenter)

This carpenter also spoke of a special project integrating green skills and respect for the land:

During the day we would teach them the skills and then in the afternoon we would go sit in the longhouse and sing and drum or listen to an elder tell stories around the fire. We had one elder take us out into the forest and teach us like this is a medicine you can harvest ... And that just builds your connection to and appreciation when you're building on this land.

Anti-racist training is needed so people respect each other and that is integrated into the job:

On a government funded job, we had ROI [return on investment] training ... everyone has to go on a Saturday, you get paid eight hours, time and a half. And it teaches you all about the residential school and the history of my people. And some of it is diversity and inclusion, but more Indigenous focus. And it's just about being sensitive to us and our culture because that's the problem with a lot of racist people, they just don't know.

Finally, training requires a long-term view as green construction is not a short-term, cheap solution; it involves workers and for everyone in the process to change their mindset. A woman working in Norway and originally from South Asia related problems she and her crew faced on a job building a wall using green cement:

Training needs to change the mindset so that people care about the end product ... I'm going to work and I'm going to make a difference, I'm going to build a wall, I'm going to make someone's apartment ... Too much is a story of who's to blame and not enough thinking about how we can work together ... green construction starts when you start to change people's mindsets.

The union as actor in training

Some women express frustration when climate-related training is reserved for managers only or is generally lacking, especially for site workers.

Employability is often cited as a motivation for unions to provide green construction training:

> ...there's a lot of money in these green jobs right now, and the union definitely wants a piece.
>
> *(North American electrician)*

Training for the installation of vehicle charging stations is one example. Union-initiated training can, however, remain an unsuccessful gamble if employers do not take advantage:

> We grabbed insulated concrete forms and made sure that was part of what we do, because we already built concrete forms. Now, that didn't end up being this job creator because contractors aren't using them, the technology came out, but the construction industry is so hesitant to change that it never got a foothold in the market.
>
> *(Carpenter)*

Accounts of unions' focus on employability-motivated green construction training also illustrate a lack of agency on the worksite to bring about change. Unions are dependent on owners, developers, and contractors to build to green standards, which are often pegged to government regulations and incentives.

Progressive politics and engagement with the union can mean increased exposure and awareness of climate change, as with a Californian electrician who works to establish links between her union youth group and the Labour Network for Sustainability (LNS) by organising workshops and gathering people for the LNS conference, using social media to increase her audience. She also described how her union's training centre, a joint venture with employers, is one of the first net zero buildings in the area.

To conclude, in terms of training, the union role is often portrayed as positive, spreading awareness among members, promoting a collective consciousness, and adopting exemplary behaviour as an organisation in a virtuous circle, attracting more climate-aware new members who can hold the unions themselves accountable.

Women – green construction and the attractiveness of the industry

Most training-related comments made by interviewees refer to recruitment and opportunities that the green transition creates for women to consider a career in the construction industry. The Californian electrician highlights that people as passionate about climate justice as herself tend to be women and therefore building green could appeal to women as a career path. This sentiment was echoed by a pipe fitter who considered a shift to green construction could make recruitment easier, though to be successful it needs to be well supported and integrated into training

Workplaces

The themes and challenges to greening the workplace raised relate to waste management, energy efficiency, the use of equipment and materials, and working conditions. Recycling and waste management are mentioned by all interviewees, occasionally with frustration at the amount of avoidable waste on site, including when company policies exist but remain unimplemented. Even in countries like Norway, where managing and sorting waste is mandatory, there are many different practices. Waste management is a complex process, entailing health and safety issues but usually undervalued and underpaid. In many workplaces, especially those labelled green such as LEED (Leadership in Energy and Environmental Design) buildings, there is also an emphasis on reducing waste by reusing materials:

> I got to help demolish and rebuild the dock and all the materials we demo'ed got thrown away. All the wood that we use for the framing and forming for the concrete … ended up just getting burned, because it was gonna cost them to put it in the dump or recycle. So, they just made a big pile and burned it.
>
> *(Pile driver.)*

With regard to bricklaying, a migrant bricklayer from Poland considered:

> Demolition is particularly sensitive, with layer-by-layer processes deemed much more effective, but dependent on different bricklaying methods, such as the use of lime mortar rather than cement.

She has even started to develop an app that could be used by workers to identify different types of waste and how to dispose of them.

Energy sources, such as wind and solar, and improving energy efficiency sprung quickly to interviewees' minds, though they had not necessarily worked on such projects. Electricians have more direct experiences, such as installation of energy-efficient lighting with daylight sensors, though this is never presented as something against climate change. Such green initiatives are not without challenges. For instance, in Los Angeles, much work on solar fields is non-union, very remote, repetitive, and physical work that employers can hire anyone with a strong back to do instead of employing expensive experienced electricians.

Several women spoke of diesel-powered equipment as causing emissions, despite electrical and propane-powered alternatives. On her trip to London during the delegation, a Canadian ironworker expressed surprise walking around work sites:

> Why is it so quiet in here? And then looking at all the cranes in the air, because we work with cranes a lot, and I'm like, man it's weird it's like Twilight Zone, 'cause we would have a lot of bip bip bip brrrrr and belching black. You folks have done electric cranes, which is fabulous.

Using environmentally friendly construction materials and technology is discussed positively, including pressed wood beams on LEED building, engineered wood acting as a carbon sink, recyclable copper pipes, lime mortar facilitating recycling of bricks, composite reinforcing bars not made solely of steel or lighter aluminium bars that can be recycled or reclaimed to reduce mining, and heat pumps. However, rather than working towards passive buildings or better operating equipment, the purchase of new equipment is encouraged, such as new heating, ventilation, and air conditioning systems:

> … people feel better, they're willing to spend more money on this big piece of equipment that they can see, rather than spending the same money on me doing work for them, that then lowers their electric bill for the next five or 10 years.
>
> *(Sheet metal worker.)*

Work organisation based on cost-driven scheduling can lead to different occupations like installation, painting, and flooring overlapping and making the job impossible. Sub-contracting and outsourcing also have adverse impacts. An example was cited of bathrooms arriving with broken parts that cannot be remodelled and for which transport was not considered part of the emissions total. Remodelling should be part of construction planning, and, when it is possible to save good past work, that needs considering seriously.

> …the new buildings, they are tearing down what it was. And I don't think that is beautiful …it's cheap … to not be aware of the history there, not putting a value on what was done well.

Drivers of change

Whilst recognising what needs improving and problems involved in the construction workplace to transition to a green economy, interviewees also identify how change might be achieved, particularly the need for union involvement, greater worker autonomy, and a more inclusive environment.

Unions as actors

A Swedish painter explained that the Building Workers International (BWI), the global federation of unions in the building, building materials, wood, forestry, and allied industries, has made climate change a major issue.

> I'm really happy the BWI has raised it because I haven't really even thought about the transition into the green economy before. So, just by the fact it's been raised will mean that we have to act on it, and we need to learn more … that's one of the big things I will take with me now from the European women's committee.

Unions are, however, not necessarily perceived by interviewees as playing an environmentally positive role; even when union leaders talk or create resolutions about climate change, this may not reach workers. An interviewee from Alaska claimed that raising the issue would not be taken seriously as union members would not see what is in it for them. Another, from Canada, suggested that union neglect of the climate crisis might be due to the specificities of certain occupations, with workers dealing with energy, like electricians, more likely to undertake climate-related training.

The priority of unions stressed by several interviewees is to organise workers and recruit members, though this does not prevent discussions on, for instance, recycling or using certain materials. One interviewee refers to disagreements within the union over jobs being always a priority, whatever they are, even working at a liquified natural gas terminal. Another gave the example of a pipeline closure, resulting in complaints to the union over job losses, whereas she did not think it was a bad thing.

> All the guys who work on those pipelines doing the welding, were up in arms, they were mad at our international president, because he wasn't doing nothing about putting pressure on it. And yet, a lot of other people like me feel we don't really need that … it's a small percentage of our membership, but there was a lot of uproar about that.
>
> *(Plumber.)*

Workers' autonomy – empowerment

A Californian electrician working in a refinery explains that her 'trade can really be a solution' through, for example, installing charging stations, solar power, occupancy sensors, photocells, and retrofitting for increased energy efficiency. She argues that talking about employment opportunities is a way to sell the green transition to older people who do not understand climate justice and just want jobs. For her, though, job satisfaction is important:

If I can do this sort of kind of intricate conduit bending or motor controls, or the things that I'm doing at the refinery somewhere else, I would love to. But this has been the only place where I've really found a lot of this work that I really enjoy.

Another example by an US carpenter of empowerment in the workplace concerns millwrights, also included in carpentry, turning to robotics, new energy, and bike rack installations.

Workers do not generally have a say on ordering materials or are not educated or paid to know what materials to use. They may not question project managers and have likely been trained to accept what has been ordered, based on initial cost. Nevertheless, there needs to be visibility when ordering materials, such as paint or caulking. As a Polish construction worker explained, it is important to create a trusting work environment so that over-ordering is not hidden, and materials can be reused. Decisions to build and how to build are at the design phase when workers' input is not often sought:

We do some design and builds, then we do get some words in on how to do it. Right now, I'm working on a healing centre for women. And they're designing the buildings, and I've been able to put my own thoughts in here and there ... They want to use a blow-in cellulose insulation [fire retardant], it's cheap, but the problem is, once it's wet, it's garbage. So, I've suggested looking at rock wool.

(Carpenter.)

A plumber observed that plumbers mostly use non-recyclable plastic, so change is in the hands of manufacturers and governments. Regarding grey water systems, decisions need taking at the design stage to include separate plumbing systems. There also needs to be more communication between engineers and construction operatives on using new materials, such as green concrete.

Not all worksites have proper facilities to ensure environmentally friendly practices, for example, the correct tap and filter system to clean paintbrushes. Breeam (Building Research Establishment Environmental Assessment Methodology, UK) and the green building rating system LEED certification may make construction more efficient and sustainable, but there are questions about how evaluations are calculated and what is green. Some interviewees note that putting a plaque on the building is about status rather than creating a sustainable facility.

Several interviewees pointed to the significance of hierarchy because responsibility for implementation of green initiatives is at management level, whereas workers are the ones on site who can check work is done properly:

Workers are the only part of this process who are always on the site, so only they can tell you that I did a good job or bad job. So, if you go to the basement and see a lot of piles of bricks sorted [and people working], so someone who sees believes they did it, but I know because I was there, I know that they only did half of it.

(Bricklayer.)

Companies may have green policies on paper but walking on site one can quickly tell if these are implemented. Nevertheless, entrusting workers comes with the caveat that these tasks need to be part of workers' job description to be carried out during paid working time.

Women – Combatting macho culture

Many women connect neglect of environmental issues with the macho culture and envision ways in which bringing in more women could change the industry for the better, as building green may mean people in charge starting to question how work can be done and by whom. The macho culture reinforces an approach whereby, as the Swedish painter explained:

If you don't care about your own health so much, the environment isn't that important to you. People who do care about their health and environment are weird and should be ignored.

She points out that, when more women enter any construction occupation, this usually results in fewer accidents, but also gives rise to smarter ways of working, as any difference in strength means women finding smarter ways to work. Both phenomena impact on cost, as fewer accidents mean lower costs and building smarter with a long-term view, foreseeing future changes in building, or maintaining the initial building, is also cost-effective.

Empowering women is thus important to bringing about social change. A Norwegian plumber shared that the women in her union pushed the leadership to organise training to strengthen women's empowerment against intimidation on the job:

When you are working with someone insecure, they will try to push you out … and the trainer gave us practical examples of techniques for how we can maintain our power in situations like this, specific things we need to work on within ourselves, like our mental capacity and emotional capacity … to make sure that we are balanced and respond instead of reacting when these things happen. And that was awesome and now we are bad bitches.

(Plumber.)

Though, she would usually refrain from using such stereotypes, a sheet metal worker acknowledges that socially women are expected to communicate more, identify social dynamics, and look out for their own safety, all useful in working collectively, so important to building green where cooperation between different occupations is essential. She also recalls her recent experience working with an apprentice who told her it was refreshing to work with her as she does not have a big ego, which is also helpful in working well collectively.

Beyond the workplace

The women interviewed do not just refer to the workplace and VET, but have wider perspectives concerning relations to nature, regulation, and the role of women and their networks in combatting climate change. Several establish connections between their work and wider societal and ecological concerns, such as linking grey water systems to the preservation of public health (plumbers), offshore construction work to marine life protection (pile driver welder), energy efficiency in housing to equality, carpentry to sustainable forestry, construction work to planning, and workplace change to systemic change. They also discuss the role of public actors, articulating their position on top-down regulation, including penalties for non-compliance, which impact their occupation, such as pre-demolition audits (bricklayers) and government incentives for charging stations (electricians). A couple of US interviewees also mentioned community benefit agreements.

When asked to reflect on the relation between barriers to the entry and retention of women and achieving low-carbon construction, some envisage similarities. For instance, advocating for green construction and increasing the number of women relies on having your voice heard, having a support system, having your credibility validated through others, and building your reputation in an adverse environment, the underlying message being that women already know how these are achieved. Addressing climate change is like addressing gender-related issues in construction: 'it's somebody stepping up and making a movement' (carpenter).

Conceptualising networks as a potential forum for climate-related discussion goes only a small step further, and several women make this connection. At a personal level, the pile driver welder explains how the London delegation constituted an opportunity for her to reflect on climate change and see her experiences at work in a different light. At a collective level, the carpentry site supervisor establishes a link between women's engagement in the delegation and their openness to advocate for green construction. Some women promote women's activism in this field, whilst others hint at the challenges such engagement could entail and perceive green issues and women's issues as separate and only adding to the weight of their existing engagement with gender issues. A plumber, in accounting for how she managed to reach a leadership position within her union, explicitly connects change in union priorities, including in relation to climate change, with the need for more women in leadership positions.

Pragmatically, the Californian electrician considers how building green appeals to women already doing this kind of work and could therefore be an organising tool. A Canadian carpenter described how attending tradeswomen's conferences has helped her find her voice and become more confident, suggesting that women are open to climate literacy and Indigenous contributions. She recounts multiple visits to the British Columbia Centre for Women in the Trades, where she met women from across the province and shared her knowledge of sustainable construction. These women were subsequently eager to share this knowledge with their own communities and offer support: 'I now have this huge team of dream women that want to back me in whatever I do'.

Reflections

This chapter shows that conversing with tradeswomen about their experience in green construction gives a lived and concrete understanding of what this means. Even though practice may be restricted in their daily work, the women have a clear idea of what is referred to and are unencumbered by policy stipulations and the language of corporate social responsibility, so often amounting to greenwashing. Their knowledge and experience can appear partial, but, as they come from a variety of occupations and regions, bringing their individual experiences together contributes to the creation of a collective narrative as, combined, their reflections and examples encompass a range of environmental solutions, from the technological to more radical systemic change. Indeed, those expressing stronger, even divergent views share core values, above all, around relations between labour and nature, and hold non-dominant, subaltern group identities, including migrant, Indigenous North American first nation, South Asian minority ethnic, and Caribbean, and/or an affinity to a collective ideological philosophy like social democracy or democratic socialism.

VET should empower workers with the knowledge and ambition to work as a broad team across occupations and with every partner in the construction process. This means fostering an inclusive culture and working for the common good. Some women look at construction from an anti-colonial perspective, changing wealthy nations' habits in the treatment of materials and waste, honouring the land, building according to what people need rather than what generates profit. For them, employing new technology is secondary to the major shifts required, the need for systemic change, greater worker autonomy, and respect. The thrust of many women's thinking was that a macho environment cannot support a sustainable society because it protects unearned privilege, inequality, and waste. Yet women construction workers also acknowledge the contradictions between sustainable work, which has positive value for communities, and work in refineries or pipelines, which pays the bills and contrasts with 'green jobs' that possibly offer worse conditions, less stability and pay.

The women's reflections are not just focussed on what needs to be done in the workplace; many are actively involved in their union and in training future generations, as well as in women's networks. They belong to and interact with companies and organisations established in multiple countries, mostly unions but also VET facilities. As such, the networks they create cut across both occupations and organisations, offering a potentially comprehensive view of green construction. Combined with an inherently critical stance toward the industry because of its exclusionary practices, these networks, through the knowledge they gather and produce, offer fertile ground for transformative change, above all, in enhancing individual and collective agency.

Thus, we return full circle to Hampton's plea for a radical transformation and Räthzel's for respect for nature, if tradeswomen are to participate fully in the transition necessary for a green economy. Through the voices of the women interviewed, we show how this change might begin to be achieved, above all, with the unions and networks of women and by challenging existing working and employment conditions and approaches to VET.

References

Baxter, K., Courage, C., & Caine, K. (2015). *Understanding Your Users: A practical guide to User Research Methods*. San Francisco: Morgan Kaufmann Publishers.

Bilginsoy, C., Bullock, D., Wells, A. T., Zullo, R., Brockman, J., & Ormiston, R. (2022). *Diversity, Equity, and Inclusion Initiatives in the Construction Trades*. Washington, DC: North America's Building Trades Union.

Bornat, J. (2008). Biographical Methods. In: Alasuutari, P.; Bickman, L. & Brannen, J. eds. *The Sage Handbook of Social Research Methods*. London, UK: Sage, 344–356.

Burchardt, T., Le Grand, J., & Piachaud, D. (2002). Degrees of Exclusion: Developing a Dynamic, Multidimensional Measure, in *Understanding Social Exclusion*, 30–43. Oxford: OUP.

Byrne, Clarke and Van Der Meer (2005). Gender and ethnic minority exclusion from skilled occupations in construction: a Western European comparison, *Construction Management and Economics*, 23(10), 1025–1034.

Cameron, M., & Rexe, D. (2022). Community-Based Access to Apprenticeship: An Indigenous Work-Integrated Learning Model. *International Journal of Work-Integrated Learning*, 23(2), 203–218.

CIRT (2024). *Climate Literacy in the Building Trades*, Climate Industry and Research Team (CIRT) of Building it Green project, Skill Plan, Canadian Building Trade Unions, Final report.

Clancy, J. and Feenstra, M. (2019). *Women, Gender Equality and the Energy Transition in the EU.* Committee on Women's Rights and Gender Equality of the EU Parliament, European Union.

Clancy, J. and Feenstra, M. (2019). *Women, Gender Equality and the Energy Transition in the EU.* Committee on Women's Rights and Gender Equality of the EU Parliament, European Union.

Clarke L., Frydendal Pederson, E., Michielsens, E., Susman B., Wall C. (2005). 'The European Social Partners for Construction: force for exclusion or inclusion?' in *European Journal of Industrial Relations*, 11(2), 151–178

Clarke, L., Michielsens E., Snijders, S., and Wall, C. (2015) *No More Softly, Softly: Review of Women in the Construction Workforce.* ProBE: University of Westminster

Clarke, L. and Sahin-Dikmen, M. (2020). Unions and the green transition in construction in Europe: Contrasting visions. *European Journal of Industrial Relations*, 26(4), 401–418.

Clarke and Sahin-Dikmen (2021). Why radical transformation is necessary for gender equality and a zero carbon European construction sector. In: Magnusdottir, G.L. and Kronsell, A., (eds.). *Gender, Intersectionality and Climate Institutions in Industrialised States*. London, Routledge, 164–180.

Clarke L., Sahin-Dikmen M. and Winch C. (2022). Vocational education and training for a greener construction sector: low road or high road approaches to apprenticeships? In *Apprenticeships for green economies and societies*, 87–96, Thessaloniki: CEDEFOP

Clarke L., Sahin-Dikmen M., Winch C., Price V., Calvert J., Bilodeau P.L., Dionne E. (2024). Differing Approaches to Embedding Low Energy Construction and Climate Literacy Into Vocational Education and Training, in *Routledge Handbook on Labour in Construction and Human Settlements: the built environment at work*, eds, Werna E. and Ofori G., London: Routledge, 76–96.

Crenshaw K (1991). Mapping the margins: intersectionality, identity politics, and violence against women of color. *Stanford Law Review* 43(6): 1241–1299.

Cruz Rios, F., Chong, W. K., and Grau, D. (2017). The need for detailed gender-specific occupational safety analysis. *Journal of Safety Research*, 62, 53–62.

Curtis, H. M., et al. (2018). Gendered safety and health risks in the construction trades. *Annals of Work Exposures and Health*, 62 (4), 404–415.

European Commission (EC) (2020). *Energy – in focus: Energy efficiency in buildings.* Brussels: European Commission.

European Construction Sector Observatory (ESCO) (2020). *Improving the Human Capital Basis.* Updated March. Brussels: European Commission.

Fielden, S.L., Davidson, M.J., Gale, A.W., and Davey, C.L. (2000) 'Women in construction: the untapped resource', *Construction Management and Economics*, 18(1), 113–121.

Goldberg, D. A., and Griffey, T. (Eds.). (2019). *Black power at work: Community control, affirmative action, and the construction industry.* Cornell University Press.

Hampton P. (2015). *Workers and Trade Unions for Climate Solidarity: tackling climate change in a neo-liberal world*, Routledge

Isingizwe, J., Eiris, R., & Gheisari, M. (2023). Racial Disparities in the Construction Domain: A Systematic Literature Review of the US Educational and Workforce Domain. *Sustainability*, 15(7), 5646.

LaTour, J. (2008). United Tradeswomen: Organising for the Guaranteed Right to Work in Any Job. In *Sisters in the Brotherhoods: Working Women Organizing for Equality in New York City*. New York: Palgrave Macmillan, 15–38.

Leung, J. (2018). *Decarbonizing US buildings.* Center for Climate and Energy Solutions.

Lundström R (2018). Greening transport in Sweden: the role of the organic intellectual in changing climate change policy. *Globalizations* 15(4): 536–549.

MacLean, N. (2006). *Freedom is not enough: The opening of the American workplace*. Harvard University Press.

McGrath J. (2021). *Report on Labour Shortages and Surpluses*, European Labour Authority

Pelenc, J. Bazile, D., Ceruti, C. (2015). Collective capability and collective agency for sustainability: A case study, *Ecological Economics* 118, 226–239.

Portelli, A. (1998). Oral history as genre. In ed. Chamberlain N. and Thompson P., *Narrative and genre*, 23–45. Routledge.

Price, V. (2002). Race, Affirmative Action, and Women's Employment in US Highway Construction. *Feminist Economics*, 8(2), 87–113.

Räthzel, N. (2021). Trade union perceptions of the labour-nature relationship. *Environmental Sociology*, 7(4), 267–278.

Räthzel N. and Uzzell D. (2011). Trade unions and climate change: The jobs versus environment dilemma, *Global Environmental Change*, 21(4), 1215–1223.

Richardson, L. and Le Grand J. (2002). *Outsider and Insider Expertise: The Response of Residents of Deprived Neighbourhoods to an Academic Definition of Social Exclusion*, CASE paper 57, Centre for Analysis of Social Exclusion, April.

Schiano-Phan, R., Guedes C., Georgiadou M.C., Clarke L., Duran-Palma, F., (2023). *Women in Construction, Wood and Forestry: a resource toolkit for gender equality at work*, European Federation of Building and Woodworkers, supported by Friederich Ebert Stiftung.

Shoemaker, S. (2023). NREL Researchers Reveal how Buildings across the United States Do- and Could–Use Energy (USDOL), September 14.

Styhre, A. (2011). 'The overworked site manager: gendered ideologies in the construction industry', *Construction Management and Economics*, 29(9), 943–955.

Suryadi, J. (2018). Examining the Labor Shortage in the Construction Industry and Possible Solutions. Presented by *Industry Members, California Polytechnic State University*, San Luis Obispo, CA.

Wagner, P., Ocelík, P., Granow A., Ylä-Anttila, T., Metz F. (2023). Challenging the insider outsider approach to advocacy: how collaboration networks and belief similarities shape strategy choices, *Policy & Politics*, 51(1), 47–70.

Wamsler, C., Osberg, G., Panagiotou, A., Smith, B., Stanbridge, P, Osika W., Mundaca L. (2022). Meaning-making in a context of climate change: supporting agency and political engagement, *Climate Policy*, Taylor & Francis.

Watts, J. (2009). 'Allowed into a man's world' meaning of work-life balance': perspectives of women civil engineers as 'minority workers' in construction', *Gender Work and Organizations*, 16(1), 37–57.

Worrall, L., Harris, K., Steward, R., Thomas, A., McDermott, P. (2010). Barriers to women in the UK construction industry, *Engineering, Construction and Architectural Management*, 17/3, 268–281.

Wright T. and Conley H. (2018). Advancing gender equality in the construction sector through public procurement: Making effective use of responsive regulation, *Economic and Industrial Democracy*, 41(4), 975–996.

9

APPLYING INTERSECTIONALITY IN CLIMATE POLICY AND PLANNING

Experiences from Gothenburg and Malmö

Nanna Rask, Angelica Lundgren and Annica Kronsell

Introduction

Climate change is considered as one of the most pressing issues of our times. In an increasingly urbanising world, cities are recognised as important arenas for driving socio-ecological change (cf. Castán Broto et al. 2019). Cities are a locus of many sustainability problems, but are also key in creating solutions, something also recognised by the international policy arena (e.g. IPCC et al. 2018, 2022, UN-Habitat 2016) and by giving cities their own Sustainable Development Goal; SDG11. Cities have in many regards been more visible than state actors in pushing for change. This chapter engages with two such cities: Gothenburg and Malmö. Both are considered 'frontrunners' in the Swedish, Nordic, and European context due to their ambitious climate objectives and their recognition of the need to incorporate justice and equality into their climate work. To do so is important because various injustices result in ineffective climate transitions/transformations.

Climate change interacts with all levels of society and requires solutions that challenge the current unsustainable patterns of business-as-usual, including current unequal power structures and injustices (Kaijser & Kronsell 2014, Magnusdottir & Kronsell 2021, Singleton et al. 2021). Examples of injustices are growing socio-economic, gender and racial inequities and widening disparities among and within communities around the world (Cuthill 2010, Murphy 2012, Vallance et al. 2011, Washington 2015, Uteng et al. 2020). The climate crisis is linked to an extensive list of acute sustainability crises which are deeply rooted in current (unjust) structures and needs to be dealt with in an integrated manner. Transformative holistic changes seem

DOI: 10.4324/9781003461005-16

imperative for inclusive and equitable climate futures. Our chapter investigates what it would take for the cities to transform in the direction of this imperative.

During our empirical investigations of the two cities' climate policy processes, we found that both cities, to an extent, often treat climate issues together with other environmental issues. Yet, they do not sufficiently include holistic perspectives and issues of justice and equality into their climate work, and it is proving difficult for civil servants to do so. The climate policy process in the two cities is centred around a technocratic and de-politicised discourse, premised on ecomodernist logics, at the expense of taking issues of power, equality, and justice seriously. Such logics tend to favour technocentric approaches to addressing sustainability issues (Hinton 2015) and reiterates the quest for continued economic growth (Turnhout et al. 2014), effectively promoting an unfeasible 'win-win' that contemporary society can be both sustainable but relatively unchanged (Eversberg et al. 2023). Secondly, a dominant bureaucratic and institutional logic that allocates problems into organisational and policy 'silos' hinders attempts to promote holistic approaches.

Even though both cities recognise the problem of silo-approaches and have taken (some) measures to reform the organisational set-up we find that there is still further need to change the deeper underlying 'ways of thinking' about sustainability issues within the organisations. We therefore argue for the need to transform these logics and to incorporate approaches and ways of thinking that are more promising for dealing with the holistic nature of climate change and for restructuring power relations. To us, policy processes and their related understandings of, for example, "expert knowledge", "the economy", "nature", and "sustainability", etc. are intrinsically value driven. As Bradshaw (2019) argues, how the world is known and how social reality and, in turn, climate issues are constructed have implications for policy and its outcomes. To expose such values, we employ a critical lens based on feminist ideas of intersectionality, together with previous research on ecomodernism and institutionalism, to be able to provide an explanation to why the two cities' organisations are struggling to include more holistic approaches to sustainability.

Intersectionality can be helpful as an analytical lens to make logics visible and to aid in (re)politicising, problematising, and disrupting seemingly apolitical understandings in climate policies and their implementation process. Using this framework, we ask what is needed for these municipalities to be inclusive of issues of justice and equality when they attempt to implement climate policy? And, how can they benefit from intersectional knowledge? Based on our findings we suggest two 'principles for transformation' that we find crucial in this process, as well as provide suggestions on how further research can develop an intersectional 'policy tool' for aiding just transformations.

The chapter is structured as follows. First, we introduce our 'critical lens'. Thereafter, we describe our empirics and methods, before presenting the results. We finish off with a concluding discussion regarding what we (academics, policymakers, civil servants, and other practitioners) can take with us from this study.

A critical lens

Our critical lens is inspired by feminist theories on intersectionality, institutionalism, and environmental critiques of ecomodern governance. Intersectionality emerged from Black female scholars lacking a voice in feminist theories (Crenshaw 1989). From this initial focus, intersectionality has broadened to encompass more forms of power and oppression. Apart from oppression of humans, our intersectional lens also includes oppressions of 'nature', inspired by schools of feminist thought that break and problematise nature-human binaries (cf. Haraway 2003; Harcourt & Nelson 2015; Rask 2022). It enlarges the analysis of relevant power relations in intersectionality and offers new perspectives on governance whereby some forms of life are enabled to thrive while others are oppressed (cf. Gruen & Weil 2010: 127). This is something we deem necessary to better understand the effects of environment and climate policy on socio-ecological inequities, and indeed its embeddedness within such unequal structures.

Intersectionality is based on the notion that knowledge is contextual and situated. People hold varying social positions where privilege and oppression are intertwined (Lutz et al. 2011). As a critical lens on climate issues, intersectionality is valuable in multiple ways (Kaijser & Kronsell 2014). Intersectional insights allow us to challenge hegemonic conceptions and question how society communicates, evaluates, and legitimises knowledge (Kelly & Licona 2017). This lens aids in (re)politicising conventional knowledge and discovering what is left unspoken in policies and policy processes (Magnusdottir & Kronsell 2021).

Based on intersectionality and feminist critiques, our lens adds institutionalist theory as an additive for analysing governance. It underscores how both formal (organisational set-up, rules of procedure) and informal structures (such as traditions, habits, attitudes, etc.) guide institutional behaviour (Scott 2017). Institutions are shaped by history, which creates historical trajectories or path dependencies, where past decisions and existing institutions provide inertia and make it difficult to adopt new and different approaches, as they challenge existing power dynamics (Goldstein et al. 2023; Jennings & Hoffman 2017). Our lens sheds light on why local governments might struggle to implement intersectional and holistic climate policies that consider social aspects and underlying power relations and helps us understand the potential roadblocks within existing institutions. Despite obstacles, institutions do evolve spurred by factors like external shocks, social movements or internal dynamics (Acemoglu et al. 2021). We focus on the potential of internal institutional dynamics.

Thus, combining the institutional approach with our critical intersectional lens, we address ecomodern governance, the dominant strategy of many governments around the world to address all types of environmental problems (Grunwald 2018) including the local governments of Gothenburg and Malmö as previous studies have also highlighted (cf. Andersson & Gyberg 2023; Hagbert et al. 2020). Ecomodernism views nature as technologically malleable and the non-human world as inanimate without intrinsic value or agency. In particular, ecomodernism holds a far too optimistic view that economic development and environmental preservation can co-exist and assumes continued growth while relying on future innovations to solve climate and other environmental issues (Kronsell & Bäckstrand 2010). For governance it means that ecomodernism has become integral to the institutional processes and organisation in policy-making (Jänicke & Jacob 2004; Mol 2002; Næss & Saglie 2019; Dryzek 2022).

Ecomodernist policy-making is coupled with a structured and bureaucratic approach to managing societal issues. It is historically derived and influenced by Max Weber's thinking (from 1921), whereby government steering is conducted in a hierarchical order where complex problems are broken down into manageable units in distinct sectors. Over time and as societies become more organised it has developed into isolated silo-structures, with each sector characterised by unique cultures and ways of working (Dryzek 2022). The complex and cross-cutting nature of climate change needs cooperation across sectors. Hence, silo-structures are not conducive to problem solving of such cross-cutting issues (Wapner 2002, Torgerson & Paehlke 2005). Instead, the siloed nature of contemporary climate governance can reinforce ecomodernism and its focus on compartmentalised, technocratic solutions that isolate issues from their broader contexts and interconnections while overlooking social dimensions and underlying power structures that shape climate policies.

Our feminist critical lens underscores the need for policies that are socially informed and justice oriented. It offers productive critique of normative assumptions in the dominant discourse of the two cities' climate policy and action. It highlights how dominant institutional practices may reproduce policies that uphold 'business-as-usual' and power relations rather than support a transformation that is inclusive of equality, equity, and social justice concerns. By integrating intersectionality, we can challenge the narrow focus of ecomodernism and advocate for more integrated and transformative policy processes.

Empirics and methods – Making visible dominant logics and power structures

Nordic cities are often seen as progressive for having ambitious climate goals (Calmfors & Hassler 2019). Gothenburg and Malmö, the second and third largest cities in Sweden, have some of the most ambitious climate programmes in place, and as such offer a relevant case to assess common practices in local climate governance. For example, they go beyond the national (Swedish) climate plans and objectives, as well as incorporating both the production and consumption perspectives on climate emissions; considering both territorial emissions and consumption-based emissions produced elsewhere. Both cities have also been chosen by the EU Commission as two of the 100 European cities to become 'carbon neutral and smart by 2030', and to lead the way for other cities to follow suit (City of Gothenburg 2022, EU Commission 2022).

In 2021, both municipalities adopted new environment and climate programmes (City of Gothenburg 2021, City of Malmö 2021). The two programmes have much in common. These documents guide the municipalities in achieving their environmental and climate goals. Due to the task descriptions of the environmental administrations, which are responsible for implementing the 'ecological dimension', the programmes primarily focus on this dimension of sustainable development. However, the 'social' and 'economic' aspects of sustainability are also discussed and there is (some) recognition that these aspects cannot be separated. Both cities aim, in various ways, to incorporate issues of justice and equality into their environment and climate policy-making and implementation process. For example, in the regulatory documents from Malmö, it is stated that climate issues are understood as overarching societal issues that are interconnected with other sustainability dimensions. Malmö aims to deal with climate change without contributing to or escalating social inequalities and tensions (City of Malmö 2021). In the case of Gothenburg, the environmental administration is currently involved in two transdisciplinary projects about inclusive and just transitions (City of Gothenburg 2023a, Urban Futures 2022). Both projects analyse potentials to, for example, connect the implementation of the environment and climate programme with other programmes in the municipality such as The City of Gothenburg's Program for an Equal City. Since 2023 there is also a task for the municipality to create a central coordination unit to work with Agenda 2030, where two dimensions have been central: (i) the idea of 'leaving-no-one-behind' and bringing issues of equality, equity and human rights to the fore throughout the municipality's work, and (ii) the idea that the goals in the Agenda are inseparable (City of Gothenburg 2023b).

The two municipalities' environment and climate programmes also mark new ways of structuring how the municipal organisations deal with these issues. Previously, the environmental administrations of both cities largely worked independently on the environmental and climate programmes. An intent with the new programmes was to better incorporate environment and climate policy across sectors and to improve cooperation with other sectors in the municipality. In Gothenburg, one way to promote such cooperation has been to give the responsibility of the strategies within the programme to other administrations and municipal companies in the municipality, as well as to organise coordination and a supportive structure to push the implementation process forward. In Malmö, the new programme adopts a so-called 'umbrella' approach, meaning an overarching approach. Instead of the environmental administration having sole responsibility for the implementation of the programme, it oversees the implementation and encourages it to become a core principle in all administrations.

Despite awareness and attempts to overcome problematic organisational silo-structures, civil servants in both cities still face difficulties in implementing the programmes across sectors. Recent evaluations of Gothenburg's achievements in reaching the goals and subgoals in the programme (cf. City of Gothenburg 2023c) reveals that the city is currently far from reaching most of its goals, such as the goal of being a close to net zero city by 2030. Current measures are not sufficient. The two municipalities' environment and climate policy implementation falls short on contributing to the transformations needed.

By applying our critical lens to these two cases we aimed to gain further understanding of how, and why, the two cities are struggling to include more holistic, radical (paradigm-shifting) approaches to sustainability. Through a critical policy analysis (cf. Bacchi & Goodwin 2016) of how Gothenburg and Malmö have carried out their environment and climate programmes, we looked for the underlying logics, ideas, and assumptions that influence the environment- and climate-related solutions that are pursued or ignored. Inspired by Bacchi (2009), Matsuda (1991), and Kaijser and Kronsell (2014), we thus used our critical lens as a 'reading tool', by asking specific questions of the material (the two programmes, interview transcripts and field notes), developed in accordance with the spirit of critical policy analysis. Both studies were conducted within the same research project and used similar questions in the policy analysis and semi-structured interviews. Thus, we write in the plural form 'we' throughout the chapter as we refer to the project's common material. Material in the case of Gothenburg was more extensive and consisted of 21 interviews and a 'meeting ethnography' (Sandler & Thedvall 2017) consisting of participant observations in relevant meetings and workshops between 2020 and 2024, while the Malmö case consisted of nine interviews and one online workshop.

In the case of Gothenburg, the interviews were pursued in 2021, after the environment and climate programme had been adopted, with the so-called 'innovation team'; a team which was active during March–September 2021 and consisted of key actors' for the successful implementation of the pro-gramme, with the purpose to both identify barriers and find possible solu-tions. The respondents consisted of 13 women and eight men in total, 14 of them working as civil servants in varying sustainability-related job positions and with various levels of seniority, among them the 'strategy coordinators' from the various administrations and municipal companies responsible for implementing the strategies within the programme. The remaining seven respondents had been included in the team as 'outside experts' (four research-ers and three consultants). The distinction between civil servant and 'outside expert' is, however, not so clear-cut. Some civil servants had research back-grounds, and some researchers were former civil servants. Roles sometimes also overlapped due to career changes during the research process. In the case of Malmö, the respondents consisted of five women and two men. Two of the civil servants were interviewed a second time at the end of 2023 to cover any potential changes or developments in Malmö's climate policy-making. Among the respondents was an elected member from the municipal council and the remainder had civil servant roles in the municipality. The partici-pants from Malmö also varied in seniority and job titles, however, all partici-pants held positions related to climate or sustainability issues.

After having acquired informed verbal consent the interviews were recorded. Interview questions focused upon i) the respondent's background and experiences, (ii) the sustainability work and context of their specific administration/municipal company, (iii) the respondent's role and connection with the environment and climate work of the municipality, (iv) if and how the respondent perceived and worked with justice and social aspects in rela-tion to sustainability, including potential barriers, and (v) information and knowledge practices within the organisation.

When we present our results, we use quotes to highlight key findings and larger trends that were identified in our analysis. The quotes, which we have translated from Swedish to English, are from the programmes, interview transcripts, or field notes from participant observations.

Results: Silo-thinking and the separation of 'environmental' and 'social' issues

Within both Gothenburg's and Malmö's environment and climate pro-grammes there are indeed some recognitions that fundamental holistic transformations are required to decarbonise societies, as well as a recogni-tion that the municipalities need to develop new ways of cross-sectoral cooperation to achieve the objectives and goals within the programmes by 2030. This is also evident in the manner in which the municipalities often-times address environment and climate issues together as (somewhat) inte-grated concerns, at least within the policies and in some discussions and presentations. Indeed, this was a motive behind including and spreading the responsibility for environmental and climate issues across various administrations and municipal companies within the two municipalities' organisations. In the case of Gothenburg this was also a motive behind combining the previous climate strategy and environment programme into one sole programme and to introducing the seven overarching key strategies to reach the three overarching goals: nature – a high biodiversity; climate – a climate footprint close to zero; and human – a healthy living environment for the citizens of Gothenburg. In the words of a civil servant who was involved in shaping the programme:

> Because we know we need to move away from the 'silos', so the idea with the strategies was for them to be a new organisational tool to increase cooperation across some of the administrations and municipal companies and so that more people in the municipality feel responsible for them and work more effectively with them and the programme.
>
> *(Respondent G12 2021)*

Our analysis shows that both cities, in many regards, have ambitious environment and climate programmes that, to an extent, also include the importance of addressing social aspects alongside environmental. Even though our critical lens makes visible several gaps within the programmes – for a more thorough analysis of the respective programmes see Rask (2022) for an analysis of Gothenburg's programme, and Lundgren (2022) for an analysis of Malmö's programme – we find that the programmes contain not solely an ecomodernist discourse, but also discourses that can be seen as more radical and holistic in character. This indicates that some of the actors involved in developing them likely have some knowledge about the systemic character of environment and climate issues and the need for transformative change. We find that the difficulties mainly arise in the *implementation process*; when the policies need to be acted upon and integrated into the organisation and across other administrations (and other actors the municipalities cooperate with). This is where the institutional logics, norms, and cultures come into play. This is also a recurring theme in our material:

> The problem is the difficulty to act and achieve results, and also the need for repercussions if the municipality does not act [referring to that there is too much focus on developing what is in the policies and not enough focus on what is actually done to reach the targets in the policies].
>
> *(Respondent G5 2021)*

There is thus a discrepancy or gap between what is articulated in the policies and what is implemented. The policy documents contain more 'holistic' and 'intersectional' thinking and less separation of 'environmental' and 'social' issues, including issues concerning justice and (in)equalities. Furthermore, Agenda 2030 is often referred to within the programmes, which itself is a policy that emphasises a holistic approach to sustainability, stressing that all goals are interconnected and are to be addressed together. It is in the implementation process that the separation and siloing is most visible in our material. One example is from Malmö where an interview participant says that responsibility for implementation still falls heavily on the environmental administration and that other administrations perceive the environmental administration in this way, as responsible for sustainability efforts (Respondent M6 2022). As similarly expressed by a civil servant in Gothenburg:

> There's this idea that the environmental administration works with environmental issues [...] that's what they do [...] and then Equal City [another program in the municipality about equality work] they work with social sustainability and gender and equality. [...] It is this parallel process, but we need to think outside of the box, I think, so that environmental issues and equality stop being parallel processes. The environment and climate issues need to become part of the main processes in the City.
>
> *(Respondent G5 2021)*

It becomes apparent that in practice these more holistic, cross-sectoral organisational measures and ideas encounter difficulties. Civil servants in both cities frequently depicted municipal work as still very much dominated by the siloed structure and often expressed, in both interviews and meetings, that such structures are not optimal for addressing environmental and climate issues. One challenge for implementation across a siloed organisational structure lies in each administration's and company's distinct focus and tasks that are set out politically, often centred around specific types of development. As expressed by two civil servants in Gothenburg:

> Many don't really see or understand that this is something that they, or we, are supposed to work with [referring to the programme]. They don't see their, or our, task in that way.
>
> *(Respondent G12 2021)*

> Many say that "oh yes, that is relevant", but also "that is not something we can work with because that is not in our task" and dismiss it [when talking about social sustainability].
>
> *(Respondent G7 2021)*

Thus, within these tasks there is room for interpretation, and these interpretations seem to be shaped by historical trajectories and path dependencies that disconnect 'environmental' issues from 'social' ones; as expressed by one civil servant at the environmental administration in Gothenburg:

> This is what we have always done, this is what we know and what we are good at [referring to environmental issues].
>
> *(Respondent G2 2021)*

In both municipalities the issue of siloed structures is often depicted as an organisational and managerial issue where the solution is deepened cooperation. As one public official in Malmö explained, currently the environmental and climate efforts largely revolve around fostering greater sector integration and reducing reliance on siloed structures (Respondent M5 2022). The same applies to Gothenburg, where many workshops and meetings concern how to better cooperate across sectors and pushing for innovation in governance and leadership within the municipality (Respondent G20 2021). However, something that does not receive as much attention is the entrenched culture of organisational silos, which makes the integration of these initiatives increasingly challenging. As described by a civil servant in Malmö:

> It is a slow process to change the culture in a municipality. It is a bureaucracy that often feels reluctant to change. [...] People are used to doing things in a certain way. [...] The administrations have their defined tasks and ways of working, they have their own cultures.
>
> *(Respondent M3 2022)*

And, relatedly, as expressed by one civil servant in Gothenburg in a meeting about the environment and climate programme:

> It is ridiculous to say that the City of Gothenburg is just one organisation. All the administrations and the municipal companies have so very different cultures.
>
> *(Respondent G7 2023)*

Culture can, in this sense, be understood as a socially negotiated system of meaning through which humans classify, construct norms and meanings (meaning-making), and make sense of the world (cf. Patterson 2014). This entrenched culture of organisational silos seems to affect how issues are constructed, viewed, and addressed. In our empirical material it becomes clear that 'silo-thinking' and the separation of issues into Weberian logic is indeed deeply entrenched in institutional logics and ways of thinking about the problems that need to be addressed, which often leads to a separation of what is perceived as 'environmental' and 'social' issues within the organisations. As further expressed by a civil servant in Gothenburg:

It is difficult for the administrations that are used to work with 'only' social issues to see the relevance for them to also work with environmental issues. To see the whole picture. The environment often gets neglected. [...] And [in the environmental administration] many are maybe trained only in the natural sciences and have those backgrounds. It is difficult for them to understand how social issues are interlinked or to think of it as something that needs to be considered.

(Respondent G2 2021)

Importantly, we find that this separation is part of a path dependency that prioritises ecomodernist discourses and 'technical' framings and solutions to environment and climate problems over more 'social', holistic, and transformative ones. As expressed by one civil servant in Gothenburg:

It's like 'as long as it's electrical it's great' and 'electrification will solve everything', and a belief in that then we don't need to change much else.

(Respondent G10 2021)

This dominance of technological solutions and problem formulations was also prevalent in the case of Malmö. An intersectional lens reveals aspects that are left unaddressed. In both Gothenburg and Malmö, social aspects receive limited attention in attempts to implement environment and climate policy. Specific social categories or differentiation among groups are often overlooked, most often leaving social groups defined in homogenous terms. The analysis suggests that only limited attention is paid to social factors and therefore falls short in, for example, deeply examining variations in engagement capacities, or the impact of pronounced segregation as noted by Magnusdottir and Kronsell (2021). Our analysis, however, revealed a varied acknowledgement of social and justice issues in the two cities' environment and climate programmes. Compared to Gothenburg, Malmö's policies show a stronger emphasis on social and justice issues. Malmö states that the city has an interconnected perspective with an outspoken aim to counteract inequalities and interconnected challenges and driving climate actions that address societal disparities. Malmö recognises the unequal impact of climate change on various groups and aims for equitable climate efforts that also reduces inequalities and enhances living conditions for all (human) residents (City of Malmö 2021). However, while Malmö's policies acknowledge the importance of equity, gender equality, and anti-discrimination in climate-related work in a general sense, it became evident in the interviews that there seems to be a gap in translating these principles into specific actions that address the diverse needs, abilities, and conditions of different groups in the city.

Furthermore, in line with ecomodernist logics, we find that nature is often described as a resource for the cities and its (human) residents, offering, for example, gains and ecosystem services. There is often a strong anthropocentric focus where the human-nature relationship is hierarchical, viewing climate and environment merely as scenery or tools for human well-being. For example, this logic is evident in both Gothenburg's and Malmö's programmes which state that their central aim is to allow people to maintain their current lifestyles, using strategies such as circular economy principles, technological advancements, and innovation to minimise their carbon footprints. The programmes highlight the need for environmental care, like preserving green spaces and biodiversity, not for nature's intrinsic value, but for the benefits to human health and future (human) generations. The focus is not on addressing the fundamental issue of viewing nature as a freely available resource separated from humanity. This is, according to our intersectional lens, a fundamental ontological concern (cf. Mansfield & Doyle 2017) as it exemplifies the dualistic view that separates humans from nature. Therefore, instead of seeing nature as a service or backdrop for humanity a just climate transformation would include placing the demands of the non-human world alongside those of humankind, rather than below them.

In sum, our material illustrates that siloing is not solely an 'organisational' or 'managerial' issue. Therefore, other than focusing on organisational structures and measures, which, as we have shown, is already receiving attention in both cities (to some extent), we find that there also seems to be a need for measures that can facilitate the inclusion of different perspectives, worldviews, and logics, such as reflexive and transformative learning practices that can explore new pathways (cf. Rohracher et al. 2023). This result will be further discussed in our concluding remarks.

Ways forward: Intersectionality as a 'policy tool' and 'transformation strategy'

In this chapter we have presented the results from a critical policy analysis of policy documents, transcripts from semi-structured interviews and 'meeting ethnography' field notes, with the aim of demonstrating how intersectionality can be used as an important tool in analysing policy processes to make visible dominant discourses and power relations. The proposed use of intersectionality as a 'reading' policy tool has been tested using two empirical cases: the environment and climate policy process in Gothenburg and Malmö, two municipalities often deemed as frontrunners in this area.

Our main contributions in this study can be seen as threefold: First, we have illustrated the dominant logics inherent within the two municipalities' environment and climate policy processes that were made visible in our empirical material with the help of our critical lens. Second, we have demonstrated if and how these logics influence the integration of justice and 'social' issues within municipal environmental and climate efforts. Thirdly, we will, in these concluding remarks, reflect on how the two municipalities can incorporate alternative perspectives, logics, and ways of thinking into their organisations, thereby enhancing their transformative potential; bringing forth what the lessons to take with us from these results could be and present our two 'principles' for transformative change.

Despite acknowledging the interconnectedness of environmental and social issues within their policies (to an extent), limitations seem to arise in mainly how they are acted upon and implemented. The critical lens reveals that the framings in the two municipalities' environment and climate policy processes follow a historical trajectory (path dependency) where social aspects are often deprioritised over technical framings and solutions. This path dependency, we find, is premised on a technocratic and de-politicised understanding of climate issues stemming from an apparent ecomodernist logic, at the expense of taking issues of power, equality, and justice seriously. Interviews and participatory observations in meetings and workshops reveal that many civil servants tasked to work with environment and climate issues view these dominant logics as barriers to fulfilling their tasks. Civil servants, and especially those in leadership positions, in both Gothenburg and Malmö seem to need to further recognise that unequal power structures, relations, and practices are embedded in the climate crisis and need recognition in proposed solutions. If not, there is a risk of reinforcing power relations along, for example, ethnicity, race, age, gender, and class lines. Environment and climate effects, as well as how such issues are governed, will likely influence people within the cities differently. If power inequalities within the city are not considered, there is a risk of increasing inequality. Based on previous research referenced in the introduction to this chapter, this likely hinders the transition/transformation towards sustainable and just societies.

Thus, importantly, we find that this separation of issues in a Weberian manner is not only prevalent in organisational structures but also in the guiding logics and ways of thinking within organisations. Compartmentalisation of environmental and climate issues from questions of social justice is thus not only sectoral, but also ontological. Such a binary or dualistic ontology allows for environmental and climate issues to become framed as a purely technocratic problem to be effectively managed, put simply. Therefore, other than focusing on organisational structures and measures, which, as we have shown, is already receiving (some) attention in both cities, we find that there also seems to be a need for measures that can facilitate the inclusion of different perspectives and logics, such as reflexive and transformative learning practices. Deeply embedded organisational practices (like silo-thinking) are not only rational responses but also culturally and socially reinforced patterns, legitimised through norms, values, and taken-for-granted assumptions.

To address the negative aspects of what we in this chapter have termed 'silo-thinking', efforts of change need to consider both the structural redesign (in the Weberian sense) and the transformation of underlying institutional logics. This dual approach can help in creating more integrated and adaptable organisations. We suspect that transforming 'ways of thinking' will likely be challenging because ingrained logics and worldviews can be sensitive and potentially conflictual.

Based on our findings and critical lens we suggest two points, or principles, that seem important in this process: the need for the organisations to (i) include time and space for reflection and reflexivity , and (ii) the need to foster an ethics of care as a guiding principle and become 'caring' organisations for both all humans and 'the environment'. We propose that bringing forth such principles could foster an attentiveness to the power relations involved in environment and climate policy processes and aid in centring questions of ethics, justice, and equality. We argue for the potential of an intersectional lens to be used as a 'policy tool' and 'transformation strategy' to promote these two (interrelated) principles and inform and enrich climate policy processes, in order to aid civil servants in these two cities and elsewhere to move forward in their work towards sustainable and just societies. Intersectionality as an analytical lens can expose underlying logics shaping current policies. In turn, enabling a more critical approach that aids to (re) politicise, problematise, and disrupt seemingly apolitical understandings of environmental and climate policies and their implementation processes. We see the need for and encourage further research to test such a tool or strategy/ies in practice, together with practitioners.

Our analysis can be seen as an initial step in making visible dominant logics, structures, and hierarchies to highlight what current deficits exist and what needs to change. Our critical lens has aided us in making visible how dominant ecomodernist logics shape understandings and perceptions in policy-making and implementation in two Swedish cities. We recommend further research to study how such logics can be changed. We see a need to investigate further: What alternative ways of knowing may these dominant logics foreclose? How can city governments include other logics, understandings, and perceptions of socio-ecological systems into policy processes? What is needed for these municipalities to include intersectional knowledge and integrate more radical (paradigm-shifting) perspectives?

This chapter has exposed the cracks in the dominant ecomodernist logics found in Gothenburg's and Malmö's environment and climate policy processes. Yet, within these cracks, we glimpse promising ground for transformation. With this chapter we offer a seed – the intersectional lens – a potentially powerful tool to cultivate policy for truly just and sustainable futures.

Acknowledgements

This chapter is a product of the Swedish Research Council for Sustainable Development (FORMAS) funded project *Intersectionality and Climate Policy Making: Ways Forward to a Socially Inclusive and Sustainable Welfare State* [grant number FR-2018/0010].

References

Acemoglu, D., Egorov, G., Sonin, K. (2021) "Institutional change and institutional persistence". In A. Bisin & G. Federico (Eds.), *The Handbook of Historical Economics*. Elsevier Science & Technology.

Andersson, M. & Gyberg, V. B. (2023) "Sustaining business as usual or enabling transformation? A discourse analysis of climate change mitigation policy in Swedish municipalities". *Environmental Policy and Governance* 2023: 1–13.

Bacchi, C. (2009) *Analysing policy: What's the problem represented to be?* Australia: Pearson Education.

Bacchi, C. & Goodwin, S. (2016) *Poststructural policy analysis: A guide to practice.* Basingstoke: Palgrave Macmillan.

Bradshaw, A. (2019) "Sustainability and Gender Equality: Exploring the 2030 Agenda for Sustainable Development." In: *Environment and Sustainability in a Globalizing World*, London: Routledge, pp. 232–245.

Calmfors, L. & Hassler, J. (2019) "Climate policies in the Nordics." *Nordic Economic Policy Review 2019: Climate Policies in the Nordics*, 7.

Castán Broto, V., Trencher, G., Iwaszuk, E., & Westman, L. (2019) "Transformative capacity and local action for urban sustainability." *Ambio* 48: 449–462.

City of Gothenburg (2021) *Göteborgs Stads miljö- och klimatprogram 2021–2030.* Available at: https://goteborg.se/wps/portal/start/kommun-och-politik/sa-arbetar-goteborgs-stad-med/hallbarhet-och-agenda-2030/program-och-planer-for-miljo-och-klimat/miljo--och-klimatprogram-for-goteborgs-stad-2021-2030 (accessed 26 April 2024).

City of Gothenburg (2022) "Göteborg blir en av de 100 klimatneutrala och smarta städerna." Available at: https://www.mynewsdesk.com/se/goteborgsstad/pressreleases/goeteborgs-blir-en-av-de-100-klimatneutrala-och-smarta-staederna-i-europa-3178405 (accessed 17 September 2023).

City of Gothenburg (2023a) "Forskarskola med fokus på rättvis lokal klimatomställning". Available at: https://goteborg.se/wps/portal/enhetssida/miljo-och-klimat-goteborg/aktuelltarkiv/aktuelltsida/76ce2463-aeb8-43e2-b57b-909bf476d444 (accessed 26 April 2024).

City of Gothenburg (2023b) "Agenda 2030 i Göteborg". Available at: https://goteborg.se/wps/portal/start/kommun-och-politik/sa-arbetar-goteborgs-stad-med/hallbarhet-och-agenda-2030/agenda-2030-i-goteborg (accessed 26 April 2024).

City of Gothenburg (2023c) *Rapport 2023:06 Uppföljning av mål och delmål i Göteborgs Stads miljö- och klimatprogram 2021–2030.* Available at: https://www4.goteborg.se/prod/Intraservice/Namndhandlingar/SamrumPortal.nsf/6865C815A54B63F9C12589CD00316450/$File/Handling%2012%20MKN%2020230620.pdf?OpenElement#:~:text=I%20rapporten%20bedöms%20graden%20av,att%20målen%20ska%20kunna%20nås (accessed 26 April 2024).

City of Malmö (2021) *Miljöprogram för Malmö stad 2021–2030*. Available at: https://malmo.se/Miljo-och-klimat/Malmo-stads-miljoprogram.html (accessed 25 January 2024).

Crenshaw, K.W. (1989) "Demarginalizing the Intersection of Race and Sex: A Black Feminist Critique of Antidiscrimination Doctrine, Feminist Theory and Antiracist Politics." *University of Chicago Legal Forum*, pp. 139–168.

Cuthill, M. (2010) "Strengthening the 'social' in sustainable development: Developing a conceptual framework for social sustainability in a rapid urban growth region in Australia." *Sustainable Development (Bradford, West Yorkshire, England)*, 18(6), pp. 362–373.

Dryzek, J.S. (2022) *The politics of the earth: Environmental discourses*. Oxford University Press.

European Commission (2022) *EU missions: 100 climate-neutral and smart cities*. Doi: 10.2777/85010

Eversberg, D., Holz, J. & Pungas, L. (2023) "The bioeconomy and its untenable growth promises: Reality checks from research." *Sustainability Science*, 18(2), pp. 569–582.

Goldstein, J. E., Neimark, B., Garvey, B. & Phelps, J. (2023) "Unlocking "lock-in" and path dependency: A review across disciplines and socio-environmental contexts" *World Development*, 161, pp. 106–116.

Gruen, L. & Weil, K. (2010) "Teaching difference: sex, gender, species." In: DeMello, M. (ed.), *Teaching the animal: human–animal studies across the disciplines*, Brooklyn: Lantern Books, pp. 127–144.

Grunwald, A. (2018) "Diverging pathways to overcoming the environmental crisis: A critique of ecomodernism from a technology assessment perspective." *Journal of Cleaner Production*, 197, pp. 1854–1862.

Hagbert, P., Wangel, J., & Broms, L. (2020) "Exploring the Potential for Just Urban Transformations in Light of Eco-Modernist Imaginaries of Sustainability". *Urban Planning* 5(4): 204–216.

Haraway, D. J. (2003) *The Companion Species Manifesto: Dogs, People, and Significant Otherness*. Chicago, IL: Prickly Paradigm Press.

Harcourt, W. and Nelson, I. L. (2015) "Are we 'green' yet? And the violence of asking such a question". In: W. Harcourt and I. L. Nelson, eds., *Practising Feminist Political Ecologies: Moving Beyond the 'Green Economy'*. London: Zed Books, pp. 131–156.

Hinton, E. (2015) "The politics of sustainable consumption." In *Routledge International Handbook of Sustainable Development*, pp. 237–249. Routledge.

IPCC, Masson-Delmotte, V., Zhai, P., Pörtner, H. O., Roberts, D., Skea, J., & Shukla, P. R. (2018) *Global Warming of 1.5°C: An IPCC Special Report on the Impacts of Global Warming of 1.5°C above Pre-industrial Levels and Related Global Greenhouse Gas Emission Pathways, in the Context of Strengthening the Global Response to the Threat of Climate Change, Sustainable Development, and Efforts to Eradicate Poverty*. Intergovernmental Panel on Climate Change.

IPCC, Masson-Delmotte, V., Zhai, P., Pörtner, H. O., Roberts, D., Skea, J., & Shukla, P. R. (2022) *Global Warming of 1.5 C: IPCC special report on impacts of global warming of 1.5 C above pre-industrial levels in context of strengthening response to climate change, sustainable development, and efforts to eradicate poverty*. Intergovernmental Panel on Climate Change.

Jänicke, M. & Jacob, K. (2004) "Lead Markets for Environmental Innovations: A New Role for the Nation State." *Global Environmental Politics*, 4(1), pp. 29–46.

Jennings, D. & Hoffman, A. (2017) "Institutional theory and the natural environment: Building research through tensions and paradoxes". In S. Clegg, C. Hardy, T. B. Knudsen, & H. V. Lindberg (Eds.), *The SAGE Handbook of Organizational Institutionalism* (pp. 759–785). SAGE Publications Ltd.

Kaijser, A. & Kronsell, A. (2014) "Climate change through the lens of intersectionality." *Environmental Politics*, 23(3), pp. 417–433.

Kelly, G. & Licona, P. (2017) "Epistemic Practices and Science Education." In: *History, Philosophy and Science Teaching: Science: Philosophy, History and Education*, Cham: Springer International Publishing, pp.139–165.

Kronsell, A. & Bäckstrand, K. (2010) "Rationalities and forms of governance: a framework for analysing the legitimacy of new modes of governance." In *Environmental Politics and Deliberative Democracy*. Edward Elgar Publishing.

Lundgren, A. (2022) The siloed umbrella: An intersectional reading of climate policy in Malmö, Sweden. *Institutionen för globala studier*, University of Gothenburg.

Lutz, H., Herrera Vivar, M. & Supik, L. (2011) *Framing intersectionality debates on a multi-faceted concept in gender studies (The feminist imagination/Europe and beyond)*. Farnham: Ashgate.

Magnusdottir, G.L. and Kronsell, A. (2021) *Gender, Intersectionality and Climate Institutions in Industrialized States* (1st ed.). Routledge: Earthscan.

Mansfield, B. & Doyle, M. (2017) "Nature: A conversation in three parts." *Annals of the Association of American Geographers*, 107(1), pp. 22–27.

Matsuda, M. (1991) "Beside My Sister, Facing the Enemy: Legal Theory out of Coalition." *Stanford Law Review*, 43(6), pp. 1183–1192.

Mol, A.P. (2002) "Ecological modernization and the global economy." *Global Environmental Politics*, 2(2), pp. 92–115.

Murphy, K. (2012) "The social pillar of sustainable development: A literature review and framework for policy analysis." *Sustainability: Science, Practice, & Policy*, 8(1), pp. 15–29.

Næss, P. and Saglie, I. (2019) *"Ecological Modernisation"*. In: *The Routledge Companion to Environmental Planning*. Routledge.

Patterson, O. (2014) "Making sense of culture". *Annual Review of Sociology*, 40, pp. 1–30.

Rask, N. (2022) "An intersectional reading of circular economy policies: Towards just and sufficiency-driven sustainabilities." *Local Environment*, 27(10–11), pp. 1287–1303.

Rohracher, H., Coenen, L. and Kordas, O. (2023) "Mission incomplete: Layered practices of monitoring and evaluation in Swedish transformative innovation policy." *Science and Public Policy*, 50(2), pp. 33–349.

Sandler, J. and Thedvall, R. (2017) *Meeting Ethnography: Meetings as key technologies of contemporary governance, development, and resistance*. Taylor & Francis.

Scott, W. R. (2017) "Institutional theory: Onward and upward." In S. Clegg, C. Hardy, T. B. Knudsen, & H. V. Lindberg (Eds.), *The SAGE Handbook of Organizational Institutionalism* (p. 853). SAGE Publications Ltd.

Singleton, B.E., Rask, N., Magnusdottir, G.L. and Kronsell, A. (2021) "Intersectionality and climate policy-making: the inclusion of social difference by three Swedish government agencies." *Environment and Planning C: Politics and Space* (April 2021): 1–21.

Torgerson, D. & Paehlke, R. (2005) "Environmental Administration: Revising the Agenda of inquiry and Practice." In: Paehlke, P. and Torgerson, D. (eds.), *Managing Leviathan: Environmental Politics and the Administrative State* (2nd ed.). Peterborough, Ontario: Broadview, pp. 3–10.

Turnhout, E., Neves, K. and De Lijster, E. (2014) "'Measurementality' in biodiversity governance: Knowledge, transparency, and the Intergovernmental Science-Policy Platform on Biodiversity and Ecosystem Services (IPBES)." *Environment and Planning A*, 46(3), pp. 581–597.

UN-Habitat (2016) *The New Urban Agenda: Quito Declaration on Sustainable Cities and Human Settlements for All.* Quito: United Nations.

Urban Futures (2022) "Styrning för jämlik och inkluderande klimatomställning i Göteborg". Available at: https://urbanfutures.se/forskning/styrning-for-jamlik-och-inkluderande-klimatomstallning-i-goteborg (accessed 26 April 2024).

Uteng, T.P., Christensen, H.R. & Levin, L. (2020) Gendering Smart Mobilities. London and New York: Routledge.

Vallance, S., Perkins, H. & Dixon, J. (2011) "What is social sustainability? A clarification of concepts." *Geoforum*, 42(3), pp. 342–348.

Wapner, P. (2002) "Ecological Displacement and Transnational Environmental Justice." *Global Dialogue*, 4(1).

Washington, H. (2015) *Demystifying Sustainability: Towards Real Solutions.* Routledge.

Interview 5

WITH MARIANNE BORGEN, TWO-TERM MAYOR OF OSLO BETWEEN 2015 AND 2023

Interviewed by Susan Buckingham, 22 January 2024 in Oslo Public Library (Deichman Bibliotek)

Background

Marianne Borgen has been, variously, a local, city, and national politician since the mid 1970s, representing the Socialist Left Party (SV). Positioned to the left of Norwegian politics, SV is committed to social and economic justice, feminism, and addressing the climate change emergency. Mayor Borgen came to the mayoralty with a strong, professional background in children's rights, and had been the Ombudsman for Children in Norway. Her approach to environmental issues was influenced by their impact of the lives of children and, under her tenure, Oslo developed impactful green transport and waste reduction programmes. Oslo was awarded European Green Capital status for 2019. Marianne Borgen resigned her position in 2023 ahead of the City elections, which were won by the Conservative Party, who appointed Anne Lindboe as mayor.

Would you define yourself as a feminist?

Yes I do!

Could you explain your background?

I was born in 1951 and grew up in the Eastern (poorer) part of Oslo – my parents had no formal education. These were the years after World War 2 during which society had to be rebuilt, such as through welfare programmes. My parents were not union members or politically active, but were community active. We grew up learning to take care of each other. I realised how divided the city was once I went to university. I grew up when women were

DOI: 10.4324/9781003461005-17

housewives; my mother worked in small shops since I was nine or ten. My parents pushed my brother and me to education as the opportunities were available, and I took optional gymnasium. No one talked at school about university, so I decided to be a social worker – to 'save the world from poverty and problems'. This needs a year of practice so I worked in an office in NE Oslo. But I was disappointed by this as it was too office bound. I was encouraged by a social work researcher to study sociology. I studied history, sociology, and German – and worked in a youth club. I graduated 1977 and went to work in the ministry with elderly people. Then I went to work in the Ombudsman's office for children (1985–95) and started to work with children's rights, which changed my career. Following this, I worked for Save the Children, campaigning for children's rights and against the abuse of children.

I was always politically active – in party membership and as a feminist. I challenged apartheid in South Africa, and was against atomic bombs. I decided to try working on the other [political] side – which I discussed with my party. I had two children and a husband who was sick, and had been a 'reserve' member of Parliament, but wanted to work locally. I was elected to Oslo city council in 1996 until October 2023.

Between 1996 and 2011 I was a city councillor while working at Save the Children. In 2011 I was elected as group leader, which was full time. Between 2011and 2015 my brief was transport and environment. As a member of a small party I had no choice of brief (which was allocated by the ruling party). I was a bit reluctant at the start, as [I] thought I didn't have the expertise. But I came to realise that it was 'great'. If you can make a city that's good for children – then it's good for everyone. Then all transport and environment questions are, are they good for children and families?

A Labour/Green/SV coalition won the election in 2015, although it was Labour Party dependent. My experience was appreciated by the Labour/Green coalition partners and I led the negotiations for building the coalition, expecting to be a city minister. Once political positions had been agreed and posts allocated/agreed, I insisted on two positions for the SV, as one is too isolated. Labour/Greens wanted me as mayor for my experience, which I accepted, although everyone was surprised.

The governing mayor (i.e. the executive mayor – the mayorship that Marianne Borgen accepted was more ceremonial) had high confidence in me, as it was known that if I promised something I'd deliver. That, and years of experience in City Hall. I had a way of cooperating which is broad, and other political parties trusted me.

When I called my husband and children to say I could be mayor, they said 'what?!'

I decided that I wouldn't stand beyond 2023, so that I could do other things with my life. But I would have lost my job anyway if I hadn't resigned as now the City is Conservative-led. (Although the majorities are never large.)

Could you provide examples of your courage – of fighting for things that you want?

[Marianne Borgen gave the example of the Munch museum in Oslo.]

The old Munch museum was in a suburban area with a lot of problems of poverty, drugs – bad for children. There were lots of negative things in the area. The big political question was where to put a new Munch museum. I fought for the original area [so as] to leverage resources into the area. The Conservatives wanted it in the centre.

In 2013, I visited a school who told me how terrible the situation was in the neighbourhood. Uncertainty [over the location of the museum] wasn't helping the area. I made a 'secret channel' with the Conservative Party. This was the most scary decision in my career – if it became public, I would have to withdraw from the city council. I negotiated changes if we agreed to build in the centre and a lot of money was spent on Munch museum designs, etc. This secret channel led to a positive outcome for the area. After negotiations, my own party was sworn to secrecy prior to the announcement of the agreement. My own party was so angry that I hadn't trusted them. However, we got libraries, [a] swimming pool, free after-school programmes for the old area. I'm still waiting for an indoor/outdoor spa – but confident it will happen. I didn't sleep that night: if my party didn't agree, I would have to leave politics. When I talk about it now, I get quite emotional. But then I rationalised it – I could always go back and work for Save the Children!

Most people say 'Wow, that was a brave thing to do'. But I can't work like that regularly – it was a one off. [All this happened before MB was mayor – and one of the reasons she was offered the mayor's position.]

Did working across boundaries/parties help with green issues?

My background is important. I have always taken a child's perspective. I use the Children's Rights Convention. We should always ask ourselves whether a policy is in the best interest of the child. This is the most important thing. And being a feminist. I had a very positive development with regard to public transport – zero emission public transport. It's women who use public transport most, who bring children to and from schools. This needs feminist and child rights' perspectives.

We won't achieve a climate friendly society if we don't solve social problems – they are interconnected. This is one of the differences between the Green and socialist parties. It is developing now, but initially the Green Party had problems seeing social injustices. You have to have the people with you. There are two pillars in [the] socialist party: reduce social inequalities; climate.

Also, Parliament and state government need to be with you: e.g. tax policy, resources for schools, care. But cities can still do a lot. In 2019, Oslo was elected as [the] European Green Capital of the year – this was the beginning, not the end of something. We got it for the waste work we have done (especially improving collection), the cleaning up of Oslo Fjord, and transport. We needed better dialogue with residents, businesses, and children. The award was an energy boost (not a 'sit back and enjoy' award). I've never seen so much activism. Being environmental/climate friendly is good for business. All building machinery should be GHG emission free – and it has helped enterprise [to] develop equipment. We are part of the Eurocities network – through which we can share good examples. We made a climate budget (on top of the financial budget) – to ensure that all parts of the community could make emissions savings. This model has been exported to New York City and elsewhere in Europe. Cities are a huge part of the problem.

Do you work with other feminist networks?

Yes – in supporting each other, and getting new ideas from each other. Oslo is a small city by world standards, but big enough to test solutions which can be interesting for others. CCS [carbon capture and storage] is being tested at a waste disposal plant – there are some economic problems but I think it will cut emissions 80–90%.

Women, mothers, and grandmothers are happy with the changes they have seen. I have seven grandchildren myself (I'm very lucky). What sort of society will we deliver to the next generations? How can I use my energy to ensure this? I used my time as mayor to talk to young people. I notice young people all over the world. I go into school classes to hear what children are worried about. They are human-made problems, and we can solve them.

Do you have grounds for hope?

What is happening in the world is something that the people can do something about. I talk about what people can do themselves, in homes, schools; but also politicians have a special responsibility. We have to be brave – to do the unpopular; this might be seen negatively in the short run, but there are longer-term advantages. An example is the smoking ban which had a lot of opposition, but now people see that it is 'so nice'. It's good for children (my own parents smoked!). Another example is toll stations to encourage people to use public transport. How did we become Number One in electric car use? National and local policy – no taxes on e-cars which didn't have to pay toll stations and can use bus lanes. Most people bought e-cars because of economic, not environmental reasons.

We should just tell people/children about scary things. To achieve a healthy society we have to do something. People will be OK. Talking to children is also about listening to them. I ask them what they would do. Children are more strict than grown-ups. I see my function as giving hope. What do we want? We can do a lot together.

But it's harder to be hopeful after the wars in Ukraine and in Palestine. It takes a lot of energy (including emotional energy). The time to turn things round is too long – for the Global South. On increasing military resources, I'm a little bit worried that resources are used for wars/military – if we could use these resources on environmental/social/climate action, we could have done so much. But when talking to grandchildren, I still try to be hopeful.

What is your thinking about feminist women's future as they are losing positions?

Now women are entering politics and now men are leaving and taking the power out with them....

There are still a lot of women out there fighting with a lot of energy. But I'm not sure.....

Who inspires you?

Gro Harlan Brundtland – I don't agree with her on many political questions but she fought a hard battle in Norwegian politics and worked hard on sustainability problems. She's a role model for people like me. She made it easier for people like me to say, 'I want power'. I grew up when women were in the background. In my party, there are strong women, including minority ethnic. But for young women to take an active part in politics – we must have a system that is functioning, e.g. fathers who are willing to take an equal part in family.

After four years with two young children I had to leave politics – there were no kindergartens and my husband had health problems, so evening meetings were difficult. Some of the same problems exist; even though gender inequality has lessened, mothers still feel the main responsibility for family. There is still a lot of pressures on families: expensive housing, the need for two incomes, high expectations of what families do (children's activities, after-school work, holidays, sports....). It is still very difficult. To be politically active with small children is hard, despite better kindergartens, activities, public transport. All this makes it harder for young women with children. It is hard to be a politician. As a member of a small political party – we have to do all the work ourselves. In my next life I will choose a big party with lots of resources!

As mayor, I tried to shift some of the evening meetings to daytime. After eight years, I hardly changed anything. It's so hard, I had objections even from the women themselves. Most women have paid work, then look after children – so they have no time for evening meetings. Employers have to give you time for council meetings – but not for party meetings.

When my children were small – meetings were 2 pm to midnight. Now they are 3–10 pm. It's hardly changed at all! Also, there are committee meetings (2–3 hours) once a month; group meetings. There are too many objections to moving council meetings to daytime, with no real gendered difference in response. Smaller municipalities have started having meetings in the daytime (especially where there is not so much business). It's exhausting combining paid work and politics – we need to have a system that makes it easier. As women get older, and have older children, they also have parents to look after. It's sad that I couldn't get more changes. The only change I did make was that committees could make their own decision. Only one committee decided to shift to meeting in the day. The argument that people [i.e. public] wouldn't come to meetings if they are in [the] day is not valid because they are streamed/recorded. These are reasons why many women don't go into politics – but it has become easier for women to get positions of leadership in political parties. That's a change I've seen in my lifetime.

We need more feminists in politics, we need more men who are feminist in politics. We are defining ourselves as a feminist party. Some of the men that I know asked whether they can be feminist – yes you can! In the 1970s it was mainly women, but that has changed. More younger men – it's easier for them now – agreed the politics but felt that the label feminist was for women, but now they can be proud to be feminist.

PART III

Methodologies

10

YOUNG PEOPLE AND OLD TREES

Posthuman intersectionality in Swedish climate litigation

Marie Widengård

Introduction

The courts are increasingly becoming the battlegrounds where individuals and groups seek to hold governments accountable for inadequate climate action. Climate change, recognised as one of the gravest threats to humanity and natural ecosystems, necessitates swift and decisive action to mitigate its risks. In this legal arena, a common theme is the argument that lax climate policies are a violation of human rights (Peel & Osofsky, 2018; Setzer & Vanhala, 2019). Moreover, the disproportionate effects of climate harm on vulnerable groups – children, youths, women, and Indigenous populations – have led to lawsuits based on discrimination, adding a layer of complexity to these legal battles.

Responding to this, scholars are advocating for the application of intersectionality to understand these multifaceted injustices (Amorim-Maia et al., 2022; Mikulewicz et al., 2023). Yet, while intersectionality in climate litigation is gaining traction (Sußner, 2023), its focus has remained predominantly anthropocentric, often sidelining how natural ecosystems like forests are implicated in legal disputes. This chapter expands the inquiry by exploring how climate litigation is shaped by multiple axes of difference, leading us towards intersectional and posthumanist visions of equality (Deckha, 2008). To this end, the concept of posthuman intersectionality, which considers the ways in which both humans and nonhumans are marginalised or privileged along the lines of social markers – including gender, race, ethnicity, nationality, class, ability, age, and species – is pivotal (Hovorka 2012).

This chapter seeks to contribute to the understanding of climate litigation through the lens of posthuman intersectionality, particularly examining the

DOI: 10.4324/9781003461005-19

Aurora case, Anton Foley and others v. Sweden (T 8304-22). Here, 636 Swedish youths demand that the court formally recognises that Sweden's climate change mitigation efforts are insufficient and violate their rights under the European Convention on Human Rights and Fundamental Freedoms (ECHR), which is incorporated into Swedish law. They claim that Sweden's ruling generations are handling climate policies in a way that exposes the young and future generations to significant risks, potentially causing widespread harm to their lives, physical and mental health, dignity, well-being, and their homes and property. Aside from the explicit age-based and intergenerational justice arguments, the Swedish case reveals how the fate of young people and old trees are tightly interwoven. The youths assert their stance against a national climate policy that they argue favours a forestry model detrimental to old-growth forests, thereby risking the carbon sink potential of forest ecosystems and violating the rights of climate protection. In essence, the chapter delves into the relationships between young people and old trees – how do their fates converge in the legal domain, and what does posthuman intersectionality add to our understanding of climate litigation?

By analysing the Aurora case from a posthuman intersectionality perspective, the chapter underscores the interconnectedness of human and nonhuman interests in the quest for climate justice. Sweden's approach to forestry, with its emphasis on young, even-aged production forests, has significant global implications due to its role-making function (Magnusdottir & Widengård, 2024). The Swedish approach, while often lauded for its high productivity, has raised questions about the true longevity of carbon sinks and biodiversity in managed forest landscapes (Beland Lindahl et al., 2017; Holmgren et al., 2022). The Aurora case is emblematic of how young citizens are not only questioning their nation's commitment to climate goals but are also contesting the environmental validity of prevailing forestry practices. Swedish forest policies and institutions are frequently criticised for being shaped by masculinist ideals, thereby marginalising care-oriented approaches and alternative forest values (Holmgren & Arora-Jonsson, 2015; Johansson et al., 2018). By highlighting these tensions, the case spotlights the complexities of climate strategies that intersect with intersectional justice and global debates on the sustainability of bioenergy, biofuels, and forest management.

The following sections will introduce the landscape of climate litigation and dissect the notion of posthuman intersectionality as it applies within this legal context. We then navigate through the Aurora case, assessing the age-based arguments and the symbolic and material connections that the young plaintiffs make between themselves and old-growth forests. Here, we examine the youths' portrayal of forests and carbon sinks as crucial nonhuman allies in their legal struggle for a fair climate future. Drawing from the case application filed with the Nacka court (T 8304-22) and available resources on the Aurora (2023) webpage, I argue that the Swedish youths' legal action transcends human rights. Claims are not merely made in defence of human rights but in advocacy for an integrated ecosystem where human and nonhuman entities coexist. In conclusion, this chapter posits that adopting a posthumanist intersectional lens could potentially reframe the interpretation of human rights claims within climate litigation. It is an invitation to consider the broader matrix of relationships that form the substrate for claims of climate justice – how intertwined the fates of humans and nonhumans are, and why acknowledging these connections is critical in legal discourses and beyond.

Climate litigation through an intersectional and posthuman lens

Climate litigation utilises legal systems to address climate change, ranging from advocating improved policies to challenging regulatory measures. This field has transitioned from relying solely on environmental laws to human rights-based approaches, acknowledging the rights of future generations and the validity of transboundary claims (Peel & Osofsky, 2018; Rodríguez-Garavito, 2022b; Savaresi & Auz, 2019). These cases pivot on diverse legal bases and the state's duty to protect its populace, invoking climate science to substantiate the damage caused by insufficient climate policies (Mayer, 2021; Rodríguez-Garavito, 2022a).

Litigants seeking climate justice frame their grievances within the context of human rights violations, with certain demographics – particularly children, youths, women, and Indigenous groups – pursuing cases on discrimination grounds, highlighting their disproportionate suffering due to climate harm. The landmark Urgenda v. Netherlands case established a precedent for holding governments accountable for inadequate climate change action, underscoring the human rights implications of failing to sufficiently reduce emissions (Rodríguez-Garavito, 2022a). The Dutch Supreme Court ruled that the likelihood of the current generation, especially the youth, facing severe impacts of climate change was evident if significant reduction in greenhouse gas emissions were not realised (Burger et al., 2022).

As of 2023, at least 34 cases have been initiated by and on behalf of children and youth under the age of 25 (United Nations Environment Programme, 2023). At the other end of the age spectrum, Swiss women over 64 years have contended that they are disproportionately affected by heat-related health problems (Klimaseniorinnen, 2024). After their case was dismissed by Swiss courts, they appealed to the European Court of Human Rights. On April 9, 2024, the court ruled that Switzerland had breached the right to respect for private and family life and their right to access the court. Although the four individual applicants did not meet the victim-status requirement, the association representing over 2,500 women was recognised as having the right to lodge a complaint (ECHR, 2024). This landmark ruling acknowledged that the European Convention of Human Rights encompasses 'a right to effective protection by the State authorities from the serious adverse effects of climate change on lives, health, well-being and quality of life' (ECHR, 2024). Conversely, a Portuguese youth-led lawsuit against 33 European countries was dismissed by the European Court due to procedural issues; the court found that they had not exhausted domestic remedies and could not sue countries other than Portugal. A French case was also rejected because the plaintiff, an ex-mayor, had relocated from the place on which his case was based (ECHR, 2024).

These instances illustrate the relevance of intersectionality, emphasising how impacts and judicial access are compounded by factors like gender and age (Crenshaw, 1991). Sußner (2023) contrasts the Swiss claimants with claimants in another gender-based case – Maria Khan et al. v. Pakistan – and posits that the Swiss women could be viewed as more privileged than vulnerable due to their status as Swiss retirees. Nonetheless, global research confirms that heat disproportionately affects women and the elderly. One Swiss plaintiff likens her situation to the melting Swiss glaciers, saying, 'I'm melting', as she notes the sweat on her face (The Guardian, December 27, 2023).

These diverse climate lawsuits[1] underscore the significance of age, gender, and location. A posthumanist legal lens further extends this scope to include nonhuman entities, calling for a justice system that acknowledges the rights and interconnectivity of all life forms, challenging conventional anthropocentric legal frameworks. Building on the foundational question of whether trees should have legal standing (Stone, 1972), this expanded perspective involves recognising collective rights and the legal personhood of natural entities (Athens, 2018; Braverman, 2018; Youatt, 2016). A ruling in a Colombian case, where youth sued to protect the Amazon from deforestation, illustrates this evolution by recognising the Amazon as an 'entity subject of rights' (Dejusticia, 2018). Colombian judges highlighted the imperative for legal systems to safeguard not only human rights but also the rights of ecosystems vital to planetary life. Such jurisprudential advancements provoke thought about how humans and nonhumans can collectively obtain rights, such as the Whanganui River and its Indigenous guardians in Aotearoa New Zealand, or the Waimea River and its community of residents, farmers and Native cultural practitioners in Hawaii (Norman, 2022; Youatt, 2016). Although posthuman perspectives often reject dualisms, posthuman intersectionality theory allows us to conceptualise certain collectives as enjoying privileges while others remain marginalised (Hovorka, 2012). This concept is vital for understanding the complexities of climate change activism, especially when considering implicit posthuman intersections, such as those forged by young Swedes with trees. The next section will delve into such alliances, underscoring their importance in the pursuit of climate justice.

The Aurora case: Young people and old trees in Swedish climate litigation

In 2022, a group of 636 young individuals filed a class action lawsuit against the Swedish state, claiming that Sweden's climate change mitigation efforts were insufficient and infringed upon their rights as defined in the European Convention on Human Rights. The plaintiffs, part of the association known as Aurora, were children and youths residing in Sweden, all born between 1996 and 2015 (para 27). They sought class action status due to their common vulnerability to the impacts of climate change and the prohibitive costs and complexities of individual lawsuits.

Before resorting to legal action, the group had approached two successive governments, voicing their concerns through letters. The response from the Social Democrats' state minister defended the Swedish climate policy, while the conservative Moderates did not respond. Left with no alternative, the youths proceeded to sue the state, seeking a decision from the Swedish judiciary. The case, filed with the Nacka district court in December 2022, was accepted for consideration in March 2023. Later, in October 2023, at the request of both parties, the district court referred the case to the Supreme Court (Ö 7177-23), which in April 2024 agreed to assess whether Swedish courts can try the case.

The analysis commences with an examination of the plaintiffs' human-focused claims before delving into their connections with Sweden's forests. It first considers the main demands and how the young plaintiffs identify themselves as an especially vulnerable and marginalised demographic. Subsequently, it examines the intertwined destinies of their lives with the trees, spotlighting the mutual dependencies. The emphasis on trees and forests is supported by their frequent mentions in the legal documents, where they feature prominently alongside other significant natural carbon sinks such as wetlands and oceans. Notably, the case file highlights the forests in two detailed sections addressing forestry, bioenergy, and climate policies in Sweden (para 124–132 and 161–167). The terms 'forest' and 'trees' are mentioned 67 and 10 times, respectively, in the case file, whereas 'ocean' appears 19 times and 'wetlands' 17 times, indicating the centrality of forests in the legal arguments. 'Climate' is cited 285 times and 'rights' 95 times, with 'youth' mentioned 23 times and 'age' eight times, predominantly in reference to the young plaintiffs (mentioned seven times) and once concerning the age of trees. These temporal and generational considerations evoke a sense of shared destiny, which I argue reflects posthuman intersections and linkages. To broaden access to the source material, in the ensuing discussion, quotations are drawn from the English translation provided by Aurora (2023), but paragraph numbers correspond to those in the official Swedish case application.

Legal claims

The young activists are urging the court to acknowledge that Sweden's approach to climate policy infringes upon their rights protected by the European Convention on Human Rights. Their chief demand is for the court to formally recognise that Sweden's efforts – or lack thereof – to reduce greenhouse gas emissions in the atmosphere fall short of the country's equitable contribution to the global initiative aimed at curbing the increase in the world's average temperature to no more than 1.5°C above the levels prior to industrialisation, as set forth in the Paris Agreement. This failure, according to the activists, contradicts the commitments made under the Paris Agreement, as well as obligations under the United Nations Framework Convention on Climate Change (UNFCCC), Swedish environmental quality objectives, national laws, and the Swedish Constitution, which all advocate for meeting the present generation's needs without compromising the ability of future generations to meet theirs. Therefore, they argue, by not implementing immediate, comprehensive, and effective strategies and actions, Sweden is not adequately mitigating the threat posed by human-induced climate change to young people. While the activists are seeking some form of compensation, they stress that it is a secondary goal. Should the court decide against treating the case as a declaratory judgment, the activists alternatively request that the court compel Sweden to fulfil its global responsibility in reducing atmospheric greenhouse gases. Potential measures could involve banning financial support or investment in fossil and biogenic fuels, such as halting infrastructure projects or investments in pension funds that contribute to fossil fuel dependency and energy-intensive activities (para 68).

That climate policy must adhere to current climate science is a demand that runs throughout the case file. Among the requests is for the state to investigate the extent of the Swedish fair share of the global measures to limit global warming with 1.5 degrees Celsius. Once these levels and responsibilities have been scientifically established, Sweden must accordingly take climate action. This entails not only reducing the currently reported greenhouse emissions but also addressing the overlooked aspects of Swedish climate policy: 'consumption-based greenhouse gas emissions abroad, greenhouse gas emissions caused by Swedish legal entities abroad and greenhouse gas emissions from LULUCF,[2] as well as by increasing the absorption of greenhouse gases by protecting and restoring ecosystems that constitute natural carbon sinks such as forests, wetlands and oceans' (para 68). Sweden's climate policy heavily relies on the forest, and I will highlight how the youths problematise this connection. But first, I focus on how the youths articulate themselves as a vulnerable and discriminated against group.

Youths facing disproportionate rights violation

The young activists claim that Sweden's current ruling generations are handling climate policies in a way that exposes the young and future generations to significant risks, potentially causing widespread harm to their lives, physical and mental health, dignity, well-being, and their homes and property (para 315). The young activists argue that the potential adverse effects of climate change infringe upon their fundamental human rights, including the right to life (Article 2), protection against inhuman and degrading treatment (Article 3), the right to private and family life (Article 8), the principle of non-discrimination (Article 14), and the right to property (Article 1 of the First Protocol to the ECHR). They highlight that, while the status of international law on these matters is a subject of debate – given that much of it constitutes 'soft law' and thus is not strictly legally binding – the principles of sustainable development and intergenerational justice, as emphasised by both international and national laws, should inform the interpretation of ECHR rights in the context of addressing climate change. For Sweden to fulfil its obligations according to the ECHR, the Swedish climate work must be 'designed in a way that does not risk exposing young and future generations to harm due to climate change in the future' (para 315). The youths highlight that:

> The right to life is not only actualized in situations where the state's action or failure to act has led to death, but also in situations where someone has been exposed to the risk of death, whereby the state has a positive obligation to protect life within the state's jurisdiction. The positive obligation includes situations where external factors pose a threat to life.
>
> *(para 240)*

The particularly vulnerable circumstances facing the youths are also described. They are at risk of experiencing the negative consequences of climate change more severely due to their young age, as they 'belong to the generation that is expected to be alive when climate change risks having more serious consequences in Sweden if sufficient and adequate measures are not taken' (p. 103). Sweden has not been spared from the effects of the climate crisis, having already experienced climate-related forest fires, severe drought, floods, and heatwaves. For instance, during the hot summer of 2018, about 700 more people died than the corresponding period of previous summers (para 81). Overall, the youths emphasise that they are 'particularly vulnerable to the climate crisis because they are at risk of suffering serious, climate-induced, physical and mental health consequences in the future' (para 96). They highlight that 'the long-term consequences of lack of climate action are even more serious than what is tangible today' and pose a specific risk to them, as 'they will probably be alive when the climate crisis risks escalating' (para 9). Citing IPCC reports, they note that from 2041–2100, the risks depend on the extent of global warming, which in turn is determined by the climate action taken or not taken in the near term (para 80).

The situation in Sweden is expected to worsen over the youths' lifetimes (para 82–96). Due to Sweden's geographical location, the average temperature increase is projected to be about twice the global average if current climate policies persist. Sweden could face more extreme weather conditions, including fewer frost days, more days of heavy precipitation, increased average precipitation, dry forest fires, floods, and longer heat waves. Prolonged heatwaves and floods would pose serious physical and mental health risks. Slower landscape changes also impact mental health. Shorter and milder winters are expected to increase tick-borne infections such as Lyme disease, anaplasmosis, herpes, babesiosis and TBE, as well as food-, rodent- and water-borne infections, with more cases of mosquito-borne infections and pollen allergies becoming increasingly common. Compared to adults, children and young people are more susceptible to climate anxiety, fear, and stress due to inadequate climate measures, which may lead to mental illnesses and affect physical health (para 83–96; p. 103). Thus, nature, manifesting in extreme weather, insects, and viruses, is here described as an immediate and growing threat, rather than an ally. Moreover, if the state delays implementing sufficient measures to mitigate the negative consequences of climate change, the youths 'risk having to shoulder an enormous burden, a burden that entails restrictions on their freedoms and rights, further down the line' (para 9).

These circumstances lead the youths to feel discriminated against. Article 14 ECHR mandates that freedoms and rights must be enjoyed without discrimination. The youths acknowledge that age is not among the explicitly listed grounds for discrimination but argue that 'the grounds for discrimination enumerated in the wording of the article (gender, race, skin colour, language, religion, etc.) are not exhaustive. Age has been considered to be a basis for discrimination' (para 252). They assert that:

> Discrimination exists if the state cannot provide objective and reasonable reasons why prima facie similar situations are treated differently. If someone who belongs to a certain age group or birth cohort, when enjoying their rights according to the ECHR, is affected to a disproportionate extent compared to other age groups or birth cohorts, it is thus a violation of Article 14 ECHR.
>
> *(para 253)*

While the younger generations are most at risk from climate change and face the most serious rights violations under the ECHR, they are also 'a social group that can be considered one of the most resource-poor. The imposition of extensive court costs can thus mean that the people most affected by the issues in the present case, including the plaintiff, are in practice prevented from having their claim heard'. Against this backdrop, the plaintiffs suggest that the state should primarily bear the legal costs for both parties, insisting that they must not bear the legal costs of the opposing party (para 401–402).

Intergenerational justice as the guiding principle

The principle of intergenerational justice orients the youths' claims. It means that 'both the quality and availability of natural resources, and the efforts to protect them, must be distributed fairly between present and future generations' (para 309). The principle originates in the Stockholm Declaration's first principle, and is clarified in the Rio Declaration's third principle, which states that the right to development must be fulfilled in a way that enables current and future generations' needs for development and a good environment to be met fairly (para 309). Highlighting this, the activists reference Sweden's "generational goal", which aims 'to hand over to the next generation a society where the major environmental problems are solved, without causing increased environmental and health problems outside Sweden's borders' (para 154).

As the below quote shows, the principle of intergenerational justice is oriented in an anthropocentric direction as these rights refer to the rights of humans *to* nature:

> The principle thus means that previous, contemporary and future generations have an equal right to the earth and its resources. Nature and its resources must therefore be preserved so that young and future generations can also enjoy them.
>
> *(para 309)*

The youths argue that 'Sweden's climate work must therefore be designed in such a way that the needs of future generations are not put at risk, and in such a way that too great a burden to reduce the anthropogenic climate impact is not placed on them' (para 313). By referring to the precautionary principle, the youths argue that even if there is scientific uncertainty about whether they will suffer adverse effects, 'this should not be used as a reason to delay effective and proportionate measures aimed to limit the potentially serious and irreversible damage caused by climate change' (para 219). They interpret the precautionary principle as requiring Sweden to shape its climate efforts in accordance with this principle, incorporating preventive measures sufficient to address the risk of the most severe and irreversible consequences of climate change becoming reality. They emphasise that the European Court of Human Rights has established in cases like Tătar v. Romania that states have an obligation to take reasonable and adequate measures to protect citizens' rights to private life, home, and health, highlighting the precautionary principle as a basis for extensive obligations. This includes the principle that scientific uncertainty should not be used as a reason to delay effective and proportionate actions aimed at preventing the risk of serious and irreversible environmental harm (para 215–223).

The young activists point out that Sweden's goal of net zero emissions by 2045 is not a viable solution to halt global warming. Instead, they call for the pursuit of negative emissions, where greenhouse gas absorption exceeds emissions. This approach underscores the need to reduce overall emissions and restore natural carbon sinks (para 120, 123, 195). The activists criticise Sweden's climate targets for excluding significant sources of emissions, such as those resulting from the land use, land-use change, and forestry (LULUCF) sector, with bioenergy emissions effectively counted as zero. Swedes' overseas consumption and from the country's international aviation and shipping activities are also not accounted for (para 161). Through their arguments, the youths emphasise the importance of forests and their role as natural carbon sinks, advocating for the preservation and restoration of forests as essential elements in the fight against climate change.

The forest as a climate ally

The Aurora case ventures into the complex debate around the best ways to manage forests for climate change mitigation and how to account for their carbon contributions. The roles of young and old forests are a point of contention among scientists. While young forests contribute significantly to carbon uptake, the long-term storage capabilities of old forests are indispensable in mitigating climate change (Pugh, 2020; Tausz & MacKenzie, 2017). The young plaintiffs take the latter view and stress the value of leaving forests undisturbed, arguing that felling forests transforms them from carbon absorbers to carbon emitters. They advocate for the preservation of natural forests, highlighting that the older and more biodiverse a forest is, the greater its capacity for carbon storage. For example, they argue that 'the longer a carbon sink is left untouched, the larger the carbon reservoir becomes'. Age is highlighted: 'The older a forest is and the richer its biological diversity, the greater the forest's carbon stock' (para 126). Legal arguments by the youths underscore the extensive clearcutting in Sweden, favouring the growth of young trees and diminishing the age and biodiversity of forests, thus reducing their carbon stock.

> When a forest is harvested for regeneration, the forest land is converted from a carbon sink to a carbon source. In other words, the large carbon stock in the ground, which previously grew larger and larger over time, instead begins to decrease. After regeneration felling, the forest land continues to be a carbon source for decades, even after replanting. The large amount of carbon, which has been stored in trees and soil, which is removed during regeneration felling of natural forest is not sequestered by the replanted trees until after decades to centuries.
>
> *(para 127)*

Here, the young activists align themselves with environmental values that champion the forest's role in storing carbon.

Furthermore, the youths challenge the categorisation of bioenergy as renewable, pointing out the lengthy payback periods for carbon reabsorption by forests used for bioenergy, which contradicts urgent climate action needs. The climate impact of the LULUCF sector is expressed as a net between carbon sink and carbon source, which means that emissions from harvested products, such as the burning of bioenergy and biofuels, are counted as zero. The youths take issue with the time perspective of this assumption.

> Biofuels are often classified as renewable because the carbon released during combustion is taken up by living biomass. The time this takes, the so-called payback period, varies depending on the type of biofuel and the fuel it replaces. For forest biofuel, the payback time varies from decades to centuries, which means that burning forest biofuel leads to increased emissions over tens to hundreds of years. Classifying biofuels with payback periods longer than a few years as renewable is thus not compatible with the clear message of science that the climate crisis requires drastic reductions in the atmosphere's greenhouse gas concentration before 2030.
>
> *(para 129)*

The youths point out that a large portion of the biomass from Swedish forestry is used for energy production and making products that do not last long, which quickly releases carbon back into the atmosphere. They find it concerning that only about 20% of the biomass is turned into long-lived wood products like furniture or construction materials, which keep carbon trapped and out of the atmosphere for many years. Most of the wood is instead burned for energy or made into items that have a short lifespan and burn quickly (para 165). They also mention that the benefit of using wood to replace other materials that are bad for the climate might not last. This is because newer, better ways to replace these harmful materials might be developed (para 130). So, they argue that the positive impact of using wood in place of other materials might decrease over time. Therefore, they believe that the best way to fight climate change is not by cutting down forests to use the wood in these ways but by keeping the forests intact and allowing them to grow. This way, forests can continue to absorb and store carbon, which is more beneficial for the climate in the long run.

> If natural forest is preserved, the entire carbon stock of the forest is preserved, in both trees and soil, while the forest may continue to be a carbon sink that binds and stores carbon in the soil every year. If natural forests are restored, for example by allowing production forests to grow into natural forests, the carbon sink is maintained and the total carbon stock increases.
>
> *(para 132)*

The legal arguments presented are framed against the backdrop of Sweden's intensive clear-cut forestry practices: nearly all unprotected forests are being transformed into commercial forests, favouring the growth of young trees.

> This transformation of natural forest into production forest has led to a sharp reduction in both the age of the trees and the biological diversity in Swedish forests, factors that reduce the forest's carbon stock.
>
> *(para 164)*

This trend reflects a contentious aspect of Swedish forest policy, which has been rigorously debated by environmental advocates. They criticise a prevailing view that younger trees, due to their rapid growth and perceived efficiency in carbon absorption, are more advantageous for climate mitigation than their older counterparts. This perspective suggests a moral valuation of trees based on their productivity and contribution to combating climate change (Palmer, 2021). The youth, aligning with environmentalists, contest this production-focused approach. They argue for preserving and restoring natural forests, positing that such actions doubly benefit climate efforts by both reducing emissions and enhancing carbon absorption. They support their stance with studies suggesting that the environmental advantages of using biomass as a substitute for fossil fuels and other materials are contingent on preserving forest ecosystems (para 167). Moreover, the youths invoke Sweden's obligation to maintain and enrich its natural habitats as a legal basis for advocating broader forest protection, referencing international and national environmental commitments like the Convention on Biological Diversity and Sweden's own environmental goals (para 305). Through these arguments, they make a compelling case for reevaluating forestry practices and prioritising forest preservation as a cornerstone of Sweden's climate strategy.

Posthuman solidarities

In analysing the Swedish Aurora case, I illustrated the symbolical and material connections between young individuals and specific nonhumans entities (forests), coining the term 'young people and old trees' to emphasise the shared experiences of discrimination and vulnerability under Sweden's current climate policies. This relationship did not involve the youths speaking on behalf of the trees. Instead, it reflected a guardianship model, in which the youths advocated for the continuous existence of the forests. However, this implied that forests deserved protection when their preservation served the interests of the youths and provided ecosystem services (cf. Harvey & Vanderheiden, 2021, p. 41). This way of presenting the relationship highlights the limits of the posthuman connections depicted in the Aurora case. It reproduces the prevailing hierarchical discourse. This discourse often places human needs above those of the forests. In essence, it shows how even well-intentioned actions can perpetuate a view of forests as secondary to human objectives. Without full insight into the youths' reasoning, this approach might be strategic, especially within the context of Swedish law and legal culture. This differs notably from Colombia, for example, where judges have the authority to recognise forests as legal entities. The instrumental framing could be a way to navigate the specific legal landscapes and cultural attitudes in Sweden. However, the case's focus on forests as vital carbon sink and allies articulates a 'collective we', merging the destinies of young individuals with those of old forests in opposition to the short-termism of current forestry practices. It suggests a form of posthuman solidarity and climate alliance (Tschakert, 2022).

As we delve deeper into the implications of the Aurora case, it becomes evident that the legal challenge extends beyond mere policy critique to question the very foundations of posthuman relationships in the law courts and beyond. The stance on forests underscores not just a legal battle over climate policies but also symbolises the evolving nature of climate litigation, where the fight for justice transcends traditional human rights boundaries to encompass a broader, multispecies perspective. This evolution is underpinned by the complex interdependencies that exist between human and nonhuman life forms.

By adopting Hovorka's (2012) perspective, we can interpret the dominant relationships as forms of othering that not only marginalise but also jeopardise the very existence of both young people and old-growth forests. It is through this posthuman intersectional lens that we begin to appreciate the full spectrum of implications stemming from our environmental policies and practices. Through this lens, we are invited to explore the intricate web of alliances and divisions that animate the legal landscape, uncovering the posthuman connections that challenge and enrich our understanding of justice.

Conclusion: Posthuman climate justice

The importance of recognising the role of climate litigation in advancing climate justice is increasingly evident as climate lawsuits proliferate globally. In this chapter, I have argued for the necessity of expanding our perspective beyond solely human-centered or gender-just concerns to include justice for a broader 'collective we', encompassing both human and nonhuman entities. Climate litigation emerges as a domain where the contours of legal advocacy stretch into the realm of the posthuman (Grear, et al., 2021). While human concerns remain central, the Aurora case illustrates a broader call to action, advocating for a justice that acknowledges the shared fate of humans and nonhumans alike. The interdependence of life forms becomes starkly apparent in the face of environmental degradation, where the loss of forests signifies not just a loss of biodiversity but a severing of the fundamental connections that sustain life on our planet (Cochrane et al., 2020).

In sum, the Aurora case not only challenges climate policies and forestry models but also invites us to reconsider the principles guiding our interaction with the natural world. It is a call to broaden our legal and moral imagination to include the myriad forms of life with which we share this planet, advocating for a legal framework that truly reflects the interconnectedness of our world.

It prompts us to transcend anthropocentric and gender justice to embrace posthuman, more-than-human, or multispecies justice and recognise 'the simultaneity of identities and categories of difference and inequalities (race, class, gender, age, ability, species, and beings) and their interlocking in structures and processes of injustice and oppression' (Tschakert et al., 2021). A posthuman, more-than-human, or multispecies justice lens can address the oversight in climate justice scholarship and activism that often neglects non-human actors and the exclusion and violence inflicted upon marginalised populations (Tschakert, 2022). An explicit consideration of humans and nonhumans relations can help unpack posthuman solidarities, and prompt reflection on 'How, where, and with whom do we position ourselves in the world?' (Tschakert, 2022, p. 279).

To dismantle multiple inequalities through climate justice litigation is no quick fix. Climate litigation is a lengthy and expensive path, though its potential benefits are evidenced by some successful cases. Many cases, however, fail, hampered by legal frameworks ill-suited to issues like climate change or youth challenging states but a loss in court can still advance a cause by elevating its public and political standing. A concerning trend is that climate activists are judicially targeted, with their rights curtailed by being labelled as state enemies or terrorists (Glazebrook & Opoku, 2018). Feminist methodologies can illuminate these shifts in rights and how they impact the interpretations and realisations of justice. Climate cases may give youths increased visibility, status and value but, paraphrasing Hovorka (2012, p. 882), any empowerment remains bounded within patriarchal, capitalist and anthropocentric structures, 'rendering them consistently and relatively marginal to privileged social groups'.

Regardless of the Aurora outcome in the Swedish courts, the struggle for climate justice will persist. A next step could involve exploring how young people and old trees can be mobilised against the privileged pairing of, for example, 'old men and young trees' which represents one of the polarising debates in Swedish climate-forestry politics. Another step could involve exploring ways to advance intersectional climate justice. This exploration would go beyond merely considering human rights *to* forests and include acknowledging the intrinsic rights *of* forests themselves. Granting forests the right to live and flourish might entail allowing them to grow old naturally. Simultaneously, we could reinterpret human rights to access forests in a way that ensures such access is contingent upon the preservation of the forests' existence (Meriläinen & Lehtinen, 2022). Posthuman alliances strengthen as both humans and nonhumans gain rights, facilitating a more equitable approach to climate policy that acknowledges cross-species vulnerabilities and strives for justice for all beings.

Acknowledgements

This chapter is based on research supported by Formas projects 2019-02023 and 2022-02102

Notes

1 More information about the various climate lawsuits can be found at the Sabin Center for Climate Change Law (https://climatecasechart.com).
2 LULUCF is the abbreviation for EU's Land Use, Land-use Change and Forestry policy which covers emissions and removals of greenhouse gases resulting from forestry, agriculture and other land uses, including harvested wood products.

References

Amorim-Maia, A. T., Anguelovski, I., Chu, E., & Connolly, J. 2022. Intersectional climate justice: A conceptual pathway for bridging adaptation planning, transformative action, and social equity. *Urban Climate*, 41, 101053.

Athens, A. K. 2018. An Indivisible and Living Whole: Do We Value Nature Enough to Grant It Personhood? *Ecology Law Quarterly*, 45, 187–226.

Aurora. 2023. Let's sue the state! [Online]. https://auroramålet.se/en/. [Accessed April 5, 2024].

Beland Lindahl, K., Sténs, A., Sandström, C., Johansson, J., Lidskog, R., Ranius, T. & Roberge, J.-M. 2017. The Swedish forestry model: More of everything? *Forest Policy and Economics*, 77, 44–55.

Braverman, I. 2018. Law's Underdog: A Call for More-than-Human Legalities. *Annual Review of Law and Social Science*, 14, 127–144.

Burger, M., Wentz, J. & Metzger, D. J. 2022. Climate Science and Human Rights Using Attribution Science to Frame Government Mitigation and Adaptation Obligations. In: Rodríguez-Garavito, C. (ed.) *Litigating the climate emergency: how human rights, courts, and legal mobilization can bolster climate action*. Cambridge: Cambridge University Press.

Cochrane, A., Fishel, S., Reid, S. & Schlosberg, D. 2020. Justice Through a Multispecies Lens. *Contemporary Political Theory*, 19, 475–512.

Crenshaw, K. W. 1991. Mapping the margins: intersectionality, identity politics, and violence against women of color. *Stanford Law Review*, 43, 1299.

Deckha, M. 2008. Intersectionality and Posthumanist Vision of Equality. *Wisconsin Journal of Law, Gender and Society*, 23, 249–268.

Dejusticia. 2018. Climate Change and Future Generations Lawsuit in Colombia: Key Excerpts from the Supreme Court's Decision. [Online]. https://www.dejusticia.org/en/climate-change-and-future-generations-lawsuit-in-colombia-key-excerpts-from-the-supreme-courts-decision/. [Accessed April 5, 2024].

ECHR. 2024. Grand Chamber rulings in the climate change cases. *Grand Chamber News*. https://www.echr.coe.int/w/grand-chamber-rulings-in-the-climate-change-cases. [Accessed April 26, 2024].

Glazebrook, T. & Opoku, E. 2018. Defending the Defenders: Environmental Protectors, Climate Change and Human Rights. *Ethics and the Environment*, 23, 83–109.

Grear, A., Boulot, E. & Vargas-Roncancio, I. D. 2021. *Posthuman Legalities: New Materialism and Law Beyond the Human*, Cheltenham, Edward Elgar.

Harvey, M. & Vanderheiden, S. 2021. 'For the trees have no tongues': eco-feedback, speech, and the silencing of nature. In: Grear, A., Boulot, E. & Vargas-Roncancio, I. D. (eds.) *Posthuman Legalities: New Materialism and Law Beyond the Human*. Cheltenham: Edward Elgar.

Holmgren, S. & Arora-Jonsson, S. 2015. The Forest Kingdom – with what values for the world? Climate change and gender equality in a contested forest policy context. *Scandinavian Journal of Forest Research* 30, 235–245.

Holmgren, S., Giurca, A., Johansson, J., Kanarp, C. S., Stenius, T. & Fischer, K. 2022. Whose transformation is this? Unpacking the 'apparatus of capture' in Sweden's bioeconomy. *Environmental Innovation and Societal Transitions*, 42, 44–57.

Hovorka, A. J. 2012. Women/chickens vs. men/cattle: Insights on gender–species intersectionality. *Geoforum*, 43, 875–884.

Johansson, M., Johansson, K. & Andersson, E. 2018. #Metoo in the Swedish forest sector: Testimonies from harassed women on sexualised forms of male control. *Scandinavian Journal of Forest Research*, 33, 419–425.

Klimaseniorinnen. 2024. Klimaseniorinnen Schweiz: Climate action [Online]. https://en.klimaseniorinnen.ch. [Accessed April 5, 2024].

Magnusdottir, G. & Widengård, M. 2024. Sweden's conflicting green leadership in the European Union. *European Politics and Society*. 26, 116–133.

Mayer, B. 2021. Climate Change Mitigation as an Obligation Under Human Rights Treaties? *American Journal of International Law*, 115, 409–451.

Meriläinen, E. & Lehtinen, A. A. 2022. Re-articulating forest politics through "rights to forest" and "rights of forest". *Geoforum*, 133, 89–100.

Mikulewicz, M., Caretta, M. A., Sultana, F. & Crawford, J. W. N. 2023. Intersectionality & Climate Justice: A call for synergy in climate change scholarship. *Environmental Politics*, 32, 1275–1286.

Norman, J. 2022. *Posthuman Legal Subjectivity: Reimagining the Human in the Anthropocene*, Oxon and New York, Routledge.

Palmer, J. 2021. Putting Forests to Work? Enrolling Vegetal Labor in the Socioecological Fix of Bioenergy Resource Making. *Annals of the American Association of Geographers*, 111, 141–156.

Peel, J. & Osofsky, H. M. 2018. A Rights Turn in Climate Change Litigation? *Transnational Environmental Law*, 7, 37–67.

Pugh, T. 2020. Are young trees or old forests more important for slowing climate change? *The Conversation*. https://theconversation.com/are-young-trees-or-old-forests-more-important-for-slowing-climate-change-139813

Rodríguez-Garavito, C. 2022a. Litigating the Climate Emergency The Global Rise of Human Rights–Based Litigation for Climate Action. In: Rodríguez-Garavito, C. (ed.) *Litigating the climate emergency: how human rights, courts, and legal mobilization can bolster climate action*. Cambridge: Cambridge University Press.

Rodríguez-Garavito, C. (ed.) 2022b. *Litigating the climate emergency: how human rights, courts, and legal mobilization can bolster climate action*, Cambridge: Cambridge University Press.

Savaresi, A. & Auz, J. 2019. Climate Change Litigation and Human Rights: Pushing the Boundaries. *Climate Law*, 2019, 244–262.

Setzer, J. & Vanhala, L. C. 2019. Climate change litigation: A review of research on courts and litigants in climate governance. *Wiley interdisciplinary reviews: Climate change*, 10, 1–19.

Stone, C. 1972. Should trees have standing? Toward legal rights for natural objects. *Southern California Law Review*, 45, 450–501.

Sußner, P. 2023. Intersectionality in Climate Litigation: The Case of KlimaSeniorinnen v. *Switzerland at the ECtHR. VerfBlog* [Online]. https://verfassungsblog.de/intersectionality-in-climate-litigation/. [Accessed April 5, 2024].

Tausz, M. & Mackenzie, R. 2017. Using forests to manage carbon: a heated debate. *The Conversation*. [Online]. https://theconversation.com/using-forests-to-manage-carbon-a-heated-debate-81363. [Accessed April 5, 2024].

The Guardian. December 27, 2023. 'We have a responsibility': the older women suing Switzerland to demand climate action.

Tschakert, P. 2022. More-than-human solidarity and multispecies justice in the climate crisis. *Environmental Politics*, 31, 277–296.

Tschakert, P., Schlosberg, D., Celermajer, D., Rickards, L., Winter, C., Thaler, M., Stewart-Harawira, M. & Verlie, B. 2021. Multispecies justice: Climate-just futures with, for and beyond humans. *Wires Climate Change*, 12, 1–10.

United Nations Environment Programme 2023. *Global Climate Litigation Report: 2023 Status Review*. Nairobi.

Youatt, R. 2016. Personhood and the Rights of Nature: The New Subjects of Contemporary Earth Politics. *International Political Sociology*, 10, 39–54.

11

PARTICIPATORY ASSESSMENT WORKSHOPS AS A GUIDING TOOL TOWARDS JUST AND INCLUSIVE ENERGY STRATEGIES

Gunnhildur Lily Magnusdottir and Anders Melin

Introduction

Energy justice overlaps with climate justice and is central for just transitions. Justice issues have in recent years gained an increased recognition at different levels of climate governance (IPCC 2022). Increased awareness, however, does not automatically lead to radical changes in strategies, political prioritisations, or institutional practices at different levels of governance (Melin et al. 2022, Magnusdottir and Kronsell 2021, 2024). This is particularly evident when it comes to energy policy-making, which is a field historically framed as a natural science topic often situated in a hegemonic masculine context (Healy and Barry 2017, Van de Graaf and Sovacool 2020: 2, Susser et al. 2022; Hultman et al. 2021). In this chapter, we focus on energy politics in Sweden. Issues of energy production and consumption, especially the question whether Sweden should continue to produce nuclear power, have been among the most controversial in Swedish politics, at least for the last 50 years (Anshelm 2000). The current liberal conservative government has the ambitious goal that Sweden should reach net zero emissions in 2045 with the help of electrification of transport and industry. More sun and wind power, as well as nuclear power, should be produced to reach that goal (Regeringen 2023).

Previous research on the understanding and recognition of justice in climate policy-making has demonstrated an institutional inertia within European energy authorities. Established institutional practices and path-dependent norms, such as the prioritisation of technological and economic approaches, for example, have resulted in limited recognition of energy justice in key

DOI: 10.4324/9781003461005-20

documents (Singleton and Magnusdottir, 2021, 2025, Magnusdottir and Kronsell 2024). We have seen similar tendencies in Swedish policy-making; for example, in our previous study on Swedish parliamentarians' motions on energy production and consumption from 2010 to 2019, where justice issues played a marginal role (Melin et al. 2022). This is unfortunate as the effects of energy production and consumption differ between different groups in society, depending on class, location, and gender as well as affecting future generations. If justice issues are not fully recognised as part of the energy debate, existing inequalities are likely to be maintained, contributing to unjust energy transition, and even increased political or societal instability.

In this chapter we contribute empirically and methodologically to the energy justice debate by analysing, with the help of feminist literature, how participatory assessment workshops can be designed to heighten awareness and understanding of energy justice and how different voices in society can be included. More specifically, we outline the main results from four workshops on justice and energy scenarios. We organised the workshops for Swedish politicians, policy-makers, and interest organisations, including labour, environmental and human rights organisations, in 2021–2022. We discuss how participatory methods in the form of workshops can be valuable in energy policy-making and make suggestions about how they can be further developed. The workshops were a part of a research project about the work on energy scenarios in Sweden and if/how justice issues were acknowledged in policy-making. The participatory workshops were conceptually grounded in literature on energy justice. The energy justice scholarship is useful for highlighting injustices in regard to the uneven effects related both to production and consumption of energy but fails to fully: "…account for intersecting dimensions of power and inequality, such as gender, race, class, indigeneity, ethnicity, sexuality, ability status, colonial history, and caste, among other statuses within the world system" (Sovacool et al. 2019:23). We therefore turn to feminist scholarship on deliberative democracy as well as feminist institutionalism when we reflect on what could have been done differently in the workshops we organised within the framework of the research project and make suggestions about how participatory assessment workshops can be further developed for inclusive energy politics. We claim that feminist scholarship, with its focus on power relations within institutions and in policy-making, can help us understand: "…what keeps us stuck in unsustainable energy cultures" (Bell et al. 2020:1) and highlight the social dimensions of energy politics needed for any discussion on energy justice.

The main empirical material that was presented and assessed at the workshops were long-term energy scenarios, built on reports from the Swedish Energy Agency (Energimyndigheten, 2016, 2021). Long-term energy scenarios are important support for decision-making on their own and an integral part of climate governance depicting different futures: "…to improve decision-practices under uncertainty" (Baard 2021:1). Energy institutions are authoritative key actors in the development of energy scenarios, especially when it comes to techno-economic knowledge, such as possible future technological development and economic consequences on states and industries. This techno-economic knowledge is a necessary starting point for making scenarios of different futures, but it risks becoming a top-down process neglecting the importance of social aspects, including questions about just transition for different groups in society. Baard (2021:3) points out that scenarios are deemed most reliable when there is a consensus for continuing on the same path. Doing "business as usual", however, risks reproducing unjust scenarios and reducing energy transitions to a matter of technology instead of fully recognising the justice aspects of energy politics. Also, an increased environmental awareness might to some extent shape energy discourses in Europe but has hitherto not led to necessary behavioural changes and: "... public opposition against energy infrastructure projects is halting transition progress in Europe and across the world" and if such factors are not taken into an account it: "...risks producing mathematically elegant but politically irrelevant scenario results" (Susser et al. 2022:2). Accordingly, energy scenarios lacking a justice understanding might be unrealistic and misleading – expecting transition at a speed not deemed just or socially acceptable (Susser et al. 2022:1).

Following this introduction, we first outline energy justice scholarship, which formed the theoretical basis of the participatory assessment workshops we designed and organised in 2021–2022. Then we discuss the importance of participatory methods for energy policy-making and present participatory-deliberative literature, which guided us when designing the workshops. We also present the feminist literature on democracy as well as feminist institutionalism which are useful for our analysis. Finally, we discuss the main results of the workshops, and make suggestions, with the help of the feminist scholarship, on how to further develop participatory assessment methods as a decision support for just energy strategies.

Energy justice and feminist scholarship as methodological tools

In this section we start by broadly discussing energy justice literature, which both forms the theoretical basis of the participatory workshops and is also a methodological tool since it articulates; "...what is morally just or right" (Sovacool et al. 2019:25) and aims to remediate injustices (Jenkins et al. 2016). Thereafter we narrow our focus and discuss methods for participatory assessments and in particular the workshops we designed.

The energy justice literature outlines several ways to stipulate how justice should be considered and assessed. The notion that access to affordable energy is a prerequisite for well-being and economic growth is a common starting point in the literature (Sovacool and Dworking 2015, 2014). Sovacool and Dworkin propose the following eight principles of justice, which were presented to our workshop participants: 1) availability, 2) affordability, 3) due process, 4) information, 5) prudence, 6) intergenerational equity, 7) intragenerational equity, and 8) responsibility (Sovacool and Dworkin 2015, 2014). This framework has inspired other scholars of energy justice, e.g. Jenkins et al.'s three tenets of energy justice – distributional, recognition, and procedural justice – which we also presented to the workshop participants (Young 1990; Jenkins et al. 2016; Walker and Day 2012). Distributional justice concerns the distribution of benefits and costs of energy production and consumption between individuals and groups in society. It applies not only to the siting of energy production, but also to the affordable access to energy and therefore to the phenomenon of fuel poverty (Jenkins et al. 2016). Recognition justice focuses on who is affected, with specific concerns regarding disadvantaged or marginalised social groups. It pays attention to whether the different perspectives of such groups are taken seriously in energy policy-making. A lack of recognition can express itself as different forms of political or cultural domination, disrespect, and degradation (Jenkins et al. 2016). Finally, procedural justice is concerned with whether individuals and groups have sufficient and equal influence on decision-making processes about energy production and consumption, thus, how to correct injustices. The engagement of local communities, impartial and full information disclosure, as well as representation in institutions are crucial to achieve procedural justice (Jack-Scott 2019; Jenkins et al. 2016). In addition, corrective justice can be relevant in the context of energy scenarios, although this tenet was only briefly presented at the participatory workshops. Corrective justice applies to situations when a person wrongly interferes with the property or well-being of individuals, and should thus compensate the victims (Miller 2020).

Participatory assessment workshops: communicative rationality, reflective mapping, and deliberative democracy

Moving on, we outline the scientific debate about participatory methods for ethical assessment of technologies, including energy scenarios, which guided us when designing the workshops. Historically, the dominant view about decisions on technology, including energy, has been that they should be made by experts and then communicated to people without scientific or technological expertise where the "public" has been perceived merely as consumers or as narrowly self-interested people who object to changes that are negative for themselves (Cotton 2014: 7–8; Melin et al. 2023). However, based on a critique of this one-way communication, participatory-deliberative methods for assessing technologies have been developed which aim to include other actors than technical experts in technology development.

Habermas' theory of communicative rationality is one central foundation for participatory-deliberative methods. According to this theory, communicative processes are rational if they fulfil certain conditions (Habermas 1984, 1987). Moreover, Habermas argues that moral justification must have its basis in dialogues between concerned individuals (Habermas 1990: 63–65). Habermas assumes that moral agents, that is, people who can act morally and possess communicative rationality, are able to jointly arrive at a moral assessment of a situation in an open dialogue with others. He describes an ideal speaking situation as one in which:

1 All subjects with competence to speak and act are allowed to participate in a discourse.
2 Everyone is allowed to question every statement.
3 Anyone is allowed to introduce any claim into the discourse.
4 Everyone is allowed to express their attitudes, wishes, and needs.
5 No speaker may be prevented by internal or external coercion from exercising his (sic) rights as described in 1, 2, 3, and 4 (Habermas 1990:89).

According to Habermas, the goal of the ideal speech situation is a shared understanding as opposed to an objective universal truth. Moreover, he contends that a moral norm can only be justified if everyone affected by it can accept its consequences (Habermas 1990:93).

An important aspect of Habermas's model is the assumption that the communicative rationality of individuals will make it possible to achieve consensus. However, the ideal of consensus has been problematised within the later debate on deliberative democracy. Many contemporary advocates argue that the theory should be more attentive to pluralism, and therefore, the ideal of consensus should be restricted to areas of compatible interests and values (Bächtiger et al. 2018: 7–8). Habermas' theory is a point of departure for our participatory workshops, but our design is, however, more in line with Matthew Cotton's model (2014) for reflective ethical mapping, which in its turn is based on the concept of reflective equilibrium. This concept has become well known within moral philosophy through the writings of John Rawls, who assumes that individuals who must make a moral decision should both consider the general moral principles they endorse as well as their moral judgements about specific cases. The idea is that the individuals should adjust these two components in relation to each other so that they become compatible. The individuals can either choose to adjust their general moral principles or their specific moral judgements to reach equilibrium between them. The concept of reflective equilibrium attempts to reconcile a top-down and a bottom-up model of moral justification by considering both general principles and specific moral judgements without presupposing that the judgements must be derived from the principles or vice versa (Rawls 1971: 47–48).

The application of the model for reflective ethical mapping has the following four stages:

1 To identify relevant topics for discussion which are affected by different technologies.
2 To stimulate the participants to discuss the moral aspects of different technologies and to make moral judgements about them.
3 To relate these moral judgements to a set of ethical principles, which represent important theoretical traditions in normative ethics.
4 To identify alternatives for political decision-making that are adapted to specific cases and based on moral judgements and principles. To agree within the group on possible action alternatives (Cotton 2014:99–100; Melin et al. 2023, 2024).

Another important aspect for the design of the workshops is literature on deliberative democracy, which has emerged as a response to a critical debate on contemporary democratic institutions. The critics contend that there is a significant gap between politicians and the public, and therefore it can be considered problematic that politicians make decisions on behalf of the public. Democratic institutions do not fully recognise socially marginalised groups, nor people in other countries or future generations. Proponents of deliberative democracy argue that legitimate forms of power presuppose dialogue, which requires deliberative rather than strategic or instrumental rationality (Smith 2003:18–26, 53–57; Melin et al. 2023). They regard deliberation as a source of legitimacy rather than the predetermined will of citizens (Smith 2003). Within the debate on deliberative democracy, consensus as a goal of deliberation is questioned as it can hinder mutual dialogue and understanding (Smith 2003: 58–59; Melin et al. 2023:9).

Feminist deliberative democracy and institutionalism

Feminist scholarship, with its focus on relational power can contribute to the development of participatory workshops. Prügl states that feminist scholarship can outweigh the shortcomings of deliberative democracy scholarship, since the feminist literature and methodologies: "…provide additional attention to hierarchies and differences and append to the democratic demand of inclusiveness, a demand for reflexivity with regard to power relations" (Prügl 2016: 26). Inclusiveness is indeed a central value for feminist scholarship, which also points out some blind spots in the deliberative democracy literature. This includes highlighting that all knowledge is situated in contexts and political struggles, and some knowledge is deemed as rational and apolitical, while approaches using feminist lenses are often deemed as political and less legitimate (Magnusdottir and Kronsell 2021, 2024). Also, established majority views and approaches of powerful actors, such as the energy sector, may be considered impartial and rational views as they are majority views (Bell et al. 2020, Prügl 2016). This might, for example, result in gender-blind energy policies since the energy sector is globally a male-dominated technological industry.

Some feminist scholars also question the model of deliberative democracy, because it excludes the expressive and aesthetic conceptions of the political that can be regarded as distinctly feminist. Expressive conceptions of the political are inspired by the work of Gilligan, who claims that women's experiences of interconnection give them another perspective on morality. Gilligan suggests that we distinguish between a masculine view of ethics focused on justice and rights and a feminist view focused on care and responsibility (Gilligan 1982). The aesthetic conception of what counts as political has, instead, its source in the writings of Judith Butler, who criticises the idea of a specific feminine form of knowledge. Butler emphasises plural rather than dichotomous differences that have their source in a performative notion of identity. Women's identities are created and maintained through discourses (Butler 1993; Squires 2008:107–109). However, the link between the expressive, the aesthetic and gender is questioned by other feminist scholars. Squires, for example, argues that women are not inherently concerned with the particular instead of the universal, or with sentiments rather than rationality. To better understand how an ethic of justice creates hierarchies and marginalises certain social groups it is necessary to move away from simple identification of justice with masculinity and caring with femininity (Squires 2008:115).

Squires argues that the labelling of care ethics as feminine ignores cultural and social differences. Instead, care ethics can be regarded as an element of subordinated cultures regardless of gender. Squires concludes that we have reason for a higher valuation of care ethics without regarding it as exclusively feminine. Instead of asking ourselves why discourses exclude women by ignoring the aesthetic, the expressive, and the concrete, we should ask ourselves how care discourses gender subjects and thus makes them associated either with the expressive and aesthetic or the rational (Squires 2008:115–116; Collins 1990).

Squires (2008:116–117) contends that there are reasons for embracing deliberative democracy also from a feminist perspective. She takes as her point of departure Young's work *Justice and the Politics of Difference*, which tries to reconcile the procedural, aesthetic, and expressive dimensions of politics. Young criticises the Enlightenment vision of a public realm of politics that expresses a universal general will and thereby relegates the body and the particular to the private realm. She argues for a transformation of the meaning of the "public" to include also group differences and emotions (Young 1990). Although Young is positive about Habermas' discourse ethics because it recognises pluralism, she rejects his distinction of the public and the private as it excludes the affective and bodily aspects of human lives (Young 1990). However, Squires (2008:119–120) points out that Young still refers to norms of just deliberation to distinguish her conception of social groups from conceptions of interest groups. She argues that social groups should be given formal recognition in the state and thus be represented in the public, but her conception of social groups is based on social relations and these groups are therefore intersecting. Young proposes a vision of group representation that differs from identity politics as it is conducive to public debate instead of the attainment of group-interests. The groups are defined based on a certain view of justice, which produces criteria for which groups need representation. Thus, demands that are solely derived from self-interests can be distinguished from demands derived from justice. Because the different social groups engage in deliberation with one another, this generates a just political debate instead of simply expressions of interest (Young 1990).

Young is critical of Habermas' ideal of impartiality because it presupposes a view of the self as dispassionate that ignores the interests, emotions, and commitments that citizens have about political situations. However, at the same time Young defends a revised version of deliberative democracy that she denotes as "communicative democracy" in defence of her notion of group representation. She contends that social groups need to enter a dialogue with one another, because it encourages representatives to express their claims by referring to justice rather than self-interest and it promotes understanding between different groups. Moreover, it forces representatives to take other groups' interests into account (Squires 2008:120–122, Young 1990).

The argumentation by Young and Squires shows that there are reasons for endorsing a theory of deliberative democracy from a feminist viewpoint, because a dialogue between different groups in society can be considered an important criterion for legitimacy. The deliberative democracy scholarship shares this emphasis on dialogue but with less focus on understanding of differences and more focus on that individuals might be able to change their mind if the quality of the arguments is high. This was not the explicit aim of our workshops, and we side with Young and argue that deliberative democratic exercises can be considered a successful process of knowledge exchange and increased understanding of differences although different views are not merged or changed. However, it cannot be overlooked that knowledge exchange may take place in a context with asymmetrical power relations favouring groups or individuals based on education, class, and professional expertise but even age, gender, location, and ethnicity.

Feminist institutionalism is also useful for further development of the workshop methods. Feminist institutionalism, inspired by historical institutionalism, focuses both on formal, and informal institutions such as norms, rules, and practices of governing authorities and very importantly the people behind the strategies, such as civil servants, involved in the policy-making process (March and Olsen 1989, 1996; Mackay et al. 2010; Holmes 2020; Magnusdottir and Kronsell 2021, 2024). This is highly relevant because established institutional practices and norms in policy-making steer the political debate and prioritisations and shape knowledge production, thus what kind of knowledge is deemed relevant in energy politics. This has to do with institutional inertia, which makes institutions, both formal and informal, resistant to change and keen to reproduce existing processes, norms, knowledge, and culture (Munk af Rosenschöld et al. 2014, Curtin 2019, Minto and Mergaert, 2018, Waylen 2014). This can lead to gender-blind strategies, limit the scope for ethical reflections on energy and might also mean that dominant economic and technical knowledge is deemed most appropriate in the work on energy scenarios. This can also lead to stickiness in policy-making, meaning that established norms and practices prevail and will be reproduced without a full understanding of their effects. Ecological modernisation is a specific example of a sticky norm or approach relevant for the workshops. Swedish environmental politics have for decades been framed in the context of "ecological modernisation" (Magnusdottir and Kronsell 2021; Hysing and Olsson 2018), where economic growth and the environment are seen to be compatible building blocks of the welfare state. This stickiness makes it difficult to change institutional practices and norms and is a part of the institutional inertia of "doing business as usual". This also means that actors involved in energy policy-making follow embedded norms and practices, according to what is appropriate within their organisation and in accordance with their professional identity (Arora-Jonsson and Sijapati 2018). Related to this discussion about the logic of appropriateness and institutional stickiness is the core concept of path dependency, which also contributes to institutional inertia. Path dependency keeps institutions and policy-makers on a certain path marked by previous decisions and thereby constrains opportunities for innovation and change (Miller 2020, Ljungholm 2017, Lowndes 2020).

We now move on to discussing the four workshops we designed and identify ways forward with the help of the feminist literature.

The design of the workshops

The empirical point of departure for this chapter is the four workshops on the work on Swedish energy scenarios and justice, held in 2021–2022. Two of them concerned national energy scenarios and were held in Stockholm, while two regional energy scenarios for Skåne (the southernmost region of Sweden) were presented in workshops held in the city of Malmö in the south of Sweden (Melin et al., 2023, 2024). The participants were representatives of governmental and regional authorities, environmental organisations, organisations for different energy sources, trade unions, and political parties.

We had also invited political youth organisations, employer's organisations, and a Sami[1] organisation. At the start, some of these were interested, but were for different reasons unable to participate, which we discuss further in our suggestions about future research and design of workshops.

Before designing our workshops, we did a trial workshop online with a group of students from the master's programme; "Leadership for Sustainability" at Malmö University. At that workshop the students expressed a lack of knowledge about energy issues and a wish for more background information, which would have required a longer time for the workshop and more preparations. This led us to invite participants that we considered to have some previous knowledge of energy-related issues or of social justice, but who could also represent different social groups and views, to the actual workshops (Melin et al. 2023, 2024).

We conducted the workshops in small groups of participants (5–7) to provide the participants with time in the discussions and to create an atmosphere of trust. At the first three workshops, there was a majority of men and at the last workshop a majority of women. Altogether the workshops were gender balanced with 14 men and 11 women participants. At the two first workshops, representatives of environmental organisations, trades unions, local authorities, and organisations for energy sources participated. We made the decision to have separate workshops with politicians, thus Workshops 3–4, as we deemed that there was a risk of imbalanced power relations between representatives of the political parties and the other participants who participated in Workshops 1–2, meaning that the politicians might dominate the discussions. We therefore planned to only invite representatives from the Swedish political parties to the third and fourth workshop. The third workshop was held with regional politicians from the region Skåne. For the fourth workshop, we initially only invited representatives from political parties at the national level. The interest in participating was, however, low among the right-wing parties, which in the end did not participate. One possible reason is that "justice" is a concept that has hitherto mostly been connected with left-wing ideologies and the redistribution of wealth and income. We therefore invited representatives of trade unions and environmental organisations to the fourth workshop. We took notes during all the workshops, and recorded parts of them to analyse the argumentation of the participants in more detail. At the end of each workshop, we conducted an oral and written evaluation of how the participants experienced the workshops and what they had learned from them and one another (Melin et al., 2023, 2024).

One aim of the workshops was that the participants should exchange knowledge and reach a better understanding of the views of other participants who represented other societal interests. In line with Young's critical feminist approach on deliberative democracy, we did not aim for a consensus or merging of views because we considered it unrealistic and possibly a hindrance to an open dialogue. Rather, the workshops could offer the participants, in dialogue with one another, an opportunity to improve their understanding of how different conceptions of energy justice affect decisions on energy politics. This is crucial, as justice is largely an unrecognised dimension of energy politics (Melin et al. 2022). The final goal of the workshops was that the participants should be able to make a more well-supported choice of energy scenarios from a justice perspective through a presentation by the members of the research team of relevant justice theories and principles and their application in the energy field. This aim is especially important for both politicians and civil servants involved in the policy-making process. Our previous research has demonstrated that justice is difficult to implement and presenting justice principles as tools in policy-making is an important step (Singleton and Magnusdottir 2021).

The workshops were between four and six hours long, depending on the commitment of the participants (Melin et al. 2023, 2024). At the workshops, we presented different scenarios for the year 2035 with different levels of total energy consumption and different energy sources. One of the scenarios was characterised by degrowth and a low total energy consumption, and the others with continued economic growth and maintained high total energy consumption. After discussing the possible outcomes of the scenarios for different social groups, we asked the participants to make an initial assessment of the energy scenarios from a justice perspective. Thereafter, we presented more general justice principles as well as more specific principles of energy justice and asked the participants to reflect on their choice of scenarios, based on a critical discussion about the principles (Melin et al. 2023, 2024).

Feminist institutionalist scholarship would also point out that what is deemed "appropriate" here is based on established norms and practices in energy politics. This became clear in the discussion about the different scenarios where the norm of ecological modernisation emerged, that environmental solutions and economic growth should be compatible. Some participants in the first three workshops also raised concerns about the degrowth scenario, although several participants in the final workshop deemed it favourable. It is interesting to note that the final workshop that was most supportive of the degrowth scenario was the one where women/left-wing parties dominated. While no conclusions can be drawn from this, it is worth further exploration.

It should still be noted that the structure of the workshops functioned well to accomplish the goal that the invited participants should improve their knowledge about other participants' views on energy justice. Many of the participants stated in their evaluations that it was valuable to enter a dialogue with participants with other perspectives. A respondent mentioned that it was both frightening and fascinating to discover that other participants had different understandings of reality (Melin et al. 2023, 2024). This is interesting and perhaps demonstrates well the complex nature of energy justice. Many of the participants expressed in their evaluations their appreciation of the opportunity to learn more about justice principles and their significance for energy policy-making. One of the respondents described the connection between energy scenarios and justice as a new and interesting perspective (Melin et al. 2023, 2024).

The workshops gave rise to many interesting discussions about justice, both intergenerational justice and global justice. At one of the workshops in Malmö, one female participant said the following:

> Somewhere it still builds on that others are worse off for us to live like this. That somewhere we must respond, and it is not possible that everyone…if everyone would fly as much as Swedes do, it would break down. It would not work. And somewhere we want to withhold others from the right to the development they could have had. They could have afforded to fly as much as we do.
>
> *(translation by the authors, Melin et al. 2023)*

In addition, the aim that the participants should attain a better-founded ethical view was reached to a certain extent. When they justified their choice of energy scenarios at the end of the workshop, many of them employed principles of energy justice. However, these principles were used mostly to support their initial ethical assessments, and not so much to question them. Moreover, in the evaluations, only a few of the participants declared that they had changed their moral view of the scenarios during the workshop, which indicates that they were not motivated by the workshop to question their initial moral assessment. Maybe this purpose would have been attained to a greater extent if we would have had more time for a critical discussion of the justice principles and choice of scenarios, for example, if it would have been possible to arrange more workshops with the same participants (Melin et al. 2023, 2024).

The workshops still functioned well for the kind of participants we recruited, who were relatively well educated, vocal, and experienced in arguing for their views. The reasons for choosing such participants were partly pragmatic, because we concluded that it would have been more difficult to recruit members from disadvantaged groups, such as low-income workers or long-term unemployed persons, who would be motivated to spend a whole day on a workshop. Moreover, for representatives from low-income groups it would also have been more difficult to attend a workshop on a workday. In addition, selecting participants from disadvantaged groups would have made it necessary to include some sessions with more information on energy politics, resulting in even longer workshops. These limitations are worth further discussion and in the following section we reflect on our choices and make suggestions for further development of the workshops.

Suggestions for future workshop design and concluding reflections

The workshops we organised were successful in the sense that the participants considered them a valuable learning experience (Melin et al. 2024) and they also gave the research team an increased insight into how the participants understand justice in energy policy-making. The design of the workshops with a clear focus on rational argumentation functioned well for the participants we recruited. The choice of participants can, however, be questioned from a feminist perspective, although the workshops were gender balanced, as we did not include members of disadvantaged groups. Therefore, as a suggestion for future research, it would be valuable to invite members of such groups not merely to make future workshops more inclusive but also to deepen the participants' understanding of the fact that all knowledge is situated in particular contexts and political struggles (Prügl 2016).

The workshops we organised functioned well since the power relations within each workshop were fairly equal – the participants all possessed some epistemic power, thus they had both knowledge of energy policy-making as well as training and experience in expressing their views in debates. For future workshops with more diverse groups of participants we therefore need to be more attentive to experiences based on other aspects than traditional knowledge founded on rationality. Feminist scholars have indeed criticised the masculinity of rational deliberation on which scholarship of deliberative democracy builds. We suggest paying more attention to views built on emotions and traditions, aspects that have hitherto been deemed less valuable in the process than those built on economic rationality. Also, in line with feminist institutionalism we think it is important to discuss established norms and institutional practices with workshop participants, since techno-economic norms are sticky in energy policy-making and might not only shape the initial positions and power of participants but also reproduce established knowledge and prioritisations in new settings. It would therefore be beneficial to introduce the core feminist institutionalist concepts of path dependency and sticky norms and practices in energy policy-making to workshop participants early in the workshop process. This would give participants tools to reflect on how possible path-dependent institutional practices, e.g. reproduction of narrow techno-economic and gender blindness in policy-making, might affect the inclusion and understanding of justice in energy policy-making. In the future it would also be interesting to conduct several workshops with the same participants as it would give them time to reflect between the workshops and possibly learn more about energy justice. However, the needs of different groups of participants should be met (e.g. constraints due to fixed working hours and the location of the workshops). These constraints could to some extent be addressed by organising the workshops in the evening, online and/or at locations close to participants that have limited flexibility or means to participate.

Finally, other material than written material and oral presentations by the research team could be useful for a more diverse group of participants, for example, those that lack prior knowledge of energy justice. Images and photos that give more room for emotional reactions to different dilemmas, could be used (Cotton 2014). The oral and written evaluations or feedback from participants which we used at the workshops, were valuable and could be used to a greater extent in future workshops. We propose starting such workshops with questions based on the participants' situated knowledge, background, and feelings in order to make the workshops less focused on technicality and economy and more attached to the everyday life of the participants. This would hopefully enable participants with limited epistemic power to share their experiences and highlight their contribution to energy justice.

Funding

The research was supported by the Swedish Energy Agency (Project No 48463-1).

Note

1 Samis are an Indigenous group in the north of Sweden.

References

Anshelm, J. (2000). *Mellan frälsning och domedag: Om kärnkraftens politiska idéhistoria i Sverige 1945–1999*. Symposium.

Arora-Jonsson, S., & Sijapati, B. B. (2018). "Disciplining gender in environmental organizations: The texts and practices of gender mainstreaming". *Gender, Work & Organization, 25*(3), p. 309–325.

Baard, P. (2021). "Knowledge, participation, and the future: Epistemic quality in energy scenario construction in *Energy Research and Social Science*, 75, p. 102019..

Bächtiger, A., Dryzek, J.S., Mansbridge, J., & Warren, M.E. (2018). Deliberative Democracy: An Introduction. In A. Bächtiger, J.S. Dryzek, J. Mansbridge & M.E. Warren (Eds.), *The Oxford Handbook of Deliberative Democracy* (pp. 1–31). Oxford University Press.

Bell, S.E., Daggett, C. and Labuski, C., (2020). "Toward feminist energy systems: Why adding women and solar panels is not enough". *Energy research & social science,* 68, p. 101557.

Butler, J. (1993). *Bodies that Matter: On the Discursive Limits of "Sex"*, New York: Routledge.

Collins, P.H. (1990). *Black Feminist Thought: Knowledge, Consciousness, and the Politics of Empowerment*. Boston: Unwin Hyman.

Cotton, M. (2014). *Ethics and Technology Assessment: A Participatory Approach*. Springer: Heidelberg.

Curtin, J. (2019). Feminist Innovations and New Institutionalism. In *Gender Innovation in Political Science* (pp. 115–133). Palgrave Macmillan.

Energimyndigheten (2016). *Fyra framtider: Energisystemet efter 2020* (ET 2016:04). Eskilstuna.

Energimyndigheten (2021). *Scenarier över Sveriges energisystem 2020* (ET 2021:6). Eskilstuna.

Gilligan, C. (1982). *In a Different Voice: Psychological Theory and Women's Development*. Cambridge, MA: Harvard University Press.

Habermas, J. (1984). *The Theory of Communicative Action, Vol. 2: Lifeworld and System: A Critique of Functional Reason*. Beacon: Boston (översättning McCarthy, T.).

Habermas. J. (1987). *The Philosophical Discourse of Modernity*. Cambridge: Polity Press.

Habermas, J. (1990). *Moral Consciousness and Communicative Action*. Cambridge: MIT Press (översättning Lenhart, C., Weber Nicholson, S.).

Healy, N. and Barry, J., (2017). Politicizing energy justice and energy system transitions: Fossil fuel divestment and a "just transition". *Energy Policy*, 108, pp.451–459.

Holmes, G. (2020). "Feminist institutionalism". *United Nations peace operations and International Relations theory*. Manchester. Manchester University Press.

Hultman, M., Kall, A. S., and Anshelm, J. (2021). *Att ställa frågan-att våga omställning: Birgitta Hambraeus och Birgitta Dahl i den svenska energi-och miljöpolitiken 1971–1991*. Arkiv Publication.

Hysing, E. and Olsson, J. (2018). *Green inside activism for sustainable development: political agency and institutional change*. Cham: Palgrave Macmillan.

IPCC Sixth Assement Report 2022, accessed in October 2022 at: https://www.ipcc.ch/assessment-report/ar6/

Jack-Scott, E. (2019). "Energy justice: a complex but vital piece to a clean energy transition". *Research Review*. Aspen Global Change Institute.

Jenkins, K., McCauley, D., Heffron, R., Stephan, H. & Rehner, R. (2016). "Energy justice: A conceptual review". *Energy Research & Social Science*. 11, 174–182.

Ljungholm, D. P. (2017). "Feminist institutionalism revisited: The gendered features of the norms, rules and routines operating within institutions". *Journal of Research in Gender Studies*, 1(7), pp. 248–254.

Lowndes, V. (2020). "How are Political Institutions Gendered?". *Political Studies*,68(3), pp. 543–564.

Mackay, F., Kenny, M., and Chappell, L. (2010). "New Institutionalism Through a Gender Lens. Towards a Feminist Institutionalism?". *International Political Science Review*, 31(5), pp. 573–588.

Magnusdottir, G.L. and Kronsell, A. (eds.) (2021). *Gender, Intersectionality and Climate Institutions in Industrialised States*. London and New York. Routledge.

Magnusdottir, G.L. and Kronsell, A. (2024). Climate institutions matter: The challenges of making gender-sensitive and inclusive climate policies. *Cooperation and Conflict*, 59(3), 361–378.

March, J.G. and Olsen, J.P. (1989). *Rediscovering Institutions: The organizational basis of politics*. New York: The Free Press.

March, J. and Olsen, J. (1996). "Institutional Perspectives on Political Institutions". *Governance*, 9(3), pp. 247–264.

Melin, A., Magnusdottir, G.L., Baard, P. (2022). Energy Politics and Justice: An Ecofeminist Ethical Analysis of the Swedish Parliamentarian Debate. *Ethics, Policy & Environment*. DOI:10.1080/21550085.2022.2115752

Melin, A., Magnusdottir, G.L., Baard, P. (2023). *Deltagande rättvisebedömningar av energiscenarier: Teori och empiriska resultat*. Malmö: Malmö universitet.

Melin, A., Magnusdottir, G.L., Baard, P. (2024). Participatory-Deliberative Ethics Assessments of Energy Scenarios: What Can They Achieve and How Should They be Designed?. *Ethics, Policy & Environment.* doi:10.1080/21550085.2024.2409025

Miller, C. (2020). "Parliamentary ethnography and feminist institutionalism: gendering institutions – but how?" *European Journal of Politics and Gender.*

Minto, R., & Mergaert, L. (2018). "Gender mainstreaming and evaluation in the EU: comparative perspectives from feminist institutionalism". *International Feminist Journal of Politics,* 20(2), 204–220.

Munck af Rosnechöld, J., Rozema, J.G. and Frye-Levine, L.A. (2014). Institutional interia and climate change: a review of the new institutionalist literature, *Wiley Interdisciplinary Rewievs: Climate Change,* 5(5), 639–648.

Prügl, Elizabeth (2016). "How do Wield Feminist Power". In Bustelo, M., Ferguson, L., & Forest, M. (Eds.). *The politics of feminist knowledge transfer: Gender training and gender expertise,* pp. 25–42. New York. Palgrave Macmillan.

Rawls (1971). *A Theory of Justice.* Cambridge, MA: Belknap Press.

Regeringen (2023). Ny klimatpolitik för att nå hela vägen till nettonollutsläpp, https://www.regeringen.se/pressmeddelanden/2023/12/regeringens-klimathandlingsplan--hela-vagen-till-nettonoll (published 202608).

Singleton, B. and Magnusdottir, G.L. (2021). "Take a ride into the danger zone?: Assessing path dependency and the possibilities for instituting change at two Swedish government agencies" in Magnusdottir, Gunnhildur Lily and Kronsell, Annica (eds.) (2021). *Gender, Intersectionality and Climate Institutions in Industrialised States.* London and New York. Routledge.

Singleton, B. and Magnusdottir, G. L. (2025). Encouraging institutional change with cultural theory. *Innovation: The European Journal of Social Science Research,* pp. 1–22.

Smith, G. (2003). *Deliberative democracy and the environment.* Routledge.

Sovacool, B.K., Dworkin, M.H. (2014). *Global Energy Justice: Problems, Principles, and Practices.* Cambridge: Cambridge University Press.

Sovacool, B.K. and Dworkin, M.H., (2015). "Energy justice: Conceptual insights and practical applications". *Applied Energy,* 142, pp.435–444.

Sovacool, B.K., Lipson, M.M. and Chard, R., (2019). "Temporality, vulnerability, and energy justice in household low carbon innovations". *Energy Policy,* 128, pp. 495–504.

Squires, J. (2008). Deliberation, Domination and Decision-making. *Theoria: A Journal of Social and Political Theory,* 117, 104–133.

Susser, D., Martin, N., Stavrakas, V., Gaschnig, H., Talens-Peiró, L., Flamos, A., Madrid-López, C. and Lilliestam, J. (2022). "Why energy models should integrate social and environmental factors: Assessing user needs, omission impacts, and real-world accuracy in the European Union" *Energy Research and Social Science,* 92.

Van de Graaf, T., & Sovacool, B. K., (2020). *Global Energy Politics.* Cambridge: Polity Press

Walker, G., and Day, R. (2012). "Fuel poverty as injustice: integrating distribution, recognition and procedure in the struggle for affordable warmth". *Energy Policy* 49, 69–75.

Waylen, G. (2014). "Informal institutions, institutional change, and gender equality'. *Political Research Quarterly,* 67(1), pp. 212–223.

Young, I.M. (1990). *Justice and the Politics of Difference.* Princeton, NJ: Princeton University Press.

12

THEATRE AND STORIES THAT RECONNECT

Embodiment Practices That Ecologise Masculinities

Paul M. Pulé, Ilaria Olimpico and Uri Noy Meir

Addressing Gendered Ecological Conundrums with Care

Industrialised modernity relies on the desire to consume, driving human societies and the planet past multiple tipping points resulting in great social and environmental upheaval (Crutzen and Stoermer 2000; Lenton et al. 2019; IPCC 2023). The current epoch now widely known as the Anthropocene has called forth a pressing 'response-ability' to address this trend, which is fundamentally a product of the separation between humanity and wider Nature[1] that has resulted in the long, slow violences of industrialised extractivism (Merchant 1980; Harari 2015; Ferrando 2019). As the social and ecological impacts on communities and ecosystems worsen, the reactive fronts of nationalist, expansionist imperialism (particularly in Global Northern nations) are lauding strongman politics to preserve post-industrial levels of profit for a privileged few at great cost to many. This long tradition of valorising prestige, status, image, fame, power, and wealth is maintained by masculinist norms of aggression, competition, blaming and shaming, that simultaneously diminish and deny empathy, intimacy, self-acceptance, acceptance of the universal rights of all, equality, and the protection and preservation of human and other-than-human life (Monbiot 2024).

With these growing problems in mind, this chapter responds to the premise that global social and ecological crises are rooted in destructive masculinities norms. We argue that engaging people in environmental discussions and sustainability policymaking requires a shift beyond current hyper-masculinist discourses, which commonly emphasise domination, competition, and control. To foster effective participatory approaches, it is crucial to reimagine

DOI: 10.4324/9781003461005-21

masculinities in ways that embrace ecological (aka. relational and caring) interdependence. Further, to effectively address the social and ecological crises that are upon us requires not only our intellectual abilities but also our willingness and ability to sense and feel our ways forward through these troubled times.

This raises the question of how to inform people and cultivate awareness about the importance and value of returning to caring and inclusive sensibilities, and how to do so in felt ways. What we mean by this is that beyond the rational/logical mind are channels of knowledge and wisdom that come from the intelligence of the body. The environmental, economic, political, social, cultural and spiritual divides that are now being exacerbated compel us to seek out and apply paradigmatic changes to create new cultures and consciousnesses. Social arts (particularly theatre) can provide pathways for transformation from a profit and power-over-driven society towards a life-sustaining culture of peace.

The chapter examines a social arts method known as *Theatre and Stories that Reconnect* (TStR), and its application at a workshop held in Milan, Italy, in June 2019. The workshop's central goal was to practice a transition from traditional forms of masculinities to the notion of ecologising masculinities to awaken relational and caring ontologies. Though experimental and not initially intended as a case study, the successes of the workshop encouraged us to publish our findings retrospectively.

TStR is a unique social arts practice that was originally developed by Ilaria Olimpico and Uri Noy Meir and draws from a series of practices that have been instrumental in shaping their respective practitioner trainings.[2] Here, we introduce TStR and illustrate how this social arts practice can bring transformational personal and structural constructs of greater care into the body, develop skills of deep listening,[3] inspire creativity, and encourage compassion; skills that are essential for ecologising masculinities that dwell in us all. TStR aims to provide a more systemic and holistic pathway that addresses personal and interpersonal complexities that are essential to understanding struggles of achieving ecological transformation through participatory theatre practices. TStR facilitates dialogues that transcend anthropocentric and dualistic frameworks; by integrating ecological metaphors, Systems Thinking, and embodied practices, TStR encourages participants to explore the intertwined dimensions of the Self, the interpersonal, and the ecological. This exploration contributes to the growing need for approaches that move beyond binary and hierarchical models, fostering transformative engagement in climate discourse and policymaking.

As the chapter proceeds, we suggest that TStR can facilitate profound transformations of individual and systemic constructs towards equitable and life-sustaining futures for all; qualities that can ecologise masculinities and are consistent with feminised climate policy.

The Story and Map of TStR as an Emerging Social Arts Practice

Social arts practices, especially those that engage theatre and performative approaches to community processes, can be an effective method of exposing social realities. These practices create spaces to reflect on lived experiences where we expand our understanding of our (inter)personal stories, our communities, and how we are inextricably connected to wider life processes. Consistent with this tradition, TStR aims to foreground the voices of Nature and otherised Others (see footnotes 1 and 4) through conversations about social and environmental justice by emphasising practices that awaken and develop relationality and wider care. Social arts are commonly associated with tangible forms of expression but can also include subtle forms of knowledge. In this sense, they are subjective. The body 'knows' in a different way to the rational mind, engaging sensory intelligence that anticipates and informs rational forms of knowing, revealing a sensory intelligence. A long-lasting personal and social transformation is possible when the body takes part in a process, enabling us to explore relationships beyond ideation or verbalisation. Through the body we tap into the processes within while also actively engaging with one's surroundings and concurrently listening to the mind. Social arts are particularly effective at deepening awareness and helping to facilitate personal and social transformation.

The combined emotional, visceral and cerebral practices that characterise TStR create place-based performative acts and moments of pause to collectively witness the Self in relational exchange with the wider world. In doing so, TStR allows us to more deeply connect with each other and the Earth. TStR calls attention to the stories that our bodies tell as the most authentic sources of knowing. The method emphasises visceral connection as the facilitator's primary tool compelling us to engage in the present moment in grounded and inclusive ways while also supporting an open/panoramic awareness of the whole space. Participants are encouraged to zoom in and out of the process at-hand through 'presencing' by being present with the sensations in the body and the internal stories that those sensations are connected to. This sensory and interconnected awareness represents a significant departure from masculinist approaches to facilitation that traditionally prioritise top-down intellectual knowledge, where so-called 'consultations' that engage 'professional facilitators' develop organisational plans in strategic and linear ways designed to enhance the aims of an organisation ahead of its impacts and accountabilities on people and planet. Accessing body knowledge is not straightforward in Westernised cultures and typically requires a process of relearning to become second nature through effective embodiment facilitation. The influence of masculinist socialisation on everybody, but particularly men, often obstructs this process. Arawana Hayashi (2021) notes that embodied presence involves integrating awareness of the Earth[4] body, the social body, and the individual body concurrently, cultivating a groundedness and connection to all three. Embodied presence is also rooted in the principles of Theory U (Scharmer, 2009), which emphasises learning and acting from the future while tending to the present as it is emerging in real time.

Consistent with these social arts offerings and based on their experience as social arts facilitators, Noy Meir and Olimpico have noted that when practising TStR, many participants experience another channel of knowing, arising from emotive and bodily awareness. This has informed their development of TStR that both acknowledges and can transition participants beyond 'stuck places' within, since embodied knowledge is taken to be both a human right and a responsibility that not only facilitates dialogue with all forms of life, but also allows individuals to engage with the uncertainty and the unease associated with not knowing. This approach helps to reveal the intuitive and creative aspects of our inner landscapes. Consequently, embodied knowledge integrates conscious and unconscious elements, the emotional and rational, and the symbolic and cognitive aspects of the Self through iterative exchanges and actions. In other words, the body does not lie.

Pausing to attune oneself to the body, a process known as 'grounding' that can be employed at the beginning, throughout, and/or at the end of a process, can foster a sense of inner calm and facilitate deeper engagement in transformational exercises. This practice is helpful as it allows participants to attain embodied knowledge that extends beyond cognitive and verbal interactions. Although body knowledge may have limitations in external communication, it is particularly useful in promoting self-discovery by revealing previously unrecognised aspects of the individual. Body knowledge brings authenticity to processes where conceptual or verbal communication may fall short. As Augusto Boal (1979) highlighted, theatre practices enable participants to use their bodies to express their realities, aligning with Paulo Freire's (2014) perspective of education as an act of freedom. Thus, body knowledge can serve as a powerful tool for transformative processes.

TStR fosters creativity and celebrates life's cycles. Facilitators and participants embrace uncertainty, integrating these experiences into the process to avoid over-rationalisation. TStR encourages the creation of maps[5] – subjective physical stories that represents inner experiential landscapes that reveal implicit (present in the Self 'in here') and explicit (present in the world 'out there') realities that enhance authenticity and clarity by guiding a process as it unfolds [see Figure 12.1 that illustrates the way that mapping is used in TStR].

Central to TStR maps are the concepts of *The Gift*, which connects individuals to their unique strengths; *The Wound*, which addresses personal and societal traumas through collective witnessing; *The Medicine*, which utilises

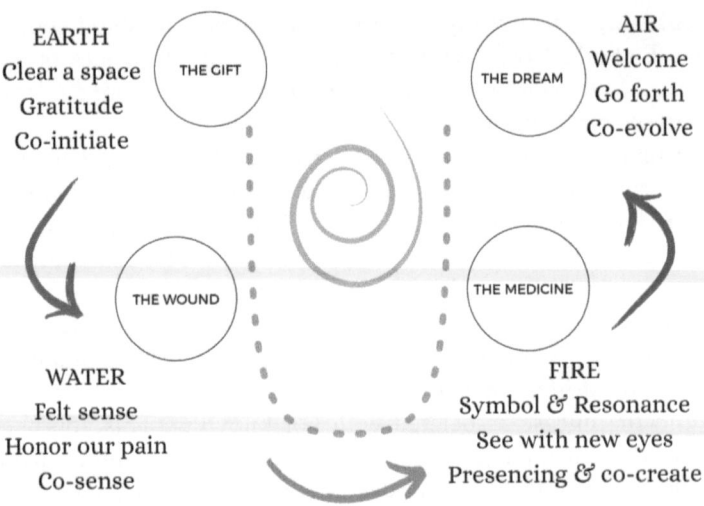

EARTH
Clear a space
Gratitude
Co-initiate

THE GIFT

AIR
Welcome
Go forth
Co-evolve

THE DREAM

THE WOUND

THE MEDICINE

WATER
Felt sense
Honor our pain
Co-sense

FIRE
Symbol & Resonance
See with new eyes
Presencing & co-create

FIGURE 12.1 Theatre and Stories that Reconnect maps.

creative play and Nature for healing; and *The Dream*, which envisions and manifests life-affirming cultural paradigms through authentic self-expression and collective action. TStR also employs the four archetypal elements of *Earth*, *Water*, *Fire*, and *Air* to guide transformative processes. *Earth* emphasises interconnectedness and gratitude, fostering a life-sustaining culture through practices like Work That Reconnects (WTR) and Theory U's ecoleadership (Macy and Johnstone 2012; Scharmer 2009). *Water* focuses on emotions and flow, embracing the felt sense[6] and honouring collective pain for healing. *Fire* symbolises transformation and courage, encouraging the unfolding of implicit meanings and fostering creativity amidst transitions. *Air* captures new awareness and movement, introducing fresh perspectives and continuous evolution. Gendlin (1978) used a metaphor of 'fresh air' as an element of relief that creates space for 'the new' to reveal itself and, from there, bring us to action.

TStR acknowledges the potential of stories to challenge the power structures that define and strengthen social narratives. To mitigate worsening social and ecological circumstances, we need to be willing to transition from 'power-over' to 'power-with' through life-sustaining cultures of peace, deep listening, creativity and empathy, where the person is recognised as a subjective story 'in process'. Stories from other social arts practices help us here. Of particular relevance, TStR takes guidance from the Spiral in the WTR as a map for facilitation (Macy and Johnstone, 2012; Macy and Brown, 2014). This Spiral encourages us to turn our attention to the inner work of cultivating courage to support outer transformational activism through four stages of the Great Turning:

- Coming from Gratitude
- Honouring our Pain for the Earth
- Seeing with New Eyes (beyond the despair of contemporary crises)
- Going Forth

This framework is mirrored in TStR, which aims to enrich Macy et al.'s Spiral through theatre, stories, and social arts practices that reconnect us to our caring selves, each other, and Nature. Like the WTR, TStR acknowledges the inseparability of personal, social, and ecological dimensions. TStR moves through a spiral-shaped map mirroring the WTR, beginning with *Earth,* where we express gratitude, clear space, and acknowledge our gifts. We then transition to *Water,* where we honour our pain for the world and our wounds, engaging in felt sensing and co-sensing of internal and external realities. With courage, we traverse *Fire,* discovering a remedy for distressing aspects of life and seeing with new eyes. Finally, we proceed to *Air,* breathing fresh air, dreaming, and co-evolving into 'Active Hope' for the future (Macy and Johnstone, 2012).

Through TStR, sharing trans-inter-cultural as well as personal stories with a focus on life-supporting skills and dreams helps remove labels and allow human beings to find and express their uniqueness. When we seek knowledge through our bodies, we create safe spaces between reality and the imagined.[7] In these ways we are better able to share our narratives as they are and as they can be, creating room for transformation and the emergence of new stories about the Self (see footnotes 1 and 4) and one's surroundings, effectively placing value on the co-arising of subjectivities. TStR therefore encourages a dynamic, inclusive, and nuanced – fuller – understanding of the world.

The deep listening required by this approach can free participants from the masculinist paradigm of problem-solving (and can instead bring our attention to what is causing the problem). In this sense deep listening complements by welcoming, receiving, pausing, and emptying out as we engage with the whole body. For example, when we strike a pose in pairs using our bodies (e.g. through a 'duet' discussed in the context of the Milan workshop below), we invite people to take a moment's pause in between movements, so they can deeply listen to themselves, their bodies and what comes to them from a posture and, in the case of the person witnessing, ensure that the person striking the pose can be truly heard [see Figure 12.2].

grounded   aware   present

Deep Listening

turns in equal time		the listener can be time keeper
storylistener		listens to in Presence, silently
storyteller		is responsible of her-his process takes care of her-himself
both		privacy and confidentiality

inspired by Focusing partnership

FIGURE 12.2 Diagram of Deep Listening as it is used in TStR. © Ilaria Olimpico and Uri Noy Meir.

We now turn to the Theatre of the Oppressed Festival workshop in Milan, Italy in June 2019 to consider one way that the social arts practice of TStR can be applied to ecologising masculinities.

A Workshop in Milan, Italy

As part of the 2019 Theatre of the Oppressed Festival in Milan, Noy Meir and Pulé (with the support of Olimpico) facilitated a workshop that aimed to expose the folly of destructive masculinities' norms and support participants to 'feel in their bodies' a transformational change; a process of becoming ecologised (in order to awaken the relational and caring self) by transitioning from industrial/breadwinner[8] and ecomodern[9] to ecological masculinities[10]; Hultman and Pulé, 2018). At this workshop, 32 participants (slightly more women than men and a small number of gender-diverse individuals) attended. Given the Milan workshop was held at the Theatre of the Oppressed Festival (organised by the Italian NGO *Casa per la Pace* in Milan), the participants represented a concentration of embodiment educators and activists who were – broadly speaking – sympathetic to community engagement through social artistry.

We did not originally design the workshop to collect case study data. Consequently, the observations that follow are anecdotal and the arguments are primarily (auto)ethnographic. They also draw from spoken participant reflections and written feedback during the workshop closing circle, with aims of improving the workshop for future use.

What We Did

A conceptual overview of three different embodiments of masculinities: Industrial/breadwinner masculinities, ecomodern masculinities and ecological masculinities (see footnote 13), participants were asked to pair up, ideally with someone they did not know, to engage in a duet exercise. Each of the three masculinities was called out and participants were asked to take 15–30 seconds to feel into the masculinity categorisation that was being called out and reflect on how its characteristics manifested in their own lives before they moved. We then instructed the participants to spontaneously move their bodies into a form reflective of the thoughts and feelings that arose. Their partner was asked to observe and note what they witnessed in both thought and feeling in response. The person striking the pose was asked to hold that pose for up to one minute, after which the pair relaxed and discussed what the person striking the pose felt while in that pose and what the observer observed. The pair then swapped roles. This shared posing and witnessing was repeated for all three masculinities. Once this iterative process was completed and everyone had embodied all three of the masculinities, the group was invited to reconvene in a closing circle in which insights, surprises, thoughts, and feelings were shared with the whole group. Participants were invited to provide written feedback on Post-it Notes if they had anonymous thoughts that they wanted to share.[11] After two hours the workshop was closed with the ritual of weaving threads between us with blue twine [see Figures 12.3–12.5].

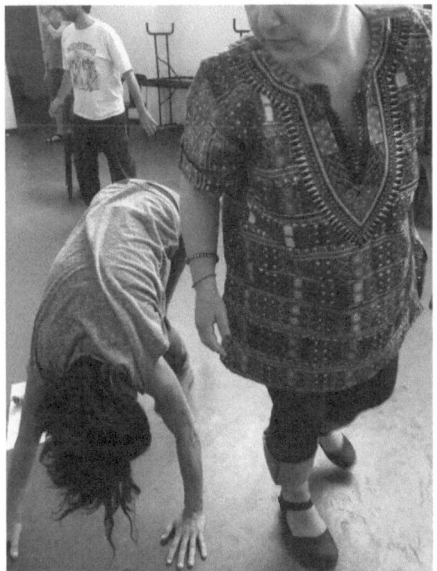

FIGURE 12.3   Image of Milan workshop participants striking embodied poses of industrial/breadwinner masculinities. Photographed by the authors.

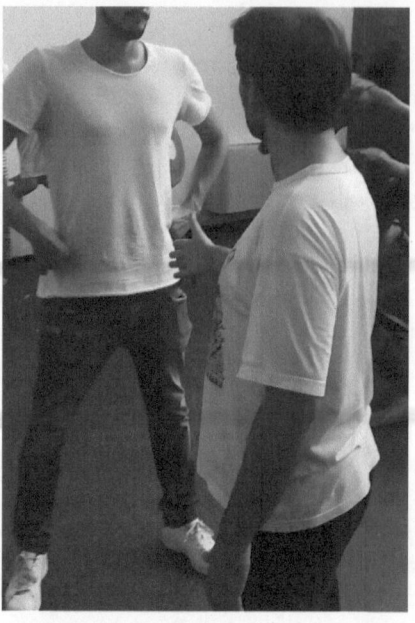

FIGURE 12.4   Image of Milan workshop participants striking embodied poses of ecomodern masculinities.Photographed by the authors.

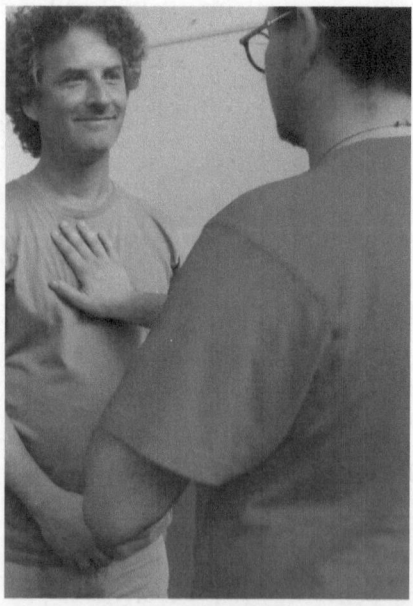

FIGURE 12.5   Image of Milan workshop participants striking embodied poses of ecological masculinities. Shared with permission from the visible participants.Photographed by the authors.

Overview of Participant (and Facilitator) Feedback

Innovations that support the conceptual development of ecological masculinities arose from the workshop. Participant feedback both during and at the end of the Milan workshop highlighted possible research and development of ecological masculinities' pedagogies going forward. From the verve and energy that participants expressed at the workshop's end, the facilitators recalled that some participants voiced desire for the work of awakening ecologisation through embodied practices to be taken forward to schools and organisations. It was evident that this style of experiential engagement could also be adapted to reach those who are the most challenging to reach who are 'intoxicated' by their own privilege, since embodiment practices can 'sneak up on' and expose destructive masculinities norms that dwell in us all. Using a subtle yet confrontational approach can engage the narcissism that commonly accompanies hegemonisation to unpick itself by exposing the costs of hegemonisation to Self and those in their immediate concern. Subtle confrontation also creates opportunities for participants to gain a felt sense of their relational and caring selves, from the personal and particulate through to the systemic and global, while minimising the risk of defensive push-back. In Milan, participants expressed an interest in exploring different types of femininities alongside the three masculinities explored. This pointed to the importance of the workshop expanding beyond deterministic conversations about masculinities alone to include gender-diverse explorations and experiments through the body as well. Conceptual developments are underway. For example, Pulé and Ourkiya (2023, pp. 213, 217) argued that:

> Masculinities and femininities are present in people regardless of how they self-identify. ... Similarly, identifying as genderqueer, two-spirited, or agender does not determine the absence nor the presence of either or both. Consequently, the ecologisation of feminisms *and* masculinities function as sub-processes within a broader, queering metanarrative ... [as] the most normative way to conceptualise gender identities... Viewed this way, ... we can ecologise all identities by passing through and reaching beyond them to arrive at the human species as 'in and of Earth'. ... To secure a truly sustainable future, we must rid ourselves of the life-destroying injustices of difference. Moving beyond traditional gender identities is integral to this forward-facing vision. ... [where] we move through, sink in, and allow ourselves to be part of Earth in its myriad intricacies; a perspective that is consistent with past and present Earth wisdoms, where a postgender 'we' surpasses an essentialised 'me' and with that, the fabricated lines of separation dissolve.

We contend that the Milan workshop is applicable to men only, women only, gender-diverse and mixed groups and can be scaled-up and adapted to different (and potentially more sceptical) audiences.[12] Given that cross-sectoral organisations and institutions, such as government agencies, not-for-profits and grassroots communities, are all important to reach with the message of ecologisation, the duet embodiment exercise discussed could be adapted to widen the reach of this work beyond sympathetic audiences, with particular applications in settings that harbour some of the most anti-ecological (and thereby counter-relational and uncaring) actors that rely most heavily on and embolden masculinist structures.

The workshop could also be adapted to become a 'Train the Trainers' experience that could focus on developing embodiment facilitation skills in social innovation spaces in support of gender equitable and ecologically just futures. Some added insights gleaned from participant feedback and facilitator reflections are:

- There is a need to make the links between pedagogical approaches to transforming masculinities and the important insights of feminism more explicit. The successes of bringing the concept of three different forms of masculinities down into the body at the Milan workshop indicate that there is scope for becoming intentional and strategic about finding the meeting places between this work, standing alongside feminist and postgender practices, to illuminate the intersecting terrain of gender-diverse realities.
- Given the strong focus on gender identity that pervaded the workshop, emphasising the similarities between pedagogies that embody ecological masculinities could strengthen the environmental aspect of ecologising masculinities.
- While the Milan workshop was a pilot and its development, adaptation, scaling up and repetition remain a 'work in progress', documenting outcomes by generating focused post-workshop evaluative data could inform future research and development of embodiment practices, such as the duet exercise discussed here, along with other social arts practices that could be explored to facilitate the ecologisation of masculinities.
- Cross-sector, in-person and online platforms, especially those that promote pro-feminist epistemologies throughout civil society, could broaden the reach and efficacy of ecological masculinities in helping to facilitate transformative change towards more sustainable futures for all of life.
- It was notable that some participants encountered intense emotional responses to the work (esp. discomfort, grief, anger, shame). This exposed the need for heightened sensitivity to the psychological impacts and thereby the wellbeing of participants in such a workshop. This is particularly relevant for those who come to such a workshop from contexts of overt social and environmental despair, and the burnout that commonly accompanies activist contexts.

- With the soft skills considerations noted above, it could also be important to prioritise medium- to longer-term sessions and support groups to ensure that participant psychological wellbeing is given consideration and factored into post-workshop follow-ups.
- It would be helpful to diversify beyond the embodied pedagogical approach studied here to include other social arts exercises that can be similarly adapted to the ecologisation of masculinities.

We suggest that there are future possibilities for ecologising masculinities through TStR.

Anecdotal Observations: What We as facilitators Saw, Thought, and How We Felt

Uri Noy Meir (Workshop Facilitator)

I arrived at the workshop with this unusual medical condition that temporarily left me with almost no voice. This physical obstacle gave me – as workshop facilitator – cause to take a step back and be more attentive to the energy in the room as the duet exercise unfolded. I was able to note subtle participant experiences with heightened acuity, noting the ways that the social body – as participants adopted various forms – expressed an array of deep and post-verbal nuances that could then arise. I realised that through my forced silence, I was better able to see how words (my own *and* those of the participants) can be an impediment; a distraction from truly listening to what is being said by a person's whole being as well as a barrier to deep connection amongst participants and between them and myself as facilitator. Facilitating the duet exercise without my voice (and with the aid of my interpreter Francesca Aloi), allowed me to *do* less and *be* more, along with the participants. I was able to confirm that embodied knowledge does indeed complement academic pursuits of knowledge when given a good measure of facilitated practice in a safe container, reminding me that intellectual knowing is not the whole story; that doing and being are additional ways of knowing that are of equal importance as we work our way from hegemonisation to ecologisation. My unusual condition allowed me to affirm that embodiment practices – like the duet – can create opportunities for us to heal from past traumas, connect with wider Nature and participate in rituals of letting go of life-destroying ontologies. I came away from the workshop filled with hope for the possibilities of social arts practices as essential pedagogies for bringing the conceptual framework of ecological masculinities down into the body where its practice and application to real-world situations could come to life.

Paul M. Pulé (Workshop Facilitator)

As a principal generator of the theory of ecological masculinities, with limited earlier experiences in the social arts, the Milan workshop was an embodiment revolution for me. It was remarkable to experience the level of engagement of the participants in a conceptual framework that I have, with others, developed (Hultman and Pulé, 2018; Pulé and Hultman, 2021). It was a surprising delight to see how easily participants could find expression for the three masculinities within their bodies. This affirmed for me that a cross-section of participants, with whom I had not previously exposed to the ecologisation of the three masculinities (but for the short introduction I gave as an 'on-ramp' into the duet exercise), could find and relate to each of the masculinities within themselves and with minimal prompting from me and the theories previously generated. As the participants progressed towards poses for ecological masculinities, the levels of openness, exchange, camaraderie, joy, and connection that reverberated throughout the room were particularly inspiring. These are qualities that participants were able to find expression for in some physical form with little effort. This affirmed for me that ecologising masculinities has exciting potential for cross-sectoral applications and does indeed dwell innately within most (if not all) of us. Further, these are qualities that I consider to be consistent with the central focus of this volume. In this sense, the duet exercise offered a pedagogical pathway that exposed, raised awareness and generated a forum for the application of feminist climate policy in a multitude of forms that could be uniquely awakened.

Conclusion

In this chapter we have argued that social arts embodiment practices, such as those employed through TStR, create opportunities for heightened body awareness (beyond the stories of the mind) that inform us about the personal and societal consequences of the three categories of masculinities. Through the practices featured here, participants experienced body knowledge of industrial/breadwinner, ecomodern, and ecological masculinities as they are present within us all. Those expressions were not simply found but were also felt in the body, giving participants an opportunity to note the visceral differences that each has on the Self intrinsically and, in holding a pose for some moments in the duet exercise while being witnessed, also noted the extrinsic implications that each form of masculinity can present to the world, replete with a wide range of differing consequences planet and people alike.

Relative to the focus of this anthology, it is important to also note that TStR is now being used with feminist policymaking.[13] Considering our current dire social and ecological circumstances, there is an urgent need for compelling, creative, and broad sweeping opportunities to awaken greater awareness of the Earth, Other, and Self-care nexus. TStR is one such opportunity to facilitate transformations towards ecologisation through the body in connection with the mind. Here, we have shown that innovative pedagogies can awaken the need to care more broadly for life on the planet and in doing so, bring it down into and out through the body. Although the long-term efficacy of these experiences is yet to be tested, the innovation of applying TStR to the complex and commonly resistant topic of transitioning from industrial and ecomodern extractivist masculinities to masculinities that are relational, caring and thereby ecological, provides a compelling beginning point for further research and applications in both the theories and the practices of ecological masculinities.

The outcomes discussed in this chapter indicate that embodiment approaches to complex social and ecological issues can be useful. We have shared the application of TStR in the context of ecologising masculinities as a gesture of hope at a time of great global social and environmental challenge.

Notes

1 Consistent with the central premises of Theatre and Stories that Reconnect (TStR), the term 'Nature' is capitalised throughout this chapter to emphasise our belief that the other-than-human world possesses agency that is worthy of our acknowledgement and respect on an equal footing to human needs and endeavours. This is a principle that was first posited by deep ecologist Arne Næss's (1973) notion of intrinsic value.

2 TStR is principally inspired by Augusto Boal's (1979) *Theatre of the Oppressed*; Macy and Johnstone's (2012) *Active Hope* combined with Macy and Brown's (2014) *Coming Back to Life*; Arawana Hayashi's (2021) *Social Presencing Theatre* that has in turn grown from Otto Scharmer's (2009) *Theory U*; Eugene Gendlin's (1978) active listening technique of *Focusing*; as well as from John Croft's (2013) *Dragon Dreaming* communitarian project management framework. Other influences include *Theatre of the Oppressed* (Boal 1979), and the *Theatre as Ritual* work of a dear friend to Noy Meir and Olimpico, their mentor and teacher Hector Aristizabal (2010) who has developed his own social arts practice referred to as 'The Medicine Next-to the Wound'. By comparison, also note *The Great Austerity Debate* (University of Cambridge, 2021), a collaborative project between academics Susan J. Smith and Mia Gray and *Menagerie Theatre Company*, which demonstrated ways that participatory theatre can translate research into public engagement. Using Forum Theatre, the project considered the impacts of austerity while inviting audiences to reshape the narrative, fostering dialogue on systemic inequality. The project highlights how creative methods can bridge the gap between academic research and its applications in civil society, revealing a precedent for TStR to expand its work on ecological masculinities into accessible, transformative public engagement through embodiment practices.

3 In TStR, deep listening is connected to practices of welcoming, receiving, pausing and emptying out in ways that engage the whole body. This strong reliance on embodiment looks beyond habitual engagement with power-over dynamics by taking us through problems that we face to solutions based on collective decision-making.

4 See fn.2. The term 'Earth' is capitalised for similar reason, in alignment with Lovelock's (1979) Gaian notion of Earth as single self-regulating organism.

5 Alfred Korzybski (1933, p. 58) noted that "a map is not the territory ... but ... has a similar structure to the territory ..." The same is true of Social Arts which reveal maps as abstractions, each offering various complex and multilayered visions of reality. There is no single correct map; they are instead as complex and as varied as each participant. Effectively, different maps of the same reality can lead to varied and unexpected insights. A map often reveals more about the user than the 'territory' it allegedly represents. TStR maps provide lenses to view reality that reveal the subjectivity and plurality of individual perspectives. For instance, feminism is not a monolithic ontology but a collection of diverse maps, each contributing unique insights into social reality. The term 'map' refers not only to the stories we tell but also the stories we are told, and the stories we live and are immersed in. Sometimes, we can be so blindly immersed in them that we are not aware that we are in fact following a map that may or may not be an exact reflection of reality. Of relevance to this chapter, we note that patriarchy is a huge story that much of humanity lives within and is determined by. Most of us work within patriarchy unconsciously, effectively normalising its implications and making it invisible. Big stories can become frameworks that we create and co-create along an old-new continuum. They can become monolithic but are in fact nothing more than big stories. When we see them in this light, it becomes more possible to transform them.

6 The 'felt sense' refers to the familiarity of body sensations and the ways we can connect the dots to various experiences in our lives even if they do not seem immediately related, enabling us to expose patterns of sensing, energy flow and various qualities of connection that represent successful or challenging relational exchanges, which can reinforce or expose needed changes in our values and actions (Goldstein 2023; Gendlin 1978).

7 Where the imagined is used here to be the ability to bring us beyond the unknown and not yet possible, to creative and intuitive aspects of life that are emergent and alive.

8 The term industrial/breadwinner masculinities refers to destructive masculinities' norms that valorise hegemonic values and behaviours for owners of the means of production. Middle- and working-class workers may also bond themselves to hegemonisation when they are broadly considered breadwinners who have been sold the false narrative that they too are free to rise to the very top of socio-political machinations; a misleading contention that fails to take onto account the many barriers that can confront many as a result of structural oppressions. This generates an aggrieved entitlement that they too ought to and can reap the benefits of post-industrial capitalism even though such a goal is unattainable for many due to structural barriers such as age, ability, gender identity, education and race (Kimmel, 2013).

9 A second category of masculinities employs techno-fixes and clean energy innovations designed to reduce the social and ecological harm associated with industrialisation while still maximising growth. They offer insufficient alternative responses to growing evidence of the life-destroying tendencies of masculinist hegemonisation.

10 Intersecting conversations about the gendered aspect of petrocultures (Daggett 2018) calls for reorienting towards innovative degrowth ethics and practices (Smith-Khanna 2021; Žďárský 2024), while postgender approaches to ecologisation aim to locate masculinities within a gender-diverse metanarrative (Pulé and Ourkiya 2023) that collectively encourages caring pedagogies to emerge (Vetterfalk and Hedenqvist 2019; Hedenqvist et al. 2021; Laurien and Öhlund 2021). This more socially and environmentally relational masculinities represents a 'third way' of manifesting masculinities through what Hultman and Pulé (2018) refer to as ecological masculinities.

11 Some notable insights from this feedback are noted in the following section.

12 The embodiment practice considered here provided a pedagogical map, that with adaptation and responsiveness to varying context, could be used as a template for empowering and transformational pedagogies with applications for activist and educators as well. Women only and gender-diverse workshop applications are yet to be trialled.

13 In 2024, Noy Meir facilitated a workshop for a subdivision of the NGO – Women Engage for a Common Future (WECF) International (Brussels Office). In a similar but separate event, Pulé raised issues of masculine hegemonisation and ecologisation with members of the Belgian parliament. Opportunities such as these call attention to the needs of marginalised people and other-than-human Others in regional and international decision-making settings such as the EU.

References

Aristizabal, H. 2010, *The Blessing Is Next to the Wound: A Story of Art, Activism, and Transformation*. Lantern Books, Woodstock, NY.

Boal, A. 1979, *Theatre of the Oppressed*, Pluto Press, London, UK.

Croft, J. 2013, Dragon Dreaming: Project Design (Version 2.09), viewed 4 August, 2024 https://dragondreaming.org/wp-content/uploads/2020/01/DragonDreaming_eBook_english_V02.09.pdf

Crutzen, P and Stoermer, E 2000, 'The Anthropocene.', *Global Change Newsletter. The International Geosphere-Biosphere Programme (IGBP): A Study of Global Change of the International Council for Science (ICSU)*, vol. 41, May, pp. 17–18.

Daggett, C 2018, 'Petro-masculinity: Fossil fuels and authoritarian desire', *Millennium: Journal of International Studies*, vol. 47, no. 1, pp. 25–44.

Ferrando, F 2019, *Philosophical posthumanism*, Bloomsbury Academic, London, UK.

Freire, P 2014, *Pedagogy of commitment*, Paradigm, London, UK.

Gendlin, E 1978, *Focusing*. New York: Bantam Books.

Goldstein, E 2023, *Making Sense of the Felt Sense*, viewed on 4 August, 2024 https://integrativepsych.co/new-blog/making-sense-of-felt-sense

Harari, Y 2015, *Sapiens: A Brief History of Humankind*, Harper Perennial, New York, NY.

Hayashi, A 2021, *Social Presencing Theatre: The Art of Making a True Move*, Presencing Institute, New York NY.

Hedenqvist, R, Pulé, P, Vetterfalk, V and Hultman, M 2021, 'When Gender Equality and Earth Care Meet: Ecological Masculinities in Practice." in G Magnusdottir and A Kronsell (eds), *Gender, Intersectionality and Climate Institutions in Industrialized States*, Routledge, London, UK, pp. 207–225

Hultman, M and Pulé, P 2018, *Ecological Masculinities: Theoretical Foundations and Practical Guidance*, Routledge, Oxon, UK.

IPCC, 2023, 'Sections', in H Lee and J Romero (eds), *Climate Change 2023: Synthesis Report. Contribution of Working Groups I, II and III to the Sixth Assessment Report of the Intergovernmental Panel on Climate Change*, IPCC, Geneva, Switzerland, pp. 35–115, doi: 10.59327/IPCC/AR6-9789291691647

Kimmel, M 2013, *Angry White Men: American Masculinity at the End of an Era*, Nation Books, New York, NY.

Korzybski, A 1933, *Science and Sanity: An Introduction to Non-Aristotelian Systems and General Semantics*, International Non-Aristotelian Library Publishing Company, Lancaster, PA.

Laurien, T and Öhlund, K 2021, *Flow Feelers – applied ecofeminism for men*, Study circle and art project through the Design+Posthumanism Network, Storbykon feransen: Oslo, viewed 3 August, 2024, https://designandposthumanism.org/2022/04/29/flow-feelers/

Lenton, T, Rockström, J, Gaffney, O, Rahmstorf, S, Richardson, K, Steffen, W, and Schellnhuber, H 2019, 'Climate tipping points – too risky to bet against: The growing threat of abrupt and irreversible climate changes must compel political and economic action on emissions', *Nature*, 27, November, viewed 3 August, https://www.nature.com/articles/d41586-019-03595-0

Lovelock, J 1979, *Gaia: A New Look at Life on Earth*. Oxford: Oxford University Press.

Macy, J, and Brown, M 2014, *Coming Back to Life: An Updated Introduction to the Work That Reconnects*, New Society Publishers, Gabriola Island, BC.

Macy, J, and Johnstone, C 2012, *Active Hope: How to Face the Mess We're in Without Going Crazy*, New World Library, Novato, CA.

Merchant, C 1980, *The Death of Nature: Women, Ecology and the Scientific Revolution*, HarperCollins, New York, NY.

Monbiot, G 2024, 'King of Extrinsics', *The Guardian*, 31 January, viewed 3 August, 2024. https://www.monbiot.com/2024/01/31/king-of-the-extrinsics/

Næss, A 1973, 'The shallow and deep, long-range ecology movement: A summary', *Inquiry*, vol. 16, no. 1–4, pp. 95–100.

Pulé, P and Hultman, M, eds 2021, *Men, Masculinities and Earth: Contending with the (m)Anthropocene*, Palgrave MacMillan, Cham, Switzerland.

Pulé, P and Ourkiya A 2023, 'Post-gender Ecological Futures: From Ecological Feminisms and Ecological Masculinities to Queered Posthuman Subjectivities', in U Mellström and B Pease (eds), *Beyond Anthropocentric Masculinities: Posthumanism, New Materialism and the Man Questione*, Routledge: London, UK, pp. 207–221.

Scharmer, C 2009, *Theory U: Leading from the Future as It Emerges*, Berrett-Koehler Publishers, Oakland, CA.

Smith-Khanna, P 2021, *Sustainable Masculinities and Degrowth: Pathways to Feminist Post Growth Societies*, Institute for Political Ecology Junior Research Fellowship, DOI:10.13140/RG.2.2.27688.65284

University of Cambridge 2021, *The Great Austerity Debate*, viewed 30 November 2024, https://www.geog.cam.ac.uk/research/projects/greatausteritydebate/

Vetterfalk, V. and Hedenqvist, R. 2019, *Men and the Climate Crisis: A Guide for Reflective Groups Among Men, with a Focus on Masculinity Norms' Impacts on the Environment and Climate. MÄN – Men for Gender Equality*, viewed 30 November 2022, https://mfj.se/assets/documents/english/Men-in-the-climate-crisis-(prototype).pdf

Žďárský, T 2024, 'What is the relevance of masculine subjectivities for the degrowth transformation?', MA Thesis. Masaryk University, Brno, Czech Republic.

13

PHOTOVOICE

A tool for countering social path dependencies in climate institutions?

*Heidi Walker, Amber J. Fletcher, Maureen G. Reed
and Nicholas Antonini*

Introduction

When climate-related hazards strike, their impacts vary considerably across and within communities. News and social media have tended to depict feminist interpretations of climate hazards as narrowly focused on emotional response (e.g. Cox et al., 2008; Öhman, Nygren and Olofsson, 2016). Yet, feminist scholarship and leadership is both broader and deeper in its understanding of how experiences are shaped by social factors like gender, race, ethnicity, socio-economic status, and age. These social factors influence peoples' experiences, responses, and adaptations to hazards like floods and wildfires (e.g. Kaijser and Kronsell, 2014; Eriksen and Simon, 2017; Vickery, 2018; Walker, Reed and Fletcher, 2021a). As climate-related hazards increase in frequency and intensity, there are growing calls to examine differential vulnerabilities and to address them through climate adaptation policy and planning (Intergovernmental Panel on Climate Change, 2022). Building truly equitable policies, however, is challenging because – like individual people – the institutions that drive responses to climate hazards are shaped by social norms that influence response. A role we may play as feminist scholars is to build, model, and promote participatory tools that support change in policy, decisions, and practice within and across institutional levels. Doing so will strengthen the alignment of feminist scholarship with feminist leadership in climate action.

In this chapter, we define institutions as "rules in use", which include both formal (e.g. laws, regulations, authorities) and informal (e.g. norms, customs, beliefs) rules, processes, and practices (modelled after Ostrom, 2005). Formal institutions, including those for climate hazard response and adaptation, are

DOI: 10.4324/9781003461005-22

shaped by historical processes, norms, and rules that create pattern-bound effects or "path dependencies" that make certain policy directions more available or acceptable than others (Mackay, Kenny and Chappell, 2010; Magnusdottir and Kronsell, 2021). Feminist intersectionality acknowledges that multiple axes of social identity and power relations intersect and contribute to these institutional path dependencies (Kaijser and Kronsell, 2014; Magnusdottir and Kronsell, 2021). Such social path dependencies can influence which impacts associated with climate-related hazards are acted upon during immediate emergency response and longer-term adaptation planning.

Feminist research increasingly recognises the intangible effects or "invisible losses" associated with climate change and climate-related hazards, such as impacts to mental and emotional well-being, social/community fabric, and connection to place (Turner et al., 2008; Tschakert et al., 2019; Walker, Reed and Fletcher, 2021a). Institutions for recovery and adaptation, in contrast, often primarily focus on the physical and economic impacts of climate-related hazards, resulting in an emphasis on technical solutions relying on external sectoral expertise (Bosomworth, 2015). Social path dependencies can also influence how various groups are represented in relation to a hazard; for example, as heroes, experts, or victims (Olofsson, Öhman and Nygren, 2016; Walker, Reed and Fletcher, 2020). Such outcomes are, in part, due to the dominant framing of climate change as a technical, scientific, and security problem (Macgregor, 2010). This framing – underpinned by masculine norms and Western rationalist ideologies – can lead to solutions that depend on male-dominated sectors (e.g. construction and engineering) and "external" expertise (e.g. academic, government) rather than local community knowledge and experiences (Enarson, 2016; Pease, 2016).

The purpose of this chapter is to consider whether photovoice – a highly participatory, action-oriented research method – is a tool that can help overcome social path dependencies and effectively integrate equity issues into climate hazard response and climate change adaptation planning at the community level. To consider the potential of photovoice to achieve these aims, we reflect on the processes and outcomes of two photovoice case studies conducted as part of a larger study that applied a feminist intersectional lens to understand the social impacts of climate hazards in Saskatchewan, Canada. These photovoice projects took place in collaboration with two rural and Indigenous communities with different geographical and demographic characteristics, and explored experiences of two climate-related hazards – wildfire and flooding.

Below, we describe the photovoice method and share the methodological procedures implemented in these two case studies. We then discuss three outcomes that point toward the potential of photovoice to enhance equity and overcome social path dependencies in institutions related to climate hazards and adaptation. These outcomes include the ability to create powerful counter-narratives to dominant framings of climate-related hazards, to reposition historically marginalised groups as active contributors to climate change solutions, and to create spaces for dialogue. We note key challenges and limitations associated with the photovoice method in this climate change context. For example, reaching policymakers is a commonly cited photovoice objective; however, we observed that reaching policymakers is a major challenge – and not always even a desirable goal – in practice. We conclude with reflections on the utility of the photovoice method for achieving inclusive reflection on, and learning about, the impacts of climate hazards and planning for future events at the local level.

Photovoice

Photovoice emerged in the 1990s under the theoretical influences of critical consciousness education, feminist theory (Coemans et al., 2019), and community-based photography, all of which share social change as a fundamental aspiration (Castleden, Garvin and Huu-ay-aht First Nation, 2008; Wang and Redwood-Jones, 2001). Methodologically, the intent behind the method is to promote local knowledge, critical thought, and change. Specifically, three common goals of photovoice are to: "(1) enable people to record and reflect on their community's strengths and concerns, (2) promote critical dialogue and knowledge about important community issues through large and small group discussion of photographs, and (3) reach policymakers" (Wang and Burris, 1997, p. 370).

Photovoice traditionally includes four components: preparation; data collection through photography; group discussion and analysis; and sharing findings (e.g. Castleden, Garvin and Huu-ay-aht First Nation, 2008; Genuis et al., 2014; Torres Slimming, Orellana and Saurez Maynas, 2014). The preparation phase often includes the development of prompts (ideally co-created with community partners/participants) and ethics and photography training. The data collection process involves participants taking photos that respond to the prompts, followed by a group analysis activity to discuss the photos and identify themes. The photo-gathering process and group analysis create spaces for critical reflection and dialogue around important community issues. Photovoice projects often include a knowledge mobilisation mechanism, consistent with the philosophical goals of participatory action research. This may include, for example, an exhibit (Wang and Redwood-Jones, 2001) or photobook (i.e. Genuis et al., 2014). Photograph-sharing activities are often meant to contribute to the goal of reaching policymakers; however, more work needs to be done regarding best practices on effecting change through photovoice (Johnston, 2016).

While photovoice typically includes these four components, it is also a highly flexible method that can be adapted to the needs and interests of those involved (Anderson et al., 2023). For example, we opted to include previously taken photographs, which is an adaptation our research partners felt would be effective for reflecting on hazard events that occurred in the past. Further, despite the often-stated photovoice goal of "reaching policymakers", our findings ultimately show that this goal may not be shared by all communities – especially when policy has historically been an instrument of colonial governance.

Photovoice case studies

Two photovoice case studies were conducted as part of our larger research initiative on the social dimensions of climate hazards in rural Saskatchewan, Canada. One of the photovoice projects included three communities in a boreal forest region of northeast Saskatchewan that experienced a major wildfire event; the other was a First Nation (Indigenous) community in a prairie region of southern Saskatchewan that has experienced several major flooding events.

La Ronge

The first photovoice project took place in the La Ronge region of northern Saskatchewan, which includes the Town of La Ronge, the Lac La Ronge Indian Band (an Indigenous community), and the northern village of Air Ronge. The total population of these three communities is approximately 6,000. In 2015, the region experienced a major wildfire that burned 138,000 hectares of forest and contributed to the largest emergency evacuation in the province's history. Some residents stayed in their community to work or volunteer; most left due to a mandatory evacuation order. The photovoice project gave community members an opportunity to reflect on and share their diverse experiences of the wildfire event four years later.

The photovoice methodology and overarching questions were developed in collaboration with a four-member local advisory committee. The three prompting questions were:

1 What experiences related to forest fire do you feel are important to share?
2 What changes, challenges, and strengths have you seen in your communities and places that are meaningful to you as a result of forest fire?
3 What is needed to build strong, prepared, and resilient communities when facing fire in the future?

While photovoice typically involves participants taking new photos relevant to the project purpose, the advisory committee chose to invite participants to either take new photos or submit photos that they had previously taken around the time of the fires. Participants were invited to submit one to three photographs, along with a short narrative to describe the meaning behind their photos. Flyer invitations were posted in public spaces and on social media. A consent form and additional information sheet containing details about the project, tasks, and ethical photography were provided to those interested in participating, with follow-up discussions with potential participants as necessary. In total, 21 people (17 women, four men) submitted 47 photos and narratives. A final "sharing circle" workshop gave participants an opportunity to share a meal, describe the meaning behind their photos, and discuss what can be learned through these experiences for dealing with fire in the future. With assistance from the local arts council and public library, the submissions were transformed into a photo exhibit, which was displayed for two months at a free local gallery space (July–August, 2019). After receiving positive public feedback on the exhibit, participants expressed interest in sharing the photos beyond their own community. As a result, the exhibit was also displayed for one month in the public library gallery in the nearby urban centre of Prince Albert (December, 2019). Opening receptions were held in each location. These exhibits became useful venues for promoting engagement among residents, decision-makers, and the wider public.

Ochapowace

The second photovoice activity was conducted in collaboration with Ochapowace Cree Nation. Ochapowace is a sovereign Indigenous nation in southeastern Saskatchewan with a population of approximately 560 residents on-reserve and more than 1,300 members in total. The community is located around the southern edge of Round Lake and has experienced repeated flooding, including major events in 2011 and 2014 that involved evacuation of community members and the loss of homes. The photovoice project was one component in a multi-year collaborative study that included three main activities: video interviews with Elders and youth to create a documentary film series; semi-structured interviews with 27 community members focusing on their flood experiences; and the photovoice activity, which occurred in late 2019. In total, eight community members (six men, two women) participated in the photovoice, which was held after a community lunch and began with a blessing by an Elder. All participants had experienced a flood, and several had been in leadership roles during flood events.

The Ochapowace photovoice project was designed in collaboration with the Nation's leadership, who assisted with organisation, participant engagement, and facilitation. During the planning phase, the researchers and community leaders decided on an unstructured approach to the photovoice, albeit centred loosely on the theme of flood losses. Rather than an individualised approach in which each participant captures and discusses their own photo(s), the leadership opted for a more collective activity. Participants brought as many photos as they wished and spread these on a table for discussion. While some participants addressed the photos they had brought, many chose to discuss other participants' photos, which evoked additional memories for them. This approach was also inclusive of participants who had no photos, but still wanted to share their experience. The photovoice discussion was recorded and transcribed for analysis by the researchers.

Photovoice outcomes

Reflecting on our two case studies, we identified three positive outcomes that arose from the use of photovoice in the context of climate hazards. These outcomes include the potential of photovoice to: 1) create powerful counter-narratives to the framings of climate change and climate-related hazards that arise from dominant climate institutions, including dominant discourses of "vulnerability"; 2) reposition historically marginalised groups as active contributors to climate change solutions; and 3) create spaces for dialogue among affected communities, policymakers, and the wider public.

Creating counter-narratives

The first outcome we observed in our case studies is that photovoice can create powerful counter-narratives to dominant framings of climate-related hazards. Framing involves selecting and giving salience to certain aspects of an issue to "promote a particular problem definition, causal interpretation, moral evaluation, and/or treatment recommendation" (Entman, 1993, p. 52). The framing of issues and events is important, as they influence which policy and planning pathways are considered worthy of implementation, where funding is allocated, who is included in decision-making, and how existing institutional practices are maintained or challenged (Swim et al., 2018).

Climate-related hazards are commonly framed as technical problems and approached with technical solutions (Adger et al., 2008; Bosomworth, 2015; Nalau and Handmer, 2015). For example, media and policy discourses have tended to construct emotion as a feminised response to wildfire – and something to be suppressed – focusing instead on material, economic, and externally derived solutions (Cox et al., 2008; Öhman, Nygren and Olofsson, 2016). We observed this in our own research where, during the recovery phase of the major wildfire event in La Ronge, emphasis was on financial solutions (e.g. utility bill credits, late bill payment waivers) and community debriefings focused on the frontline fire operations and the economic impacts of the fires (Walker, Reed and Fletcher, 2020, 2021b). Less public attention was given to intangible losses community members experienced, such as those to mental and emotional well-being and feelings of disconnection to land and community.

The La Ronge photovoice project provided an opportunity for community members to share their personal experiences of the 2015 fires, pushing back against the dominant framing of climate-related hazards as primarily a technical problem dependent on the expertise of external agencies. Many of the submitted photos spoke to the losses experienced. These included losses associated with physical places – cabins and important sites on the land (e.g. favourite picnic places) that had burned in the fires. The photos and narratives, however, showed that losses associated with damage to physical places extended far beyond financial impacts to include impacts to the emotional attachments, memories associated with those places, and even sense of self. For example, along with a photo of the burnt forest where their cabin once stood, one male participant wrote, "This place that looked so familiar is changed, even your life is changed [...] it's a feeling of being uprooted, of displacement. It's life changing for you and the forest. You don't know the place you called home anymore, you don't even know yourself anymore." Other photos and narratives touched on impacts to mental and emotional well-being experienced during and after the fire event; for instance, the grieving of the loss of an important community event and the emotional fatigue and lingering anxiety associated with evacuation. We observed no significant differences in themes raised by men and women; rather, the photovoice activity provided a powerful opportunity for all participants to explore their emotional and affective experiences of fire.

While many photos touched on the tangible and intangible losses experienced with the wildfire event, they just as often contained themes of strength and resilience. Several photos, for example, depicted the first new growth following the fires, symbolising healing, the process of adapting to changed landscapes, or the old and new community ties that had been strengthened through the collective experience of fire. Ultimately, the photovoice project became an opportunity to push back against the dominant framing of wildfire as primarily a technical problem in want of technical and financial solutions. By putting these photos out into the world through a public exhibit, the photovoice project – even if in a small way – created a powerful counternarrative that demonstrated the importance of including a wider range of physical, emotional, and mental consequences when recovering from a major wildfire event and planning for future events.

While the La Ronge participants challenged technocratic responses to wildfire, participants in Ochapowace rejected a different, but similarly dominant, climate change frame: the framing of Indigenous communities as uniformly "vulnerable". Such framings have the problematic tendency to justify intervention by external agencies into Indigenous peoples' lives and communities, thus "perpetuating the legacies of paternalistic and top-down measures that are symptomatic of dysfunctional Indigenous-settler relations, rather than achieving progress towards equity" (Veland et al., 2013, p. 325).

At Ochapowace, participants' narratives exemplified the necessary balance between acknowledging structural inequity while, at the same time, recognising the agency and capacity that can be a source of strength for communities experiencing a disaster (Gauer, Schaepe and Welch, 2021).

Climate change researchers have documented the disproportionately negative impact of climate hazards on Indigenous people and communities (e.g. Scharbach and Waldram, 2016; Martin et al., 2017). Colonial violence and institutionalised racism, both historical and contemporary, have created deeply rooted structural inequities that continue to shape Indigenous people's experiences of disaster (Cameron, 2012; Whyte, 2017). As part of future climate response and adaptation, the unequal impacts of climate hazards must be acknowledged, and the structural causes of these inequities redressed. Indeed, Ochapowace members talked about their losses, which were exacerbated in part by a long history of government underfunding for housing that led to less flood-resilient homes on the reserve. They questioned the lack of effective assistance from external authorities during and after the event, including private companies and government agencies. They noted, as one example, how insurance companies requested that residents produce receipts to receive insurance coverage, but those receipts had been destroyed in the flood along with their other possessions.

Residents shared the emotional pain of seeing their belongings thrown in the trash, steeped with sewage. They shared the physical drain of sandbagging and the feeling of not being able to carry on, while nonetheless pushing through their exhaustion. Participants also spoke about the health impacts of living in mould-infested homes. One said, "The mould in my house, it is still inside the wood and whatnot. My kids were still getting sick from the furnace blowing air in the house like a week after. Yes, they were sick and congested and it was pretty brutal." Even five years after the most recent flood, some residents reported that they continued to live in homes contaminated with mould.

However, these stories of loss and harm were overshadowed by a stronger emphasis on community solidarity, mutuality, and compassion. Many of the submitted photos showed community members working together to place sandbags, organise the response, or prepare and share food. Ochapowace participants repeatedly commented on how the Nation's residents and leadership (e.g. Chief and Council) came together to create an organised, cohesive response effort, which ultimately strengthened community ties. One participant said:

> You know, we came together to fight this problem, not once but twice. And you know, it takes people like that to come together and put all of their things behind them and think of other people, and that is what a lot of people did here, was just go out and help. You know, like the people said, we were tired, we were hungry, we were stressed out and people are crying – it was not nice – but I don't think anybody ever gave up.

Participants showed gratitude toward the Nation's leadership and Council members who helped them during the floods. The photovoice discussion was also marked by moments of humour and positivity. One participant shared a beautiful moment when the northern lights (aurora borealis) appeared and seemed to surround one of the community members who was sandbagging. Another laughed about the time a canoe tipped while bringing supplies to an isolated community member, plunging the community responders into cold water.

Overall, these stories combined to create a powerful counter-narrative to dominant discourses of Indigenous communities' vulnerability. Community members experienced vulnerability and loss, but they preferred to highlight their resilience and collectivity as a community. In this way, the photovoice activity facilitated a resistant discourse that brings nuance to the vulnerability framing, adding a dimension of agency and a strengths-based framing that should be supported and enhanced in future flood response.

Repositioning communities as active responders and adaptors

How people are represented has implications for whose knowledge, experiences, and expertise are perceived as important and legitimate (or not) in climate hazard response and adaptation (Cameron, 2012; Haalboom and Natcher, 2012). Media framings are an important form of representation by which "individuals and groups are subjected – most often unconsciously or automatically – by cultural codes and hegemonic practices that serve to shape how they are identified by society, often manifesting in inequalities based upon gender, class, race, as well as other categories of social difference like disability or geographical location" (Eriksen et al., 2015, p. 528). Some studies, for example, have found that media tends to focus on stereotypically masculine roles during emergency response and recovery, while representing women as either virtuous supporters or not at all (e.g. Eriksen, 2014; Olofsson, Öhman and Nygren, 2016). Other studies have also noted both a general lack of media attention to Indigenous communities' experiences of major hazard events and disempowering framings of Indigenous groups (particularly youth) during evacuation (Christianson and McGee, 2019; Scharbach and Waldram, 2016). Similarly, our previous work found that media discourses reinforced a victim-hero dualism during the La Ronge wildfire event, representing northern Saskatchewan and Indigenous communities, as well as women, as in need of protection, while centring the roles of external and historically male-dominated organisations such as the military and wildfire management agencies (Walker, Reed and Fletcher, 2020).

Many of the La Ronge photovoice submissions countered this framing by depicting the many ways in which local community members – men and women, Indigenous and non-Indigenous people – actively contributed to the wildfire response. One participant, for example, submitted a photo of a note reading, "Kitchen staff, you rock!" that was left for Lac La Ronge Indian Band (LLRIB) volunteers working to feed evacuees and work crews in the community. In the accompanying narrative, she wrote:

> [LLRIB] played such a significant role during the fires and I don't think they got enough recognition for it. They were feeding workers and contractors [...] As more and more people started coming to [the LLRIB response location], nobody even blinked an eye – nobody said, 'whose budget is this coming out of?'; 'why are they showing up here?'; 'they are not our people, we can't feed them'. No, that wasn't even uttered, nobody even batted an eye. Our First Nation said you have nobody feeding you, come and eat.

Other photos and narratives illustrated, for example, local volunteers who went from home to home feeding animals that had to be left behind when their families were evacuated, people working together to set up pumps and hoses to protect their neighbours' homes, and community residents who supported fellow evacuees after leaving their communities.

Ochapowace residents similarly emphasised the powerful community response to the flood events in 2011 and 2014. They described the importance of looking out for Elders and those who lived alone – including the innovative approach of using a canoe to bring bread and other supplies to their neighbours. Beyond immediate response, Ochapowace members described longer-term adaptation measures that had been implemented because of previous floods. After the flood of 2011, one participant had built a berm (a raised barrier) to protect his home from the nearby creek, which overflowed again in 2014 but with reduced damage due to the berm.

Ochapowace participants clearly positioned themselves as adaptors. Near the end of the group discussion, one participant queried the others, asking: "When it boils down to it, have we really learned anything from it to prepare ourselves for the next one; have we done that, are we doing that, or what?" Other members responded by noting the purchase of generators and the establishment of the local skating rink as an emergency operations centre, in addition to other emergency response planning efforts at the community level. In Ochapowace, community members were not merely victims of flooding, but active responders who initiated both short-term coping and longer-term adaptation efforts.

The recognition of communities' adaptive agency adds crucial nuance to dominant framings of disaster. Such framings can help disrupt the "procedural vulnerability" that is created when Indigenous knowledge is ignored in favour of Western approaches (Veland et al., 2013). At the same time, however, agential framings must not excuse the ongoing effects of colonisation and structural racism and must not justify the extinguishment of Indigenous rights claims. Further, agential framings should not diminish or downplay the responsibility of larger, relatively well-funded institutions to support community-led disaster response efforts. Agential framings instead reinforce the importance of community-driven, collaborative approaches to both emergency response and climate change planning, which recognise both local strengths and the need for publicly provided safety nets – particularly in the face of future extreme events exacerbated by climate change.

Creating spaces for dialogue

Participatory research methods, including photovoice, can create focal points and platforms for dialogue among people with diverse values and experiences, and effectively facilitate knowledge transfer through the process of developing and sharing the final products (Masterson, Mahajan and Tengö, 2018; Zurba et al., 2019). Common goals of photovoice include fostering critical reflection, dialogue among participants, and learning and social change by sharing the photographs with key decision-makers and the broader community through forums, displays, and exhibits (Wang and Burris, 1997; Krieg and Roberts, 2007). Importantly, both case studies demonstrated that successfully creating spaces for dialogue requires adapting the photovoice method and mobilisation strategies to fit the values and priorities of participating communities – even if those adaptations do not necessarily align with existing scholarly guidance.

The La Ronge photovoice sharing circle workshop and two community exhibits successfully fostered dialogue and reflection among the photovoice participants, the local community, the wider public, and government representatives. The local advisory committee's preference to invite participants to submit photos from around the time of the fires, instead of taking new photos as is typical of the photovoice method, enhanced the ability of participants to share the stories that were most meaningful to them. The degree to which dialogue and reflection translated to learning, action, or policy change, however, varied across scales. The final sharing circle workshop provided participants an opportunity to share their photos with each other. Attention was given to creating a safe space for discussion. The workshop began by sharing a meal together and gently introducing the sharing circle activity, including establishing ground rules (e.g. encouraging respectful dialogue, confidentiality, and inviting participants to step away from the circle at any time). The sharing circle workshop took place in several rounds. Each participant had the opportunity to share the stories behind their photographs and reflect on what the shared stories meant for establishing future community resilience and preparedness. Even though the workshop took place four years after the fire event, the discussion became intensely emotional at times.

Afterwards, several participants mentioned that the photovoice project and sharing circle felt like a healing experience, giving them a chance to share their experiences in ways they were not able during public community forums immediately following the fires.

The two public exhibits enabled sharing and dialogue among the La Ronge photovoice participants, fellow community members, and the wider public. Comments left in the guestbooks at each site provide at least some evidence that the exhibits fostered reflection, learning, and action beyond the participants themselves. For instance, after viewing the La Ronge exhibit, one community member wrote "I'm going home to test my fire pump...." Several comments from the Prince Albert exhibit noted that the photos exhibit gave them a deeper understanding of wildfire, its impacts, and the diverse ways that people experienced the fires. For example, viewers wrote, "great display – helps us a little to understand the human impact" and "amazing way to get a glimpse into the experiences of those living through forest fire". Local wildfire management contacts were informed about the photovoice project as it developed and staff from both local and regional wildfire management agencies were invited to attend the exhibit opening receptions. While there was some attendance among these groups, there was no evidence that the photographs and narratives influenced wildfire management policy or practice.

While some photovoice studies employ an individualistic approach in which photography is positioned as an emancipatory or even skill-building activity, Ochapowace community leaders and participants chose a more collective approach using existing photographs. Discussion focused on the photos holistically, resulting in an overarching emphasis on shared experience. The photovoice activity largely omitted the emancipatory photography element of photovoice, focusing instead on existing rather than induced photos.

Although photovoice is often promoted as a route to social and policy change that extends beyond the participants themselves, Ochapowace participants prioritised depth over breadth of mobilisation. Participants, as well as the Elders who advised the project, were not particularly interested in scientific or technical reports stemming from the activity, nor did they express a desire to see the results shared with external agencies or policymakers. Instead, Ochapowace participants were interested in seeing the photos displayed in the form of a book or scrapbook that would allow them to share their story of the floods with their children and grandchildren.

Despite these differences from the reported practice, the photovoice activity successfully created a space for dialogue within the community and an opportunity for intergenerational knowledge transfer. It also provided attendees with the opportunity to share their experience as individuals and as a community, to reflect on strengths, and to express gratitude to the Nation's leaders and others who led the flood response. The activity did not facilitate knowledge mobilisation to policymakers outside the Nation – nor was that its intent. Indeed, given the problematic history of settler-Indigenous relations in Canada, the idealised notion of photovoice as a method of broad knowledge mobilisation beyond communities may in fact contradict communities' own desire to pass knowledge internally, rather than externally.

Challenges and limitations of photovoice

One challenge of photovoice relates to ethical issues, including privacy around who or what is being photographed and ensuring ethical considerations are specific to a particular project, which may require adjustment over time. Wang and Redwood-Jones (2001) addressed how different elements of the research process (e.g. consent mechanisms, being sensitive to disclosing embarrassing/incriminating information, providing ethics training) are practical measures to honour privacy laws. Scholars suggest that as photovoice methods evolve, so too should the questions of ethics (Creighton et al., 2017; Rosemberg and Evans-Agnew 2020). We argue that the method need not rigidly adhere to typical photovoice methodology outlined in the scholarly literature, but should be flexible to align with the values, priorities, and objectives of the communities involved. We agree, therefore, that photovoice requires committed attention to ethical issues that are both anticipated prior to the research process, but also to those which emerge on a case-by-case basis (Tarrant and Hughes, 2020).

In the La Ronge case, for example, many participants found submitting previously taken photos more conducive to reflecting on a past hazard event. This led to a situation where some participants wished to submit a photograph originally taken by someone else along with their own reflective narrative. In these cases, we extended the informed consent process to the original photographers and only included photos where the original photographer consented. Appropriate credit was given to the original photographers in the final photovoice displays.

Some scholars have argued that, in the digital world, photos are virtually permanent in some form, which may have implications long after the project is complete (Creighton et al., 2017; Rosemberg and Evans-Agnew, 2020; Tarrant and Hughes, 2020). In the case of Ochapowace, participants wished to pass a record of the photos and flood stories to future generations; therefore, the permanence of photographs aligns with the community's own objectives. Generally, however, photovoice researchers – especially those focusing on hazard experiences – should carefully consider their method, since photos may create a permanent record of people during their hardest moments. Both projects described in this chapter were conducted several years after the hazard events, and although the discussions still evoked emotion, participants had arguably had more time to process their experiences and thus to consider the photos submitted and their long-term implications.

A second key challenge of photovoice relates to the ability of the method to influence decision-making. As the Ochapowace case study showed, influencing policy is not always a desired outcome of photovoice; rather, the goals of any photovoice project should be decided in collaboration with the communities involved. Our experiences show that when policy influence is desired, it can be difficult to apply photovoice in a way that achieves this outcome. Even where photovoice is effective in creating spaces for dialogue, path dependencies and entrenched epistemic power relations in governance institutions may determine what data truly "count". Qualitative methodologies like photovoice, which do not align with positivist paradigms, might be less valued as evidence to inform policymaking (Greenhalgh and Russell, 2009). For example, it may be easier to suggest that for dollars spent, specific outcomes will be realised than it is to suggest that community empowerment will improve "general" responses to hazard events. Deeply embedded path dependencies may ultimately be more influential than new evidence and insights from community-based research. Due to such challenges, Johnston (2016, p. 807) suggests that photovoice may best be understood as "policy informing rather than policy changing." It can be an effective tool for informing community policymakers of pertinent local issues, but policy change likely requires "a variety of material, social and cultural factors" beyond the scope of a single photovoice project (p. 807).

Notwithstanding these limitations, photovoice may still be effective in catalysing change in policies, programmes, and activities if photovoice collaborators (e.g. researchers and communities) consider more diverse outputs and outlets – when reaching policymakers is a desired goal. Social science researchers often claim that only fundamental or transformative change is sufficient, yet, when working in the policy arena, whole-scale changes are unlikely. Rose et al. (2020) offer concrete advice for making tailored recommendations that may appear incremental, but can build on prior changes by academics and other "voices" in policy debates to move towards large-scale change. We have found that alternative outlets might showcase the outcomes of the research to a wider audience and thereby gain greater interest and impact. For example, practitioner magazines outside of our research field (e.g. *Policy Options*) or news outlets like The Conversation offer researchers an outlet to showcase their research to a broad audience in a timely way (e.g. see Walker, Reed, and Fletcher 2020). This broader reach might help to demonstrate why policymakers should care about the work communities and feminist, community-engaged researchers do.

Conclusions

Forging equitable, inclusive, and locally appropriate emergency response and climate adaptation processes require tools that illuminate the experiences of residents at diverse social locations and the contexts that shape those experiences. Our experiences with photovoice suggest that it is one such tool. In their favour, photovoice methodologies are excellent at illustrating multiple and intersecting social dimensions, inequalities, and collective actions. Importantly, as described in this chapter, photovoice can overcome the sometimes ossifying trope of marginalisation, offering a strengths-based approach that showcases local agency and action.

Despite the potential of photovoice to illuminate more strengths-based and locally appropriate hazard response and climate adaptation pathways, institutions for emergency response and climate adaptation often still favour quantitative, technical knowledge and tools. Part of our role as feminist researchers is to continue to promote the value of, and build capacity for implementing, participatory tools such as photovoice in community and policy spheres. It is also to design and implement such tools collaboratively in ways that align with the values and priorities of the communities involved. At the same time, we must recognise the need to employ multiple strategies to forge new pathways towards equitable and just responses to climate change. If applied in the absence of other tools, photovoice will not likely have a policy impact. No single method will. We need to bring our tools together to support incremental and transformative change, including through broader feminist approaches to regional, national, and international climate change decision-making. Together, these small- and large-scale strategies can help work towards a more equitable and just climate emergency response.

Acknowledgements

We deeply appreciate everyone who shared their powerful and inspiring stories through the photovoice projects. The La Ronge photovoice project was completed in collaboration with a local advisory committee and with support from the La Ronge Arts Council and the Alex Robertson Public Library. The Lac La Ronge Indian Band was a key partner and provided direction for this project. The second photovoice project was developed and carried out in collaboration with the Ochapowace Cree Nation. This chapter draws on research supported by the Social Sciences and Humanities Research Council.

References

Adger, W.N., Dessai, S., Goulden, M., Hulme, M., Lorenzoni, I., Nelson, D.R., Naess, L.O., Wolf, J., and Wreford, A. (2008) 'Are there social limits to adaptation to climate change?' *Climatic Change*, 93(3–4), 335–354. https://doi.org/10.1007/s10584-008-9520-z

Anderson, K., Elder-Robinson, E., Howard, K., and Garvey, G. (2023) 'A systematic methods review of photovoice research with Indigenous young people', *International Journal of Qualitative Methods*, 22, 1–37. https://doi.org/10.1177/16094069231172076

Bosomworth, K. (2015) 'Climate change adaptation in public policy: frames, fire management, and frame reflection', *Environment and Planning C*, 33, 1450–1466. https://doi.org/10.1177/0263774X15614138

Cameron, E.S. (2012) 'Securing Indigenous politics: A critique of the vulnerability and adaptation approach to the human dimensions of climate change in the Canadian Arctic', *Global Environmental Change*, 22(1), 103–114. https://doi.org/10.1016/j.gloenvcha.2011.11.004

Castleden, H., Garvin, T., and Huu-ay-aht First Nation. (2008) 'Modifying photo-voice for community-based participatory Indigenous research', *Social Science & Medicine*, 66, 1393–1405. https://doi.org/10.1016/j.socscimed.2007.11.030

Christianson, A.C. and McGee, T.K. (2019) 'Wildfire evacuation experiences of band members of Whitefish Lake First Nation 459, Alberta, Canada', *Natural Hazards*, 98, 9–29. https://doi.org/10.1007/s11069-018-3556-9

Coemans, S., Raymakers, A.-L., Vandenabeele, J., and Hannes, K. (2019) 'Evaluating the extent to which social researchers apply feminist and empowerment frameworks in photovoice studies with female participants: a literature review', *QualitativeSocialWork*,18(1),37–59.https://doi.org/10.1177/1473325017699263

Cox, R.S., Long, B.C., Jones, M.I., and Handler, R.J. (2008) 'Sequestering of suffering: Critical discourse analysis of natural disaster media coverage', *Journal of Health Psychology*, 13(4), 469–480. https://doi.org/10.1177/1359105308088518

Creighton, G., Oliffe, J.L., Ferlatte, O., Bottorff, J., Broom, A., and Jenkins, E.K. (2017) 'Photovoice ethics: critical reflections from men's mental health research', *Qualitative Health Research*, 28(3), 446–455. https://doi.org/10.1177/1049732317729137

Enarson, E. (2016) 'Men, masculinities, and disaster: an action research agenda', in Enarson, E. and Pease, B. (eds.) *Men, masculinities, and disaster*. London and New York: Routledge, pp. 219–233.

Entman, R.M. (1993) 'Framing: toward clarification of a fractured paradigm', *Journal of Communication*, 43(4), 51–58.

Eriksen, C. (2014) 'Gender and wildfire: landscapes of uncertainty.' New York: Routledge.

Eriksen, S.H., Inderberg, T.H., O'Brien, K., and Sygna, L. (eds) (2015) *Climate change adaptation and development: transforming policies and practices*. London and New York: Routledge, pp. 1–18.

Eriksen, C., and Simon, G. (2017) 'The affluence–vulnerability interface: intersecting scales of risk, privilege and disaster', *Environment and Planning A*, 49(2), 293–313. https://doi.org/10.1177/0308518X16669511

Gauer, V.H., Schaepe, D.M., & Welch, J.R. (2021) 'Supporting Indigenous adaptation in a changing climate', *Elementa: Science of the Anthropocene*, 9(1), 00164. https://doi.org/10.1525/elementa.2020.00164

Genuis, S.K., Willows, N., Alexander First Nation, and Jardine, C. (2014) 'Through the lens of our cameras: children's lived experience with food security in a Canadian Indigenous community', *Child: Care, Health, and Development*, 41(4), 600–610. https://doi.org/10.1111/cch.12182

Greenhalgh, T. and Russell, J. (2009) 'Evidence-based policymaking: a critique', *Perspectives in Biology and Medicine*, 52(2), 304–318. https://doi.org/10.1353/pbm.0.0085

Haalboom, B.J. and Natcher, D.C. (2012) 'The power and peril of "vulnerability": lending a cautious eye to community labels in climate change research,' *Arctic*, 65(3), 319–327.

Intergovernmental Panel on Climate Change. (2022) *Climate change 2022: impacts, adaptation, and vulnerability.* Cambridge and New York: Cambridge University Press. Contribution of Working Group II to the Sixth Assessment Report of the Intergovernmental Panel on Climate Change. https://doi.org/10.1017/9781009325844

Johnston, G. (2016) 'Champions for social change: photovoice ethics in practice and 'false hopes' for policy and social change', *Global Public Health*, 11, 799–811. https://doi.org/10.1080/17441692.2016.1170176

Kaijser, A. and Kronsell, A. (2014) 'Climate change through the lens of intersectionality', *Environmental Politics*, 23(3), 417–433. https://doi.org/10.1080/09644016.2013.835203

Krieg, B. and Roberts, R. (2007) 'Photovoice: insights into marginalisation through a community lens in Saskatchewan, Canada', in Kindon, S., Pain, R., and Kesby, M. (eds.) *Participatory action research approaches and methods: connecting people, participation, and place.* Abingdon and New York: Routledge, pp.150–159.

Macgregor, S. (2010) 'A stranger silence still: the need for feminist social research on climate change', *Sociological Review*, 57, 124–140. https://doi.org/10.1111/j.1467-954X.2010.01889.x

Mackay, F., Kenny, M. and Chappell, L. (2010) 'New institutionalism through a gender lens. Towards a feminist institutionalism?' *International Political Science Review*, 31(5), 573–588. https://doi.org/10.1177/0192512110388788

Magnusdottir, G. and Kronsell, A. (eds.) (2021) *Gender, Intersectionality and climate institutions in industrialised states.* Abingdon and New York: Routledge.

Martin, D.E., Thompson, S., Ballard, M. and Linton, J. (2017) 'Two-eyed seeing in research and its absence in policy: Little Saskatchewan First Nation Elders' experiences of the 2011 flood and forced displacement', *International Indigenous Policy Journal*, 8(4), 1–25. https://doi.org/10.18584/iipj.2017.8.4.6

Masterson, V.A., Mahajan, S.L. and Tengö, M. (2018) 'Photovoice for mobilizing insights on human well-being in complex social-ecological systems: case studies from Kenya and South Africa', *Ecology and Society*, 23(3), 13. https://doi.org/10.5751/ES-10259-230313

Nalau, J. and Handmer, J. (2015) 'When is transformation a viable policy alternative?' *Environmental Science & Policy*, 54, 349–356. https://doi.org/10.1016/j.envsci.2015.07.022

Öhman, S., Nygren, K.G. and Olofsson, A. (2016) 'The (un)intended consequences of crisis communication in news media: a critical analysis', *Critical Discourse Studies*, 13(5), 515–530. https://doi.org/10.1080/17405904.2016.1174138

Olofsson, A., Öhman, S. and Nygren, K.G. (2016) 'An intersectional risk approach for environmental sociology', *Environmental Sociology*, 2(4), 346–354. https://doi.org/10.1080/23251042.2016.1246086

Ostrom, E. (2005) *Understanding institutional diversity.* New Jersey: Princeton University Press.

Pease, B. (2016) Masculinism, climate change and "man-made" disasters, in Enarson, E. and Pease, B. (eds.) *Men, masculinities, and disaster.* Abingdon and New York: Routledge, pp. 21–33.

Rose, D.C., Mukherjee, N., Simmons, B.I., Tew, E.R., Robertson, R.J., Vadrot, A.B.M., Doubleday, R. and Sutherland, W.J. (2020) 'Policy windows for the environment: tips for improving the uptake of scientific knowledge.' *Environmental Science & Policy*, 113, 47–54. https://doi.org/10.1016/j.envsci.2017.07.013

Rosemberg, M-A.S. and Evans-Agnew, R. (2020) 'Ethics in photovoice: a response to Teti', *International Journal of Qualitative Methods*, 19. https://doi.org/10.1177/1609406920922734

Scharbach, J. and Waldram, J.B. (2016) 'Asking for a disaster: being "at risk" in the emergency evacuation of a Northern Canadian Aboriginal community', *Human Organization*, 75(1), 59–70. https://www.jstor.org/stable/44127063

Swim, J.K., Vescio, T.K., Dahl, J.L. and Zawadzki, S.J. (2018) 'Gendered discourse about climate change policies', *Global Environmental Change*, 48, 216–225. https://doi.org/10.1016/j.gloenvcha.2017.12.005

Tarrant, A and Hughes, K. (2020) 'The ethics of technology choice: photovoice methodology with men living in low-income contexts', *Sociological Research Online*, 25(2), 289–306. https://doi.org/10.1177/1360780419878714

Torres Slimming, P.A., Orellana, E.R., Saurez Maynas, J. (2014) 'Structural determinants of Indigenous health: a photovoice study in the Peruvian Amazon. *AlterNative*, 10(2), 123–133. https://doi.org/10.1177/117718011401000203

Tschakert, P., Ellis, N.R., Anderson, C., Kelly, A. and Obeng, J. (2019) 'One thousand ways to experience loss: a systematic analysis of climate-related intangible harm from around the world', *Global Environmental Change*, 55, 58–72. https://doi.org/10.1016/j.gloenvcha.2018.11.006

Turner, N.J., Gregory, R., Brooks, C., Failing, L. and Satterfield, T. (2008) 'From invisibility to transparency: identifying the implications,' *Ecology and Society*, 13(2), 7.

Veland, S., Howitt, R., Dominey-Howes, D., Thomalla, F. and Houston, D. (2013) 'Procedural vulnerability: understanding environmental change in a remote Indigenous community. *Global Environmental Change*, 23(1), 314–326. https://doi.org/10.1016/j.gloenvcha.2012.10.009

Vickery, J. (2018) 'Using an intersectional approach to advance understanding of homeless persons' vulnerability to disaster', *Environmental Sociology*, 4(1), 136–147. https://doi.org/10.1080/23251042.2017.1408549

Walker, H.M., Reed, M.G. and Fletcher, A.J. (2020) 'Wildfire in the news media: an intersectional critical frame analysis', *Geoforum*, 114, 128–137. https://doi.org/10.1016/j.geoforum.2020.06.008

Walker, H.M., Reed M.G. and Fletcher, A.J. (2021a) 'Applying intersectionality to climate hazards: a theoretically informed study of wildfire in northern Saskatchewan', *Climate Policy*, 21(2), 171–185. https://doi.org/10.1080/14693062.2020.1824892

Walker, H., Reed, M.G. and Fletcher, A.J. (2021b) 'Pathways for inclusive wildfire response and adaptation in northern Saskatchewan', in Magnusdottir, G. and Kronsell, A. (eds.) *Gender, intersectionality and climate institutions in industrialised States*. Abingdon and New York: Routledge.

Wang, C. and Burris, M.A. (1997) 'Photovoice: concept, methodology, and use for participatory needs assessment,' *Health Education & Behaviour*, 24(3), 369–387.

Wang, C. and Redwood-Jones, Y.A. (2001) 'Photovoice ethics: perspectives from Flint photovoice', *Health Education & Behaviour*, 28(5), 529–644. https://doi.org/10.1177/109019810102800504

Whyte, K. (2017) 'Indigenous climate change studies: Indigenizing futures, decoloniz- ing the Anthropocene', *English Language Notes*, 55(1–2), 153–162. https://doi. org/10.1215/00138282-55.1-2.153

Zurba, M., Maclean, K., Woodward, E. and Islam, D. (2019) 'Amplifying Indigenous community participation in place-based research through boundary work', *Progress in Human Geography*, 43(6), 1020–1043. https://doi.org/10.1177/0309132518807758

14

FEMINIST CLIMATE APPROACHES: HOW, WHY AND WHAT?

Why we need feminist climate approaches more than ever, what would they look like and how do we get there?

Martin Hultman, Karen Morrow, Gunnhildur Lily Magnusdottir and Susan Buckingham

It's a long road that has no turning

Our concluding thoughts are shaped by the nature of change for both good and ill. The past, as amply demonstrated in this anthology, has shown we can make progress on climate justice; the present shows how vulnerable even seemingly secure gains are to new political circumstances, and the future offers prospects for both despair and hope. When we (the editors and many of the contributors) met in person to plan this volume at Malmö University, Sweden, in May 2022, the mood in the room was one of excitement and confidence. It was at the end of the COVID-19 pandemic and the opportunity to reflect and rebuild was full of possibility. We brought to the table fresh memories of the enforced respite from modernity that, while it came at a huge human cost, also gave us an unprecedented experience of living in a world with significantly lower pollution, including greenhouse gas emissions, and noise from cars, ships, and airplanes. We were also energised by the high-profile mass mobilisations of a plethora of progressive social movements, including Black Lives Matter, #MeToo, Fridays For Futures, Extinction Rebellion, and many more, giving voice to and hope for meaningful change on seemingly intractable issues. For many it was the first in-person meeting for almost two years and finding ourselves in community with colleagues again was wonderful, remedying the brutal separation endured during the pandemic.

We found cause for cautious optimism in the election defeats of some of the most misogynist far-right leaders who came to power in 2016–2017 (Donald Trump, Scott Morrison and some months later Jair Bolsonaro) and in the fact that other countries still had progressive feminist leaders (such as Jacinda

DOI: 10.4324/9781003461005-23

Ardern, Sanna Marin, Nicola Sturgeon, Katrin Jakobsdottir) some of whom were inspired by Sweden's feminist foreign policy (FFP) practice.

However, we also found ourselves in a polycrisis, with states confronted by the economic aftermath of the pandemic, burgeoning climate, and biodiversity crises, as well as multiple conflicts around the world, which have led to increased militarisation and where energy has been securitised. In Europe, the war in Ukraine, has indeed securitised all policies, both at the national and regional level, and reinforced existing gender-blind patterns in policymaking (see Allwood in Chapter 3).

So, in the face of this polycrisis, and turn to anti-environmental, anti-social justice populism, including the dismantling of Swedish FFP by a conservative government in Autumn 2022 (see Rosén-Sundström and Elgström in Chapter 1), can a feminist approach to dealing with the climate and other environmental emergencies offer us a practical way forward?

Feminist climate leadership

Feminist governance principles are designed to be transformative, reflective, caring, responsible, transparent, non-violent, inclusive, courageous and with zero tolerance for discrimination and the abuse of power (Action Aid, n.d.; Fair Share of Women Leaders, n.d.; Government of the Netherlands, 2023). Climate leaders who self-identified as feminist interviewed for this volume referred to many of these qualities and gave examples of how they pursued them in practice when describing what distinguished their approach (see Table 14.1). Arguably they had achieved some transformational changes (although it is difficult to speculate on the longevity of these once the people have left office).

Reeves and MacArthur, reflecting on interviewing Catherine McKenna, highlighted the importance of building strong and diverse networks for policy leadership, as illustrated by her domestic work on women in political office (Run like a Girl) or internationally focused women in climate leadership (Women Kicking it on/leading on Climate). Additionally, the level of opposition to her advocacy on policy issues like carbon pricing and/or the coal phase-out was strong and significant and the vitriol she dealt with on

TABLE 14.1 Common qualities identified by feminist climate leaders

- Mentoring/supporting women, young people, indigenous groups
- Inclusivity, collaboration (e.g. with grassroots groups)
- Being courageous/brave
- Cooperation/alliances
- Networks
- Challenging colonialism, patriarchy; confronting privileged interests
- Holistic/intersectional (climate crisis and social injustice are interlinked)
- Non-violent/active listening

these issues in her official capacity is well documented in Canada. As Raney has argued, women's climate leadership represents a double threat to those who cling to misogynist views. Not only do such women defy patriarchal expectations of leadership, they also espouse policies that pick at the delicate threads of heterosexual masculinity, which are closely entwined with industrialised capitalism and climate denialism. (Raney, 2019, online)

McKenna's policy alliances reached across the aisle to opponents, and across international borders in order to militate against charges of job theft and economic irrelevance.

Marianne Borgen's ability to work across political divides was also notable in her interview and was one reason for her election to the mayor's position in Oslo (see Interview 5). Nevertheless, the process of cross-party negotiation was nerve-wracking and showed a degree of courage. Both Borgen and McKenna stressed the importance of climate communication to reach out to non-specialist audiences and for a need to speak clearly of benefits, as well as honestly in terms of costs.

In order to move toward more feminist climate policy, the fundamental interconnections between the varied aspects of climate impacts (on health, society, identity, economy) need to be more widely recognised, just as the leadership and voice of women in this space needs to be.

Tellingly, in the book's developed world context, Marama Davidson was the only ethnic minority/racialised politician interviewed for the book: a self-declared brown women leader, a *mana wāhine* [Māori feminist]. She explained to Reeves and MacArthur how, for her, the starting point for navigating climate policy is to understand that colonisation violently impacted on the status of Māori women and that this has led to ongoing and intergenerational impacts in the stronghold of *whanau* [family]. She contrasts Māori feminism with 'white feminism', the former taking a much more holistic approach, a whole of *whanau* well-being, which includes men, all genders, children, and the community. It is the collective nature of Māoridom that provides Marama with her inspiration, her strength and her resilience. Marama states that she can claim nothing individually and that this reflects another feature of the holistic collective Māori voice. Her motivation is derived from the mandate and support of flax roots (grassroots) leaders, their messaging, clarity, focus, approach, and strategy (Interview 2 in this volume).

The importance of grassroots networks and support was also very clear from Ada Colau's interview, in which she stressed the importance of participatory budgeting, and working with communities in local transport planning. As with Marianne Borgen, Colau identified the need to communicate with children, while also invoking the power of communities when confronting big business – something that she was able to do with Airbnb to control their power in Barcelona (Interview 3 in this volume). Guðmundur Ingi Guðbrandsson, former environmental – and social and labour – minister in

Iceland, argues along similar lines and stresses how important it is for politicians to listen to the grassroots and social movements as they often are the first ones to frame new mindsets and have bold ideas, which politicians need to implement. Guðbrandsson, who, before his political career, was the chairperson of the largest environmental organisation in Iceland, for example, mentions that the idea to put carbon neutrality on the agenda of his party, the Left Green Movement, originally came from the environmental grassroots in Iceland (Interview 4 in this volume).

As a *mana wāhine*, Marama Davidson stresses that we need to understand the link between the role of women and the role of climate destruction; the link between our violence to *Papatūānuku* [Earth mother], and the patriarchal drivers of exploiting women and inflicting violence on women. Significantly, in the context of this book, she sees a flawed theory of change as a major barrier to addressing climate change and global warming and refers to a political system that is set up to favour big power and big money influence. She stresses the importance of connectedness; amongst *whanau*, amongst grassroots groups and leaders, with indigenous peoples world-wide, and with projects like the misinformation project in Wellington NZ. She advocates for a theory of change that is grounded in connected progressive grassroots movements, which is certainly echoed by Ada Colau.

Marianne Borgen, Guðmundur Ingi Guðbrandsson, and Marama Davidson all stated how all policy is related to the climate. For example, Borgen, with her background in children's rights, adopted the practice of reviewing all environmental decisions through the eyes of how they would affect children. Guðmundur Ingi Guðbrandsson also suggests a cross-sectoral approach in climate policymaking, involving all policies, including social policy. He also emphasises the importance of the care discourse as a guiding tool in policymaking, for diminishing human suffering, and increasing respect for nature. Another quality that has been revealed by the interviews is the ethical value of strategic alliances which ensure that networking is informed by feminist values, rather than for consolidating leadership for its own sake.

Threats to climate ambition

The optimism with which the editors and writers embarked on the project of this book has, however, diminished as many feminist leaders, including those who we interviewed, have left or been voted out of office, to be replaced, in some cases, by more autocratic, right-wing politicians (Angiolollo et al., 2024). The opportunity to, as the World Economic Forum put it, 'build back better' has been revealed to be hollow (WEF, 2020), as Baruah and Burke's chapter on Canada's economic response to the pandemic reveals. In this, they show how redistributive policies were abandoned as soon as the pandemic was declared over. The pandemic served to illuminate and accelerate a more fundamental long-term societal erosion that is now proving decisive in the direction of travel in many developed nations. This was observed by Foa et al. in the first global overview of how the pandemic had impacted on public attitudes, as characterised by falling support for:

> … core democratic beliefs and principles, including less liberal attitudes with respect to basic civil rights and liberties and weaker preference for democratic government.
>
> *(Foa et al. 2022)*

The parlous situation that democracy now finds itself in provides fertile ground for right-wing populism and associated ample opportunities for eroding legal protections for vulnerable people and groups, while facilitating further illegitimate power grabs by already powerful petrostates. Meanwhile, one example of particular concern for those of us who have worked on this volume is the online persecution of the influential Centre for Feminist Foreign Policy, which has had to disable its online presence in response to persistent cyberbullying and cybercrime (CFFP 2024).

Both gender and climate backlash flourish in this toxic milieu and thus it is no surprise that the intersection between them is already attracting particularly aggressive attention. Likewise, there is a gender dimension (Waldman et al. 2018) to more generalised online denigration of climate science (Huber et al. 2021). Silencing critical voices, by denying basic human rights to freedom of expression and access to information, and attacking core elements of democracy, is a depressingly familiar autocratic strategy, finding renewed vigour online, and in the context of this anthology it is a prime example of a particularly telling form of gender-based violence (Arimatsu, 2019).

Oscillations between liberal and illiberal governments during the last ten years are becoming more extreme, alongside an underpinning shift to the right in the politics of many states in the developed world. (Angiolollo et al. 2024, Pető 2021) This anthology has made it clear that gains in feminist politics, gender rights, and ecological policies, while they benefit society as a whole and the environment in the long term, will be met with fierce resistance in the short term. This is, however, unsurprising when almost every society and government on our planet are drenched in fossil fuels.

Alternatives to this 'drenching' can be found in the fields of material feminism, posthumanism, relationism, and the more-than-human turn, that all take into account how alliances with other creatures can be included in feminist climate policy work. This is explored by Marie Widengard in her analysis of using a legal approach to challenge climate harm (Chapter 10), and by the interviews with Marama Davidson and Guðmundur Ingi Guðbrandsson in this volume.

A key struggle now might be over the European Green Deal (EGD) and its accompanying laws. Without grassroots – including feminist – activity, the

EGD was unlikely to have been adopted in the first place. However flawed we might understand it to be in terms of feminist engagement, it is still an outcome of these progressive years and the influence of grassroots involvement (Bauhardt, 2022). But while it is important to defend and work with the EGD, as Chapter 3 in this volume proposes, it is also important to recognise the substantial revision that has taken place through political negotiations. This limits the scope of what the EGD can achieve.

Even though we admit that it looks tougher than ever to mitigate greenhouse emissions and stop heating our planet (and the sadness and grief that comes with such knowledge are important emotions to recognise and share), very recent history also shows that achieving dramatic changes in societies' ways of dealing with our ongoing climate catastrophe is not linear or totally predictable. However, while there is much room for dramatic progressive changes, these are not in evidence in the recent annual global Conferences of the Parties to the UNFCCC.

Global climate-related agreements

Gender Backlash at COP29

As in previous climate summits, COP29 began with the host state (Azerbaijan) appointing an all-male organising committee and it only added (a minority of) female members after a public backlash (Mooney and Bryan, 2024). Dispiriting as this was, much more concerning was the unexpected gender backlash to the extension of the UNFCCC's Gender Action Plan (GAP), which has been in place for more than a decade. The GAP's very future was called into question by a bizarre alliance of the Vatican and a number of conservative petrostates (Saudi Arabia, Iran, Russia, Egypt), who objected to the use of the term 'gender' as encapsulating women in all their diversity and to the application of intersectionality (Garric 2024). While the state parties ultimately agreed to renew the GAP, the significance of weaponising the term gender in the context of climate change's recognised disproportionate impact on women and girls, and in the chronic under-utilisation of female agency to address it, must not be downplayed. These issues must be kept in the public eye with renegotiating the GAP at COP30 in mind.

Beijing +30 promise and Peril

2025 marks the thirtieth anniversary of the Fourth World Conference on Women and the adoption of the Beijing Declaration and Platform for Action (BPfA, UN 1996). In preparation for the Beijing+30 review schedules at UN Commission on the Status of Women's sixty-ninth session, reviews of progress are underway in each of the five global regions (UNECE/UN Women

2024). The BPfA essentially functions as a global bill of rights for women, and its twelve focus areas address multiple issues of relevance to the gender-climate nexus. While, significantly, the BPfA was one of the first international documents to identify the gender-climate nexus, the importance of improving its climate change credentials is clearly apparent in its prominence and persistent presence in the report of the Expert Group on Beijing +30. This was charged with evaluating progress overall and identifying gaps and challenges that need to be addressed (CSW69 2024). The Expert Group report also identifies the significance of gender backlash in its multiple guises as one of the greatest challenges that we face going forward. Its conclusion succinctly states that the Beijing Declaration and the BPfA:

> ... continue to stand as a visionary blueprint for equality and justice. While there have been critical areas of progress, there remain myriad gaps in implementation. These have become more difficult to close in the context of multiple, overlapping crises, exacerbated by [a] powerful backlash on gains made on gender equality and human rights.
>
> *(CSW69 2024, 21)*

Likewise, the European Union's thirty-year review of BPfA reveals a concern with potential backlash against climate activism, and resistance to green policies by fossil fuel dependent industries and the rise of climate sceptic politicians. It acknowledges advances at the European Commission level regarding the statement of commitments to a socially just and inclusive transition to address the climate emergency declared in 2019, and a rise in the proportion of women in environmental decision-making roles. However, concurrently, it notes a lack of concrete measures and of mandatory commitments at the national scale. (EIGE, 2025) With these concerns in mind, Beijing+30 will be crucially important as a call to safeguard and improve upon the imperfect progress that has been made thus far and build on it for a better future. Not just for women, but for all.

Concluding thoughts – a call to research and to action

The chapters in the global section of the book highlight what can be achieved with feminists leading, for example, Feminist Foreign Policy in Sweden (Chapter 1 in this volume), and feminist leaders in some oceans negotiations, which have led to some tentative progressive shifts (Chapter 4). The robust statistical link that has been made between women's leadership and greenhouse gas emission reduction (Chapter 2) demonstrates how clearly gender equality and action on climate are intertwined, while a critical consideration of EU policy demonstrates a potential for change in EU (Chapter 3). Initiatives at the intra-national and local scale show what can be achieved, whether in a declared 'emergency' (COVID), or through doing transport, energy and climate planning differently (See Chapters 7, 6, 9, and 11 respectively). Ways of generating grassroots evidence, and mobilising action from generally unrepresented groups, while as yet small scale and vulnerable, illustrate how challenges to the global hegemony are mounting (Chapters 12 and 13). This, together with material from the interviews, shows that feminist climate leadership can and does make a difference, but that it is vulnerable to greater sexist/misogynist/patriarchal forces at work. It is a start to at least expose this, and, given the latest evidence on global heating which indicates that the world has now breached the 1.5 degrees of heating (Copernicus Climate Change Service (C3S) and the World Meteorological Organization (WMO), 2025), it is too important a link to be allowed to fall by the wayside.

A crucial task is to make the links between gender and climate backlash impossible to ignore. Ways need to be found to document and expose climate misogyny and failure by states and big tech to address it and the ways in which this is silencing women's voices, and aggravating existing under-representation in societal decision-making for (in the developed world) just over 50% of humanity (World Bank 2022). Mainstream political parties, conventional media and big tech are more often than not complicit in structural misogyny and this must be made to cost them votes and business.

The UN and its constituent bodies respectively provide significant opportunities to expose retrogressive practices and to press states to make good on their promises in international climate law, human rights and gender equality law, and at the intersection between gender and climate change. In the latter context, identifying and publicising ways to better exploit the opportunities that the UN system offers, to identify cross-cutting links and activate mutual support between its gender and climate change machinery, and in its gender-informed environmental human rights agenda, offers significant levers for progress (UNHRC 2023). Likewise, international institutional settings provide opportunities to shed light on state practices, both by placing issues firmly in the public domain and inviting comparison between good and bad practice and positive and negative actors. The current parlous state of both the climate and action to address it requires urgent progress at the global level. This means that we need to grasp every available opportunity to make obstructing the necessary radical changes that need to be made as politically unacceptable as it is ecologically objectionable.

References

Action Aid (n.d.) ActionAid's Ten Principles of Feminist Leadership | ActionAid International

Angiolollo, F. Lundstedt, M. Nord, M. and Lindbery, S.I. (2024). State of the World 2023: democracy winning and losing at the ballot. *Democratization*, *31*(8), 1597–1621, doi:10.1080/13510347.2024,2341453

Arimatsu, L.(2019). Silencing women in the digital age. *Cambridge International Law Journal*, *8*(2), 187–217.

Bauhardt, C. (2022). Ecofeminist political economy: Critical reflections on the green new deal. In *Post-Capitalist Futures: Paradigms, Politics, and Prospects* (pp. 87–95). Singapore: Springer Singapore.

Centre for Feminist Foreign Policy (2024) Maintenance Page – CFFP (Accessed 08 January 2025).

Copernicus Climate Change Service (C3S) and the World Meteorological Organization (WMO), (2025). European State of the Climate (ESOTC) report doi:10.24381/14j9-s541

CSW69 (2024). Beijing +30: Progress, Gaps and Challenges Report of the Expert Group. https://www.unwomen.org/sites/default/files/2024-12/csw-69-expert-group-meeting-report-en.pdf

European Institute for Gender Equality (EIGE) (2025). *Review of the implementation of the Beijing Platform for Action in the EU – Beijing +30* Lithuania: EIGE.

Fair Share of Women Leaders (n.d.) FAIR SHARE – Fair Share of Women Leaders.

Foa, R.S., Romero-Vidal, X., Klassen, A.J., Fuenzalida Concha, J., Quednau, M. and Fenner, L.S. (2022). "*The Great Reset: Public Opinion, Populism, and the Pandemic.*" Cambridge, United Kingdom: Centre for the Future of Democracy.

Garric, A. (2024). At COP29, Saudi Arabia, Iran, Russia, Egypt and the Vatican oppose gender equality measures. Le Monde, 22 November 2024.

Government of the Netherlands (2023). Shaping Feminist Foreign Policy Conference 2023 | Ministry of Foreign Affairs | Government.nl.

Huber, R. A., Greussing, E., & Eberl, J. M. (2021). From populism to climate scepticism: the role of institutional trust and attitudes towards science. *Environmental Politics*, 31(7), 1115–1138. https://doi.org/10.1080/09644016.2021.1978200

Mooney, A. and Bryan, K. (2024). Baku backtracks after backlash over all-male UN COP29 committee. *Financial Times*, 19 January 2024.

Pető, A., (2021). Gender and illiberalism. In Sajó, A., Uitz, R., & Holmes, S. (eds.).*Routledge handbook of illiberalism* (pp. 313–325). Routledge.

Raney, T. (2019). The Problem with Canadian Politics is written across Catherine McKenna's Window. Huffington Post. October 25, 2019. [Online] Available: https://www.huffpost.com/archive/ca/entry/catherine-mckenna-office_ca_5db31a7de4b05df62ebe6170 (accessed 8 May 2024).

UNECE/UN Women (2024). Reviewing 30 Years of Beijing Commitments to Accelerate Gender Equality in the ECE Region. https://unece.org/sites/default/files/2024-09/concept-note_beijing30_rrm.pdf

United Nations (1996). Report of the Fourth World Conference on Women, Beijing, 4–15 September 1995, chap. I, resolution 1, annexes I and II.

United Nations Human Rights Council (2023). A/HRC/52/33 Women, girls and the right to a clean, healthy and sustainable environment: Report of the Special Rapporteur on the issue of human rights obligations relating to the enjoyment of a safe, clean, healthy and sustainable environment, David R. Boyd.

Waldman, S., Heikkinen, N. and E&E News (2018). As Climate Scientists Speak Out, Sexist Attacks Are on the Rise | Scientific American. (Accessed 08 Jan 2025).

WEF (2020). COVID-19: How to build back better with climate action | World Economic Forum.

World Bank (2022). Population, female (% of total population) | Data.

INDEX

Pages followed by "n" refer to notes.

activism 3, 157, 185, 195, 221, 230, 240
age 10, 99, 117, 138, 141–142, 164, 227, 230, 234, 240, 251, 277
Agenda 21 89
Agenda 2030 204
Ardern, Jacinda 4, 122, 127–128, 176, 296–297

Beijing Declaration 66, 73, 89, 96, 301–302
Borgen, Marianne 4, 218–223, 298
Brundtland, Gro Harlan 222

care 51, 83, 94, 96, 98, 108, 160, 166–168, 177, 250; care economy 169; child care 167–168; elder care 168; ethics of care 250
caring 13, 100, 213, 261, 297
carbon: neutral 122, 178, 204, 299; pricing 45, 297; tax 44
Carson, Rachel 178
Centre for Feminist Foreign Policy 7, 83, 300
civil society organisations 78, 80, 82–83
climate: action 37, 49, 76–77, 222, 233, 237, 277; anxiety 234; backlash 303; budget 221; crisis 1–2, 192, 200, 212, 233, 297; emergency 1, 8, 45, 122, 152–153, 157, 218, 302; footprint 54–56, 60–67; litigation 227–229, 239–240; movement 125

Colau, Ada 4, 152–158, 298–299
colonialism 74, 83, 122, 124, 297
colonisation 37, 124, 286, 298
Commission of the Limits of the Continental Shelf 93
Committee of the Regions 78, 82
Council of the European Union 77, 79–85
COP: COP22 40; COP26 4, 38, 90; COP27 38; COP28 89; COP29 71
COVID-19 6, 11, 28–30, 127, 160–172, 296–297, 299–300, 302

Davidson, Marama 122–129, 298–300
De Landro Clarke, Wanda-Lee 93
deliberative democracy 245, 248–251, 253, 256
disability 117, 138, 142
disabled people 124, 143, 179
division of labour 51, 145, 166

eco-feminism 75
ecomodernism 5, 200, 203, 207, 210, 212–213, 252, 254, 273
EIGE 107–109, 113, 149
ethnicity 10, 37, 99–100, 107–108, 138, 141–142, 165, 227, 245, 251, 277
European Commission 77–85, 156
European Court of Human Rights 229–230, 232, 234–235
European Green Deal 9, 77, 79, 82–85, 105–106, 149, 169, 300–301

European Parliament 72, 77–85
European Union 9, 71–85, 136
Extinction Rebellion 3, 296

family policies 27
feminism 5, 19, 30–34
feminist: approaches 6–7; governance 7;
    Green New Deal 169;
    institutionalism 3, 9, 21–24, 32–33,
    77–78, 245, 251–252, 254, 256;
    leaders 4; leadership 6–7;
    Participatory Action Research 97
Feminist Climate Policy 9, 21, 34,
    71–85, 90
feminist foreign policy 5–6, 9–10,
    19–34, 40, 65, 71–85, 297, 302
Feminist Political Ecology (FPE) 98, 100
Figueras, Christina 4, 39
fossil fuels 1–3, 13, 37, 39, 42–43, 113,
    123, 180, 182–183, 300, 302
Fridays for Future 3, 178, 296

gender: action plan 39, 76, 91, 95–96;
    based analysis 40; based violence
    28–30, 300; disaggregated data 97;
    identity 270; justice 6; policy 98–99;
    stereotypes 3, 65, 99, 115
gender equality 5–6, 9–10, 19–34, 48,
    57, 75, 80, 89, 91, 93–94, 99,
    106–109, 111–115, 118, 136,
    160–162, 165, 167, 176, 222, 302;
    gender equality index 60; gender
    equality strategy 96, 108, 117
gender focal point 26, 95–96
gender mainstreaming 9–10, 66, 71–85,
    94–97, 105–107, 111–114, 117–119,
    149n2, 176, 180
grass-roots movement 7, 105, 128–129,
    178, 298–300, 303
Greens 80, 83–84, 122–124, 129, 178,
    219–220
Gudbrandson, Gudmundur Ingi
    176–180, 298–300

Hidalgo Anne 4
human rights 39, 123, 162, 228–229,
    239, 241, 263, 300, 303

IPCC 1–2, 5, 233
implementation theory 9, 20–24, 32
Indigenous 125, 129, 196, 229, 245;
    community 187, 278, 280, 283–285;
    knowledge 286; leaders 45, 123;
    nation 281, 285; people 37, 108,

119, 122–124, 187–188, 227,
    283–285; rights 287; settler relations
    289; women 39, 73
International Council for the
    Exploration of the Sea (ICES) 97,
    100
International Intergovernmental
    Organisations (INGOs) 91, 93,
    95–98
International Seabed Authority (ISA) 93,
    95
International Tribunal for the Law of
    the Sea (ITLOS) 93–94
intersectionality 3, 7–9, 11–12, 39, 64,
    74, 83, 99–100, 106, 124–125, 135,
    138, 146, 201–203, 208, 210–213,
    227–241, 278, 301

Jakobsdottir, Katrin 4, 176, 297
justice 3, 5–6, 12, 200–201, 206, 208,
    218, 230, 239–240, 245, 247, 251,
    263–266; and climate justice 3, 7–8,
    12, 85, 125, 189, 228–229, 240,
    244, 296; and energy justice 12, 246;
    and gender justice 3; and social
    justice 6–7, 39, 46, 161–163, 212;
    and intergenerational justice 235,
    247, 255

Kelly, Elsa 93

Leave it in the Ground 3
Lee, Rena 94
Leopold, Aldo 178
LGBTQI+ 73, 157
life history interviews 185
Lima Work Programme on Gender 89

macho culture 194
Marin, Sanna 4, 297
masculinities 2, 13, 111–112, 250, 256,
    260–261, 266–270, 272–273; and
    petro-masculinities 2–3
McKenna, Catherine 37–47, 297–298
methodologies 7, 9, 12, 21, 24–25, 50,
    55–56, 78, 161, 185, 240, 246, 280
mentor 39, 96, 297
MeToo 3, 109, 115–116, 120
mobility 136; and Gender Smart
    Mobility 10–11, 135–151; and green
    mobility 10
Montseny, Federica 157–158, 158n4
Morgon, Jennifer 39
misogyny 3, 13, 303

networks 11, 40, 45, 91, 126–127, 143, 156, 183–184, 186, 195, 197, 221, 297, 299
non-governmental organisations (NGOs) 2, 91, 93, 99, 172, 177

organisational culture 96

pandemic *see* COVID-19
parental leave 177, 182
Paris Agreement 1, 4, 39, 48, 149n1, 169, 171–172, 177, 232; and Paris climate talks 38
participation 12–13, 22, 96–97, 100, 143, 156
participatory approaches 260–261, 278; and participant observation 206; and participatory assessment workshops 12, 244–245; and participatory budgeting 155–156; and participatory research methods 247, 287; and participatory tools 247–248, 277
path dependencies 3, 13, 21, 33, 202, 208, 210, 212, 252, 256, 277–291
patriarchy 8, 124, 183, 240, 297, 299, 303; and patriarchal norms 108
Petro-states 2, 300; and Provinces 37
polluter pays principle 180
polycrisis 297
practice theory 9, 22–24, 32
precautionary principle 235

queering 269
quotas 158

race 10, 83, 160, 164–165, 227, 240, 245, 277, 285
racism 8, 10, 74, 284
renewables 42, 123, 183, 237
rights: of ecosystems 230; of women 22, 25
right wing populism 5, 300
Rio Declaration 49, 235
Robinson, Mary 39, 46
role models 65, 222

sexism 3, 13, 74, 109
sexual: abuse 90; harassment 3, 99, 109–110, 114
sexuality 142, 245
social movements 3, 7, 177, 296, 299
STEM 94, 99
Sturgeon, Nicola 4, 176, 297
sustainable development goals (SDGs) 89, 94, 105, 176, 200

Thunberg, Greta 178
trans 124, 165
transformative 39, 43, 49, 72–73, 75–76, 78, 80, 82, 92, 100–101, 112, 166, 197, 200–203, 206, 210–213, 261–264, 290, 297
Tubiana, Laurence 39

UN Convention on the Law of the Sea (UNCLOS) 89, 92–93
UN Decade of Ocean Science for Sustainable Development 10, 89, 91, 95, 100
UN Sustainable Development Goals 10
UNFCCC 2, 4, 39, 45, 89–90, 93–94, 100, 232, 301; and Conference of the Parties 2, 39–40, 45, 49, 301; *see also* COPs; and Gender Action Plan 301
UN Women 28

Wallstrom, Margot 19, 22, 24–26, 31, 72
We Stay on the Ground 3
well being 4, 90, 228, 233, 278, 282; well being economy 5, 177
Well Being Government 4, 6, 11, 90, 176
women's political representation 49–63, 55–67

young: activists 232, 236; environmentalists 39; people 44–45, 221, 227–241
youth 3, 12, 229; youth organisations 253